PARTITION OF CELL PARTICLES AND MACROMOLECULES

PARTITION OF CELL PARTICLES AND MACROMOLECULES

THIRD EDITION

Separation and Purification of Biomolecules, Cell Organelles, Membranes, and Cells in Aqueous Polymer Two-Phase Systems and Their Use in Biochemical Analysis and Biotechnology

PER-ÅKE ALBERTSSON

Department of Biochemistry
University of Lund, LUND, Sweden

A Wiley-Interscience Publication

JOHN WILEY & SONS

New York Chichester Brisbane Toronto Singapore

Library of Congress Cataloging in Publication Data:

Albertsson, Per-Åke.
Partition of cell particles and macromolecules.

"A Wiley-Interscience publication"
Includes bibliographies and index.
1. Separation (Technology) 2. Biopolymers—Separation. 3. Phase rule and equilibrium. 4. Cell fractionation. I. Title.

QH324.9.S4A423 1986 574.1′92′028 85-26335
ISBN 0-471-82820-3

Printed in the United States of America

10 9 8 7 6 5 4 3 2 1

To

Alexandra, Charlotte, Daniel, Elin, Elisabet, Erik, Ingrid, Johan, Jonatan, Kristian, Rikard, and Torun.

PREFACE

This third edition of my book *Partition of Cell Particles and Macromolecules* is an up-to-date revision of the second edition, published in 1971. It presents a method for separation and purification of cells, cell organelles, membrane vesicles, biopolymers such as proteins and nucleic acids, and viruses. The method involves partition between two aqueous, yet immiscible, liquid phases. I have retained from previous editions the description of the basic theory of partition, the properties of the aqueous two-phase systems, and the general partition behavior of particles and molecules. Since the last edition there has been a growing number of applications of phase partition of interest to scientists in biochemistry, molecular biology, cell biology, and biotechnology, and new methods such as affinity partition have been introduced. The method has also recently been applied on a large scale in biotechnical processes such as enzyme purification in industry. All the new advances and applications of the method are described in this new edition. The work described has been carried out both in my laboratory and in several other laboratories.

I wish to thank all my co-workers at the Department of Biochemistry, University of Lund, particularly Elisabet Gersbro for typing the manuscript and Agneta Persson for drawing many of the figures.

PER-ÅKE ALBERTSSON

Lund
February 1985

CONTENTS

PARTITION OF CELL PARTICLES AND MACROMOLECULES

1 INTRODUCTION

Progress in biochemistry and cell biology depends to a great extent on the development of efficient separation methods. This holds both for soluble substances, such as proteins and nucleic acids, and for suspended particles, such as cell organelles and whole cells. Much interest is devoted to the study of complex particles obtained by disintegration of cells or cell organelles, such as mitochondria, chloroplasts, or various cellular membranes. These procedures yield very complicated mixtures of particles differing in size, form, and chemical composition. The particles are also very fragile and they may aggregate, dissociate, or generally change their state with time. This is also true of suspensions of whole cells or organelles. In many fields of biological research there is a pressing demand for mild and efficient fractionation methods.

PARTITION IS A SURFACE DEPENDENT METHOD

Since a multitude of components is present in mixtures such as those mentioned, one cannot expect to solve a separation problem by one type of method alone. It is necessary to combine different methods, which utilize different properties of the particles. Centrifugation methods, for example, which separate according to size and density of particles, should be complemented by methods in which other properties, such as surface properties, comprise the separation parameter. One of these methods is distribution in a liquid–liquid two-phase system. This book deals with the application of liquid–liquid partition to macromolecules and particles using phase systems obtained by mixing water with different polymers. Both phases of these systems are aqueous, and they are, therefore, suitable for particles and macromolecules from biological material. The particles distribute mainly according to their surface properties, and partition can, therefore, be used in combination with centrifugation methods to improve separation in a two-dimensional manner.

Aqueous two-phase partition was introduced in 1956–1958 with applications for both cell particles and proteins (Albertsson, 1956, 1958a). Since then it has been applied to a large number of different materials, such as plant and animal cells, microorganisms, virus, chloroplasts, mitochondria, membrane vesicles, proteins, and nucleic acids.

The basis for separation by a two-phase system is the selective distribution of substances between the phases. For soluble substances, distribution takes place mainly between the two bulk phases, and the partition is characterized by the partition coefficient

$$K = \frac{C_t}{C_b} \tag{1}$$

where C_t and C_b are the concentrations of the partitioned substance in moles per liter of top and bottom phase, respectively. Ideally, the partition coefficient is independent of concentration and also independent of the volume ratio of the phases. It is mainly a function of the properties of the two phases, the partitioned substance, and temperature.

The interface between the phases should, however, also be considered. It has a certain capacity for adsorption of the partitioned substance. This does not usually play any significant role with respect to soluble substances, but when suspended particles are present the interface may adsorb relatively large quantities of material. Therefore, in the separation of cell particles there are, in fact, three "phases" to consider: the upper phase, the interface, and the lower phase. It is the selective distribution among these phases which forms the basis for separation of particles by a two-phase system.

THE PHASES MUST BE AQUEOUS

Obviously, the choice of a suitable phase system is the key step in all partition work. Special problems arise when a phase system has to be selected for biogenic particles and macromolecules. The phase system should be mild, that is, consideration must be given to the water content, ionic composition, osmotic pressure, ability to elute out substances from the particles, denaturing effects, and so forth. The larger and more complicated the particle, the more limited is the choice of environment. Whereas a protein may tolerate wide ranges of pH or ionic strength, a cell organelle may have a very strict requirement with respect to pH and salt composition. Further, to be separable, the substances should differ in their partition characteristics, that is, they should not have the same affinity for the two phases.

Several of the factors mentioned above rule out most of the conventional phase systems containing an organic solvent. These are unsuitable both because of the denaturing effects of organic solvents and because, in a water–organic system, biogenic particles and macromolecules almost always segregate completely to the aqueous phase, thereby precluding separation. In addition, the liquid–liquid interface of conventional phase systems has a rather strong interfacial tension which might damage fragile cell structures.

Aqueous–aqueous systems have been used to overcome the above deficiencies. They consist essentially of two immiscible aqueous solutions of different polymers. It is a general phenomenon that mixtures of solutions of unlike polymers in a given solvent result in phase separation. A mixture of aqueous solutions of polymers results in phases with a water content in the range between 85 and 99%. See Chapter 2 for a detailed treatment of polymer phase systems. They are very mild towards various biological activities. One can also obtain a reproducible partition of soluble macromolecules such as proteins and nucleic acids of high molecular weight. The interfacial tension is extremely low, between 0.0001 and 0.1 dyne/cm compared with 1–20 dyne/cm for conventional systems, which, therefore, allows reproducible adsorption of even delicate cell particles, such as chloroplasts and mitochondria, without structural damage. In fact, experience thus far is that the polymers stabilize, rather than damage, the particle structures and the biological activities.

THE MECHANISM OF PARTITION IS COMPLICATED

The mechanism governing partition is largely unknown. Qualitatively it can be described as follows. When a particle is suspended in a phase it interacts with the surrounding molecules in a complicated manner. Various bonds, such as hydrogen, ionic, and hydrophobic, are probably involved, together with other weak forces. Their relative contributions are difficult to estimate; however, their net effect is likely to be different in the two phases. If the energy needed to move a particle from one phase to the other is ΔE, one would expect, at equilibrium, a relation between the partition coefficient and ΔE which can be expressed as

$$\frac{C_1}{C_2} = e^{\Delta E/kT} \tag{2}$$

where k is the Boltzmann constant and T the absolute temperature. Obviously, ΔE must depend on the size of the partitioned particle or molecule since the larger it is, the greater the number of atoms that are exposed and can interact with the surrounding phase. Brønsted therefore suggested the following formula for partition:

$$\frac{C_1}{C_2} = e^{\lambda M/kT} \tag{3}$$

where M is the molecular weight and λ is a factor which depends on properties other than molecular weight. For a spherical particle, M should be replaced by A, the surface area of the particle. Thus,

$$\frac{C_1}{C_2} = e^{\lambda A/kT} \tag{4}$$

and λ in this case is a factor which depends on properties other than surface area, for example, the surface properties as expressed by the surface free energy per unit area (surface tension). Both size and surface properties are, therefore, of great importance in determining partition.

One would also expect the net charge Z of a particle to play a role. If, for example, there is an electrical potential difference $U_1 - U_2$ between the phases an energy term $Z(U_1 - U_2)$ has to be included and the relation would then be

$$\frac{C_1}{C_2} = \exp\frac{\lambda_1 A + Z(U_1 - U_2)}{kT} \tag{5}$$

where λ_1 then depends on factors other than size and net charge.

In this manner we may formally divide the overall effect determining partition into a number of different factors, such as size, hydrophobicity, surface charge, and, probably also, conformation of the particle or macromolecule, which in turn determines the size and number of groups exposed to the surroundings. The theory of partition is treated in Chapter 3.

PARTITION HAS A GREAT POTENTIAL FOR BEING SELECTIVE

The main point of the Brønsted partition theory is the exponential relation between the partition coefficient and properties that enter into the λ factor, for example, size and charge. Small changes in such factors will cause relatively large changes in the partition coefficient. The theory, therefore, predicts a high degree of selectivity.

This compares well with analogous relations for other methods. In centrifugation, sedimentation coefficient is proportional to size; in free electrophoresis, mobility is proportional to charge; and in gel electrophoresis, mobility is proportional to log *M*. Therefore, the theory of partition predicts that this method will give a much higher resolution in separation according to size as compared with centrifugation, or according to charge as compared with electrophoresis.

More important is that factors other than charge, size, or density also determine partition. By binding hydrophobic groups to the polymers, partition can be made dependent upon the hydrophobicity of the particle surface. By binding biospecific ligands to the polymers, a highly specific affinity partition of particles bearing receptors for these ligands can be achieved. Affinity partition also can be applied to proteins and nucleic acids. The possibilities for modifying the phase systems for a given purpose are almost unlimited.

SINGLE AND MULTISTEP PROCEDURES

From Eq. (3) we also learn that even a single partition step may give rather efficient separations of large molecules. If, for example, there is a certain difference in λ for two substances, their separation will be more efficient the larger the size of their molecules. An almost complete separation of two types of macromolecules or particles, therefore, can often be obtained by a single partition. An example is the separation of native from denatured DNA, or bacterial spores from vegetative cells. Such applications are described in Chapter 5.

A mixture which is only partly separated by a single partition step can be fully resolved by a multistage procedure, such as countercurrent distribution or partition chromatography. Particles which distribute between one of the phases and the interface also can be separated in this manner by liquid–interface countercurrent distribution. In this technique each type of particle gives rise to a peak in the distribution diagram which can therefore be used to analyse suspensions of cells and cell organelles such as chloroplasts and mitochondria. See Chapters 6, 7, and 8.

LARGE SCALE AND BIOTECHNICAL APPLICATIONS

Isolation and purification of biopolymers and particles on a large scale are also of increasing importance in industry for example, in such processes as the production of pure enzymes for various technical applications, and the culture and isolation of viruses for vaccines. In these fields there is a need not only for mild and efficient separation methods but also for such methods which can be applied economically to tons of material. Liquid–liquid distribution methods are of special interest here since they can be scaled up rather easily. Using standard liquid separators large volumes of phases can be separated in a short time. The examples published so far indicate a very favorable economy compared to other procedures as shown in Chapter 9.

Bioconversion in phase systems has also been used where an enzymatic conversion of a substrate to product is allowed to take place concomitant with selective extraction of the desired product. These methods are particularly attractive for continuous processes and are described in Chapter 9.

BINDING STUDIES

Binding of one molecule to another also can be studied by partition. A complex between two components usually has a partition coefficient which differs from the coefficient of the two components when partitioned separately. The dissociation constant can be determined and partition can also be used for analytical purposes. Of particular importance is that macromolecule–macromolecule interactions, such as between two enzymes, antigen and antibody, or between protein and nucleic acid, can be studied, as seen in Chapter 10. Even rather weak complexes between enzymes can be detected by partition.

THREE-, FOUR-, AND POLYPHASE SYSTEMS

If three mutually incompatible polymers are mixed with water, a three-phase system can be obtained. If four polymers are used, a four-phase system can be obtained, and if five polymers, a five-phase system, and so forth. A solid phase also may be included in such a mixture. Systems with several phases and interfaces can be used for separation purposes, see Chapter 11. Of particular interest is that they allow an almost instant separation of a mixture into several fractions.

REFERENCES

Albertsson, P.-Å. (1956). *Nature,* **177,** 771–774.
Albertsson, P.-Å. (1958a). *Biochim. Biophys. Acta,* **27,** 378–395.
Albertsson, P.-Å. (1958b) *Nature,* **182,** 709–711.

2 AQUEOUS POLYMER-PHASE SYSTEMS

TWO-PHASE SYSTEMS

Mixtures of different polymer solutions often give rise to two or more liquid phases. For example, a mixture of dextran and polyethylene glycol dissolved in water is turbid above certain concentrations of the polymers. On standing, two liquid layers are formed. These two phases are in equilibrium (Albertsson, 1958). The upper phase is enriched in polyethylene glycol while the lower is enriched in dextran. Both phases are aqueous (Fig. 2.1). Dextran and polyethylene glycol are each fully water soluble, yet the two polymers are incompatible and separate into two phases.

Phase diagrams describing the composition of each phase and the concentration range where phase separation occurs are shown in Figure 12.6. With these two polymers, therefore, one can easily obtain phases which are immiscible yet both aqueous, seemingly a paradox.

The example of dextran and polyethylene glycol is not unique. Phase separation in polymer mixtures is a very common phenomenon. In fact, miscibility is an exception rather than the rule for polymer mixtures. Table 2.1 gives examples of different aqueous polymer-phase systems.

Such liquid-phase separation in mixtures containing one or more colloids was first reported in the literature long ago. Beijerinck in 1896 observed that if aqueous solutions of gelatin and agar, or gelatin and soluble starch (but not agar and soluble starch), were mixed, a turbid mixture which separated into two liquid

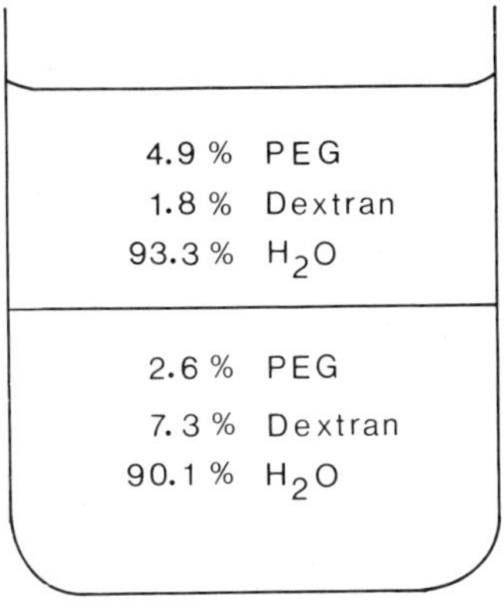

FIGURE 2.1. Composition of the two phases formed by a mixture of 5% w/w dextran (Dextran 500) and 3.5% w/w polyethylene glycol (PEG 6000). See also Figure 12.6.

TABLE 2.1 Aqueous Phase Systems

Liquid Polymer Two-phase Systems		
P	Q or L	Reference
	A. Polymer–Polymer–Water Systems	
	1. Nonionic polymer (P) *Nonionic polymer* (Q)–*Water*	
Polypropylene glycol	Methoxypolyethylene glycol	
	Polyethylene glycol	Albertsson (1958)
	Polyvinyl alcohol	
	Polyvinylpyrrolidone	Albertsson (1958)
	Hydroxypropyldextran	Albertsson (1958)
	Dextran	Albertsson (1958)
Polyethylene glycol	Polyvinyl alcohol	Albertsson (1958)
	Polyvinylpyrrolidone	Albertsson (1958)
	Dextran	Albertsson (1958)
	Ficoll	
Polyvinyl alcohol	Methylcellulose	Dobry (1938)
	Hydroxypropyldextran	
	Dextran	Albertsson (1958)
Polyvinylpyrrolidone	Methylcellulose	Dobry (1939)
	Dextran	Albertsson (1958)
Methyl-cellulose	Hydroxypropyldextran	
	Dextran	Albertsson (1958)
Ethylhydroxyethylcellulose	Dextran	Albertsson (1958)
Hydroxypropyldextran	Dextran	
Ficoll	Dextran	

Table 2.1 *(continued)*

Liquid Polymer Two-phase Systems		
P	Q or L	Reference
	2. Polyelectrolyte (P)*–Nonionic polymer* (Q)*–Water*	
Na dextran sulfate	Polypropylene glycol	
	Methoxypolyethylene glycol NaCl	
	Polyethylene glycol NaCl	
	Polyvinyl alcohol NaCl	
	Polyvinylpyrrolidone NaCl	
	Methylcellulose NaCl	
	Ethylhydroxyethylcellulose NaCl	
	Hydroxypropyldextran NaCl	
	Dextran NaCl	
Na carboxymethyldextran	Methoxypolyethylene glycol NaCl	
	Polyethylene glycol NaCl	
	Polyvinyl alcohol NaCl	
	Polyvinylpyrrolidone NaCl	
	Methylcellulose NaCl	
	Ethylhydroxyethylcellulose NaCl	
	Hydroxypropyldextran NaCl	
Na carboxymethylcellulose	Polypropylene glycol NaCl	
	Methoxypolyethylene glycol NaCl	
	Polyethylene glycol NaCl	
	Polyvinyl alcohol NaCl	
	Polyvinylpyrrolidone NaCl	
	Methylcellulose NaCl	
	Ethylhydroxyethylcellulose NaCl	
	Hydroxypropyldextran NaCl	
DEAE dextran·HCl	Polypropylene glycol NaCl	
	Polyethylene glycol Li_2SO_4	
	Methylcellulose	
	Polyvinyl alcohol	
	3. Polyelectrolyte (P)*–Polyelectrolyte* (Q)*–Water*[a]	
Na dextran sulfate	Na carboxymethyldextran	
Na dextran sulfate	Na carboxymethylcellulose	
Na carboxymethyldextran	Na carboxymethylcellulose	
	4. Polyelectrolyte (P)*–Polyelectrolyte* (Q)*–Water*[b]	
Na dextran sulfate	DEAE dextran·HCl NaCl	
	B. Polymer* (P)*–Low-Molecular-Weight Component* (L)*–Water Systems	
	1.	
Polypropylene glycol	Potassium phosphate	Albertsson (1958)
Methoxypolyethylene glycol	Potassium phosphate	Albertsson (1958)

Table 2.1 *(continued)*

Liquid Polymer Two-phase Systems		
P	Q or L	Reference
Polyethylene glycol	Potassium phosphate	Albertsson (1958)
Polyvinylpyrrolidone	Potassium phosphate	Albertsson (1958)
Polypropylene glycol	Glucose	Albertsson (1958)
Polypropylene glycol	Glycerol	Albertsson (1958)
Polyvinyl alcohol	Butylcellosolve	Albertsson (1958)
Polyvinylpyrrolidone	Butylcellosolve	Albertsson (1958)
Dextran	Butylcellosolve	Albertsson (1958)
Dextran	Propyl alcohol	Albertsson (1958)
	2.	
Na dextran sulfate	Sodium chloride (0°C)	

[a]Both electrolytes have acid groups.
[b]P has acid groups, Q has basic groups.

layers was obtained (Beijerinck, 1896, 1910). The bottom layer contained most of the agar (or starch) and the top layer most of the gelatin (Beijerinck, 1896, 1910).

In a systematic investigation Dobry and Boyer-Kawenoki (1947) studied the miscibility of a large number of pairs of different polymers soluble in organic solvents. In most cases they found demixing and phase separation. Thus out of 35 pairs of polymers tested, only four gave homogeneous solutions. Experiments on water-soluble polymers indicated the same phenomenon (Dobry, 1948).

Both nonionic and ionic polymers may be used. In some cases both polymers are found mainly in one phase while the other phase is relatively poor in polymers. This occurs when there is attraction between the two polymers, for example, between positively and negatively charged dextran derivatives. Such phase separation is called complex coacervation (Bungenberg de Jong, 1949).

Liquid-phase systems also may be obtained by mixing one polymer with two low-molecular components. For example, polyethylene glycol, ammonium (or magnesium) sulfate and water can form a two-phase system (Albertsson, 1958). The bottom phase is rich in salt and the upper phase is rich in polyethylene glycol. Another example is dextran–propyl alcohol–water.

The phases formed are not always liquid. Depending on the type of polymer used a phase may be more or less solid or gel-like. Dextran forms liquid phase systems with polyethylene glycols with molecular weights above 1000. When lower molecular weight polyethyelene glycols are used dextran forms a solid-gel phase.

In summary, three alternative phenomena are observed when mixing two different polymer solutions:

1. *Incompatibility*: phase separation occurs and the two polymers are collected in different phases.
2. *Complex coacervation*: phase separation occurs and the two polymers are collected together in one phase while the other phase consists almost entirely of solvent.
3. *Complete miscibility*: a homogeneous solution is obtained, and as mentioned above, this is the exception rather than the rule in polymer mixtures.

Examples of a number of different aqueous-phase systems are given in Table 2.1. They have been classified in two main groups, A and B. Those systems containing two different polymers belong to group A, while those containing only one polymer belong to group B. It is convenient to divide group A further into four sub-groups, according to whether or not the polymers are charged. Thus, group A1 contains only nonionic polymers; group A2 contains one nonionic polymer and one polyelectrolyte (positively or negatively charged); group A3 contains two polyelectrolytes with the same sign of their net charges, and group A4 contains two polyelectrolytes with opposite net charges.

Group B may be divided into two subgroups. Thus, group B1 contains systems with a nonionic polymer and a low molecular component, and group B2 contains systems with a polyelcctrolyte and a salt. Any requirement concerning the presence of salt is listed in the second column of Table 2.1.

All systems belonging to the groups A1, A2, and A3 show polymer incompatibility, that is, the two polymers go to different phases. Even very closely related and highly hydrophilic compounds may show phase separation when mixed. Such examples are dextran–methylcellulose, dextran sulfate–dextran and the sodium salts of dextran sulfate–carboxymethyldextran. The systems which belong to group A4 either show complex coacervation, that is, both polymers collect predominantly in one of the phases, while their concentration in the other phase is relatively low; or incompatibility, that is, the two polymers go to different phases, depending upon the salt concentration. Thus, when sodium dextran sulfate is mixed with the chloride salt of diethylaminoethyldextran at low NaCl concentrations, a phase containing both polymers is obtained; but if the same mixture is made with a high NaCl concentration, the first polymer collects in the bottom phase and the second polymer in the top phase. At intermediate NaCl concentrations the two polymers are miscible.

Phase separation in mixtures containing only nonionic polymers occurs independently of the salt concentration and pH. The systems containing a polyelectrolyte, however, depend very much on the concentration of salts present and the pH. Sometimes phase separation occurs only above certain values of the ionic strength or within certain pH intervals. There is no sharp boundary between the different groups listed in Table 2.1. All possible intermediate systems may be encountered, such as those containing polymers with only a few charged groups. As the degree of substitution is progressively reduced, polyelectrolytes will tend to behave as nonionic polymers.

Those polymer–polymer combinations, which are possible with the polymers listed here and not found in Table 2.1, either give homogeneous mixtures or a solid precipitate of one of the polymers. Several polymer-phase systems with organic solvents have been published by Dobry and Boyer-Kawenoki (1947).

Some of the polymers used here may be regarded as representatives of a whole group of closely related polymers. Phase separation is obtained if, for example, in the systems of Table 2.1 dextran is replaced by any of the following mutually compatible polymers: levan, glycogen, and soluble starch. Ionic derivatives of these compounds can replace dextran sulfate and carboxymethyldextran. In the same way, methylcellulose and ethylhydroxyethylcellulose may be replaced by other water-soluble cellulose derivatives such as ethylcellulose and hydroxyethylcellulose.

Proteins and nucleic acids may also serve as phase forming polymers. For example, serum albumin and polyethylene glycol are incompatible, and two phases are obtained on mixing their aqueous solutions above certain concentrations. Often the protein-rich phase is more or less solid-gel-like, but a clear liquid phase may also be obtained. See phase diagram described by Edmond and Ogston (1958).

POLYPHASE SYSTEMS

When three different aqueous polymers, all mutually incompatible, are mixed with water a three-phase system is obtained (Albertsson, 1971). Such a system is obtained, for example, with dextran, Ficoll® (a polysucrose), and polyethylene glycol. Dextran is enriched in the lower, Ficoll in the middle, and polyethylene glycol in the upper phase. If four polymers are used a four-phase system is obtained and so on (Table 2.2).

Polyphase systems may also be obtained from certain highly heterogeneous polymer mixtures containing the same type of polymer (Albertsson, 1971). An

TABLE 2.2 Examples of Some Polyphase Systems[a,b]

Three-phase Systems
Dextran (6)–HPD (6)–PEG (6)
Dextran (8)–Ficoll (8)–PEG (4)
Dextran (7.5)–HPD (7)–Ficoll (11)
Dextran–PEG–PPG
Four-phase Systems
Dextran (5.5)–HPD (6)–Ficoll (10.5)–PEG (5.5)
Dextran (5)–HPD; A (5)–HPD; B (5)–HPD; C (5)
Five-phase Systems
DS–Dextran–Ficoll–HPD–PEG
Dextran (4)–HPD; a (4)–HPD; b (4)–HPD; c (4)–HPD; d (4)
More than Five-phases
Highly polydisperse and heterogeneous dextran sulfate[a]
Eighteen-phase Systems
Dextran sulfate (10)–Dextran (2)–HPDa (2)–HPDb (2)–HPDc (2)–HPDd (2)

[a]Figures in brackets are examples of % w/w compositions.
[b]Dextran is ''Dextran 500'' or D 48; PEG is Polyethylene glycol 6000; PPG is Polypropylene glycol 424; DS is Nadextran sulfate 500; HPD is hydroxypropyl dextran made from Dextran 500. A, B, C, a, b, c, and d denote different preparations with different degrees of substitution.
[c]Dextran sulfate 2000 lot No. 6668, 0.4 g dissolved in 1.6 mL 0.4 *M* NaCl.

example is a preparation of high-molecular-weight dextran sulfate, which gives as many as nine liquid phases when dissolved in 0.3 *M* NaCl at a total dextran sulfate concentration of 20% w/w. The phase separation in this case is most probably due to a combination of high molecular weight, high polydispersity, and an uneven substitution of the dextran molecules by sulfate groups, so that the polymer solution contains an entire range of molecules with different degrees of substitution.

A five-phase system is obtained with dextran, together with four hydroxypropyldextran preparations: HPD-a, HPD-b, HPD-c and HPD-d, each having different degrees of substitution. The difference in the number of hydroxypropyl groups on the dextran backbone makes these sufficiently different so that they are all mutually incompatible and give rise to phase separation. Thus a mixture containing 4% of each of the polymers gives rise to a five-phase system (Fig. 2.2).

By dissolving dextran, the four hydroxypropyldextrans, and a certain dextran

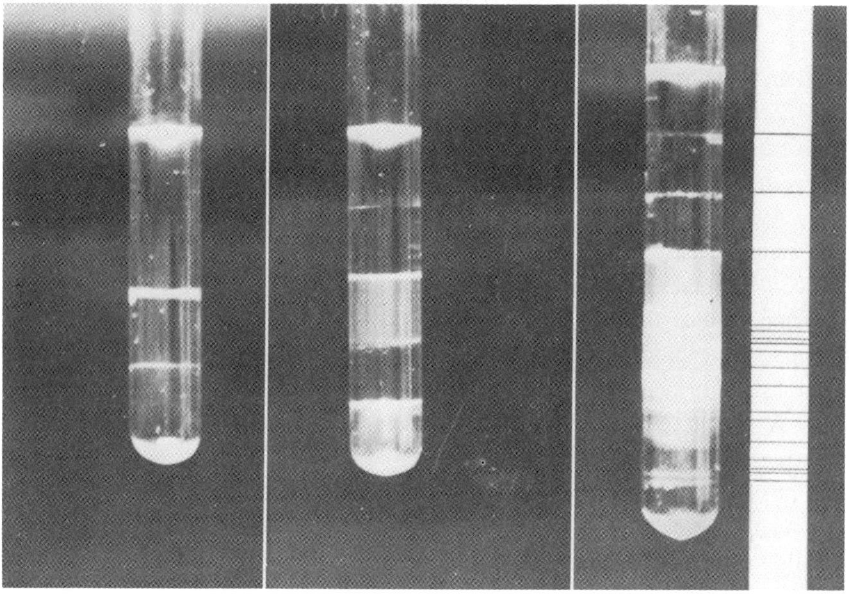

FIGURE 2.2. Polyphase systems. (Left) three-phase system (7% Dextran 500; 7% Ficoll; 6% polyethylene glycol, PEG 6000, 80% H_2O). (Middle) four-phase system (4% Dextran 500; 4% of each of four hydroxypropyldextrans with different degree of substitution, HPD-a, HPD-b, HPD-c, and HPD-d; 80% H_2O). (Right) 18-phase system (10% Dextran sulfate 2000, batch 6668, 2% Dextran 500; 2% of each of HPD-a, HPD-b, HPD-c, and HPD-d; 80% 0.4 *M* NaCl).

sulfate preparation in 0.3 *M* NaCl, a phase system with as many as eighteen liquid phases has been obtained (Fig. 2.2).

WHY DEMIXING IN POLYMER MIXTURES?

The frequent phase separation in polymer mixtures is due to the high molecular weight of the polymers combined with the interaction between the segments of the polymers.

The phenomenon has been treated theoretically by applying theories on the thermodynamic properties of polymers in solution (Koningsveld, 1968; Scott, 1949; Tompa, 1956). See also the books by Flory (1953), Huggins (1958), and Morawetz (1975). The explanation for phase separation in polymer mixtures can be summarized as follows:

Two factors determine the result on mixing two substances. One is the gain in entropy which occurs on mixing the molecules, and the other is the interaction between the molecules. The gain in entropy on mixing two substances is related to the number of molecules involved in the mixing process. As a first approximation, the entropy of mixing is, therefore, the same for small and large molecules if defined on a molar basis. The interaction energy between molecules, however, increases with the size of the molecules since it is the sum of the interaction between each small segment of the molecules. Therefore, for very large molecules the interaction energy per mole will tend to dominate over the entropy of mixing per mole. Thus it will be mainly the type of interaction between the molecules which determines the result of mixing two polymers.

Suppose then that the interaction between two unlike polymer molecules is repulsive in character, that is, molecules prefer to be surrounded by their own kind instead of being mixed. In this case, the system will have its energetically most favorable state when the two polymers are separated. The result of mixing solutions of two such polymers is, therefore, incompatibility, and there arises one phase which contains the one polymer and another phase with the second polymer. This is the most common result obtained when mixing two polymer solutions; as mentioned above, it can occur in mixtures containing nonionic polymers, polyelectrolytes, or both.

If, on the other hand, the interaction is attractive in character, that is, unlike polymer molecules attract each other, then there is a tendency for these to collect together, and their separation into a common phase is favored. This is known as complex coacervation. The attractive forces necessary for this to occur must be great, however, such as those between oppositely charged polyelectrolytes.

Finally, in the absence of comparatively strong attractive or repulsive forces, complete miscibility may result.

In principle there is no difference between the mechanism behind phase separation in a polymer mixture and that in a mixture of low-molecular weight substances. We may, for example, compare the incompatibility of two polymers with the immiscibility of water and benzene, and complex coacervation with the formation of an insoluble inorganic salt like $BaSO_4$. The point about polymer mixtures is that, owing to the large size of the molecules, phase separation takes place even for very closely related polymers and when they are present only in low concentrations of a few per cent.

THE HYDROPHOBIC LADDER

We may compare the aqueous polymer phases with conventional solvents as is done in Figure 2.3. To the left in this figure a number of solvents are listed according to their hydrophobic–hydrophilic nature; at the bottom a salt solution,

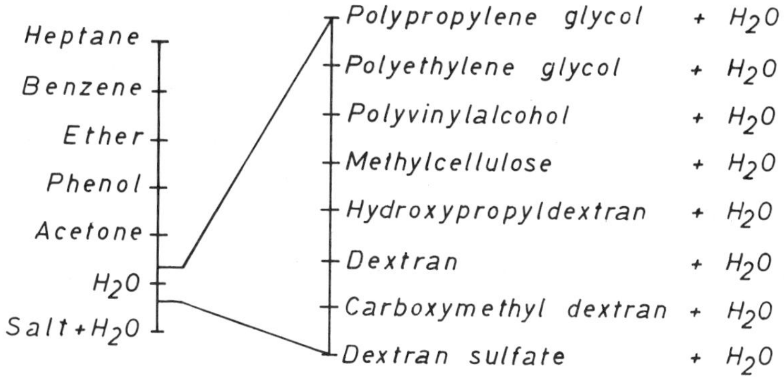

FIGURE 2.3. Hydrophobic ladder. To the left, a number of solvents have been selected from a spectrum of solvents with increasing hydrophobicity. Aqueous solutions of polymers to the right are mutually immiscible, but since they all contain mainly water they fall within a narrow part of the solvent spectrum to the left.

then water, acetone and so forth, with increasing hydrophobicity, up to heptane. We may consider these as selected solvents from a continuous solvent spectrum with increasing hydrophobic character. If we want to place the polymer solutions in this spectrum it is obvious that, since they all contain mainly water, they should fall within a very narrow part of the spectrum. To the right in Figure 2.3, a number of polymer solutions are listed according to the hydrophilic–hydrophobic nature of the polymers. The order may be doubtful here, but the main point is that we have a number of immiscible liquids which are all very close to each other in the solvent spectrum. This means that phase systems formed by these solvents can be expected to be selective in separating substances which themselves fall within the same aqueous part of the solvent spectrum, as, for example, particles and macromolecules of biological origin.

THE PHASE DIAGRAM

In a mixture of two polymers and water, a two-phase system will arise only when the constituents are present in a certain range of proportions. The constituent compositions at which phase separation occurs may be represented in a phase diagram. Figures 2.4 and 2.5 show such a diagram for a system polymer P–polymer Q–solvent. The concentration of polymer P is plotted as the abscissa and the concentration of polymer Q as the ordinate; the concentrations are expressed as percentage. The curved line separating two areas is called a *binodial*. All mixtures which have compositions represented by points above the line give

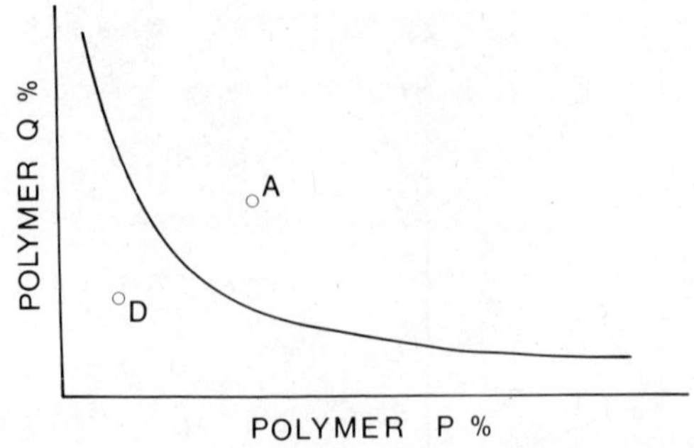

FIGURE 2.4. By mixing two polymers P and Q in water, phase separation occurs above certain concentrations of the two polymers, while other mixtures give a homogeneous solution. Thus, mixtures represented by points above the curved line, such as point A, give two liquid phases, while mixtures represented by points below the curved line, such as point D, give one liquid phase. The curved line is called a binodial.

rise to phase separation, while mixtures represented by points below the tie line do not. Thus a composition represented by point A in Figure 2.4 gives a two-phase system while a composition represented by point D gives a homogeneous solution.

In order to describe the two-phase system in more detail, we have also to account for the compositions of the two phases which are in equilibrium. Suppose point A in Figure 2.5 represents the composition of the total system (percentage of polymer P and Q per total weight mixture). The compositions of the bottom and the top phases obtained with this system will then be represented by points B and C respectively. In the same way, the system with the total composition of A′ will have a bottom phase composition of B′ and a top phase composition

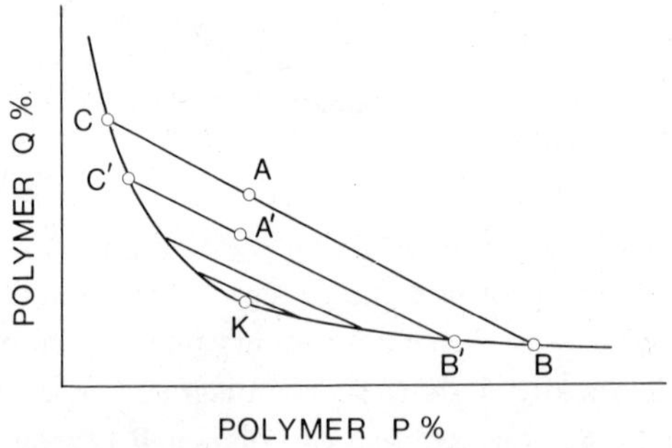

FIGURE 2.5. Tie lines, for example the lines BC or B′C′, connect the points representing the composition of two phases in equilibrium. Thus, point B represents the composition of the bottom phase, and point C the composition of the top phase of a system with a total composition represented by point A. K is the critical point.

of C′. Like all other points representing the composition of pure phases, points C, B, C′, and B′ lie on the binodial. Pairs of points like B and C are called nodes and the lines joining them are called tie lines. Point A, representing the total composition, lies on the tie line joining B and C. Any total composition represented by points on the same tie line will give rise to phase systems with the same phase compositions, but with different volumes of the two phases. If composition is expressed in per cent weight per weight (w/w), the weight ratio bottom phase/top phase is equal to the ratio between the lines AC and AB. This follows from the fact that the weight m_t of polymer P in the top phase plus the weight m_b of polymer P in the bottom phase should be equal to the total weight m_0 of polymer P.

$$m_t + m_b = m_0 \tag{1}$$

but

$$100\, m_t = V_t d_t C_t \tag{2}$$

where V_t is the volume, d_t the density, and C_t the concentration of polymer P in per cent (w/w) of the top phase and

$$100\, m_b = V_b d_b C_b \tag{3}$$

where V_b is the volume, d_b the density, and C_b the concentration of polymer P in per cent (w/w) of the bottom phase and

$$100\, m_0 = (V_t d_t + V_b d_b) C_0 \tag{4}$$

where C_0 is the total concentration of polymer P in per cent (w/w). Substituting Eqs. (2), (3), and (4) in (1) we obtain

$$V_t d_t C_t + V_b d_b C_b = (V_t d_t + V_b d_b) C_0 \tag{5}$$

or

$$\frac{V_t d_t}{V_b d_b} = \frac{C_b - C_0}{C_0 - C_t} \tag{6}$$

but from the diagram,

$$\frac{C_b - C_0}{C_0 - C_t} = \frac{\overline{AB}}{\overline{AC}}$$

hence

$$\frac{V_t d_t}{V_b d_b} = \frac{\overline{AB}}{\overline{AC}} \quad \text{and} \quad \frac{V_t}{V_b} = \frac{d_b}{d_t}\frac{\overline{AB}}{\overline{AC}}$$

The densities of the polymer phases are not very different from that of water (usually in the range 1.00–1.1) and the ratio of the volumes of the two phases may be obtained therefore approximately from the distances AB and AC on the tie line.

As may be seen in Figure 2.5, the more the composition of the phase system approaches point K the less is the difference between the two phases. K is the critical point (also called the plait point) and the composition represented by point K is called the critical composition. At the critical point the compositions and the volumes of the two phases theoretically become equal. Thus a very small change in the total composition from below to above point K means a change from a one-phase system to a two-phase system with very nearly equal volumes of the two phases formed. All compositions at the binodial are critical in the sense that a small change in composition gives rise to a drastic change in the system, namely from a one-phase system to a two-phase system or vice versa, but it is only at the critical point that the volumes of the phases become equal. Figure 2.6 illustrates this fact as seen from the division of the lines through A′and B′.

It is evident from Figure 2.7 that near the critical point the property of a two-

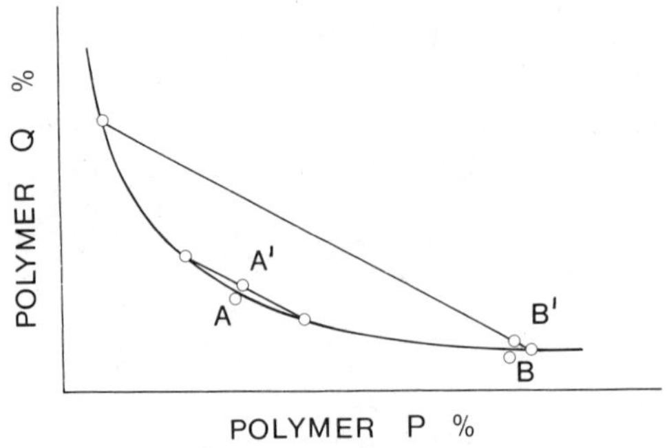

FIGURE 2.6. Phase diagram showing the volume ratios obtained for systems near and far away from the critical point, when the composition is changed so that the system shifts from a one-phase to a two-phase system. Point B represents a one-phase system, which by a slight change in composition to point B′ is converted to a two-phase system with one small and one large phase. A corresponding shift in composition from A to A′ results in a two-phase system with about equal volumes of the phases.

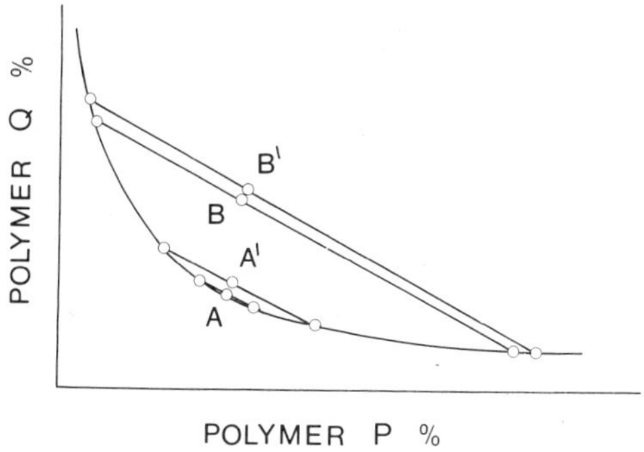

FIGURE 2.7. Phase diagram with tie lines showing the sensitivity of a system near the critical composition. A change in the composition of a system from point A to A′, which takes place near the critical point, causes a relatively much larger change in the difference between the two phases than a change from B to B′, which takes place far away from the critical region.

phase system is most sensitive to changes in its total composition. Compare, for example, the change from A to A′ and the change from B to B′. The change from A to A′, which occurs near the critical point, causes a relatively larger change in the difference between the compositions of the two phases than does the B–B′ change.

The same holds for a change in temperature. The two curves in Figure 2.8 thus represent binodials at two different temperatures. A change in temperature obviously causes a relatively larger change in the difference between the two phases when the composition is near the critical point.

Phase systems near a critical point, therefore, require more precise experimental control than others and should, if possible, be avoided.

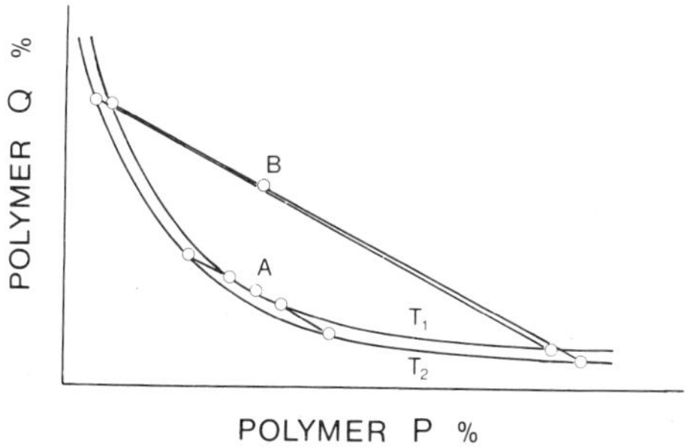

FIGURE 2.8. Phase diagram of a system at two different temperatures T_1 and T_2. A change in the temperature causes a relatively much larger change in the difference between the two phases for a system (A) near the critical composition than for a system (B) removed from the critical composition.

These considerations apply for ternary systems, that is, composed of three monodisperse components. Polymers are, however, usually polydisperse, composed of a large number of molecular species with different molecular weights. A system of two polymers and water is, therefore, not a three-component but a multicomponent system. This means that some of the statements given above do not hold strictly for a polymer-phase system. We should expect that phase separation from a mixture containing a polydisperse polymer leads to a fractionation of this polymer. Thus in a dextran–polyethylene glycol system, for example, the dextran of the bottom phase has a molecular weight distribution other than the dextran of the top phase, the dextran of the bottom phase is, therefore, not identical with that of the top phase. The statement that ''all points on a tie line give the same phase compositions'' no longer holds if we are dealing with polydisperse polymers. If in a system of monodisperse polymers we construct the binodial from the mixtures which just give turbidity (see next section) or from the compositions of the phases, two identical lines should be obtained. With polydisperse polymers, however, the two lines may not be identical. The practical significance of these deviations from a strict ternary system will be discussed later (see Figs. 2.17 and 2.18).

A phase diagram of polydisperse polymers has, however, a defined critical point. Likewise, the condition that the three points, representing the composition of the total mixture, the top phase, and the bottom phase should lie on a straight line, the tie line, also holds for a system of polydisperse components. This follows from Eqs. (1)–(4), which can also be applied to polydisperse systems.

How the Binodial, the Tie Lines, and the Critical Point May be Determined Experimentally

A binodial may be determined experimentally in one of the following ways:

1. A few grams of a concentrated solution of polymer (P) are put into a test tube. A solution of known concentration of polymer (Q) is then added dropwise to the test tube. First, a homogeneous mixture is obtained, but after a certain amount of polymer Q has been added, one further drop will cause turbidity and a two-phase system will arise. The composition of this mixture is noted. One gram of water is then added and the mixture becomes clear again. More solution of polymer Q is then added dropwise until turbidity and a two-phase system is again obtained. The composition of this mixture is noted and more water is added to get a one-phase system, and so on. In this way a series of compositions, close to the binodial, are obtained and, if the concentration of polymer P is plotted against that of Q for these compositions, a line as that of Figure 2.4 is obtained.

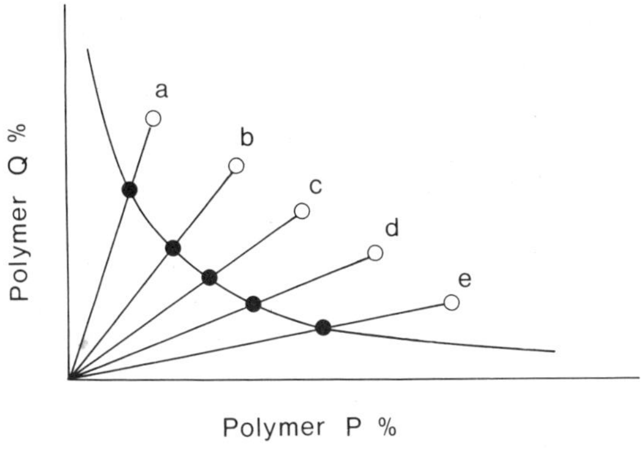

FIGURE 2.9. Determination of the binodial. Mixtures represented by points a, b, c, d, and e are prepared. Water is added until the mixtures turn clear and become one phase systems. The composition of the shift (filled symbols) lies on the binodial.

When the polymers are polydisperse, as is usually the case, there is not a sharp change from a clear solution to a turbid mixture, particularly when one of the polymers is present at a low concentration. By adding polymer Q a very weak turbidity appears first, probably because only the largest molecules of polymer P separate into a phase. Then, as more polymer Q is added, more polymer P separates out and the turbidity increases. For very polydisperse polymers this method is, therefore, rather tedious.

2. A number of phase systems having different compositions, represented by, for example, the points *a*, *b*, *c*, *d* and *e* in Figure 2.9 are first weighed into test tubes. Water is then added until the mixture becomes clear, that is, a one-phase system is obtained. The compositions where this change from a two-phase to a one-phase system occurs lie on the binodial.

3. The compositions of the phases of a number of different systems are determined, and a curve, the binodial, is drawn through the points representing these compositions.

The tie lines are obtained by analyzing the composition of each phase. This gives points *b* and *c* of Figure 2.5 and the tie line is that joining these points. Alternatively, only one phase may be analyzed. If the total composition and the ratio between the weights of the two phases are known, the composition of the other phase may be obtained by calculation according to Eq. (6).

The critical point is determined in one of the two following ways.

1. By trial and error, a mixture is made such that, after the addition of a drop of the one-polymer solution, conversion from a one-phase to a two-phase system with equal volumes of the phases occurs.

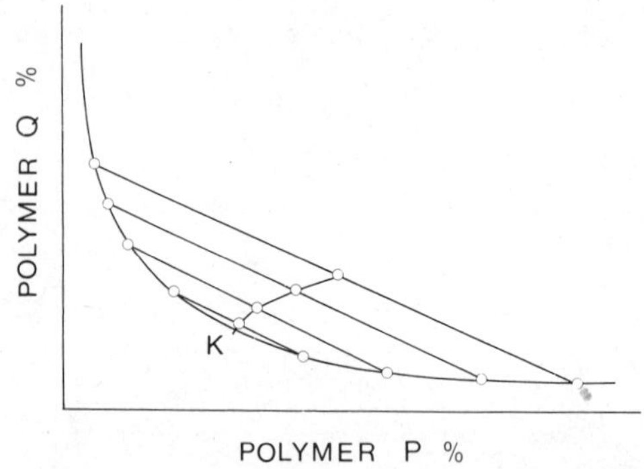

FIGURE 2.10. Phase diagram showing how the critical point K may be obtained by extrapolation. A line is drawn through the middle points of the tie lines and intersects the curved line, that is, the binodial, at the critical point.

2. If tie lines have been determined near the critical point, a line is drawn through the middle points of these and extrapolated to the binodial. The critical point lies, approximately, where this line intersects the binodial, see Figure 2.10.

Phase diagrams of a number of phase systems are found in Chapter 12.

PROPERTIES OF PHASE SYSTEMS

The properties of the phase systems depend on many factors, such as, type of polymer, its molecular weight, temperature, and the presence of low-molecular-weight compounds.

Molecular Weight of the Polymers

In Figure 2.11 the effect of the molecular weight of dextran on the phase diagram of dextran–PEG–H_2O is shown. The higher the molecular weight of the dextran, the lower the concentration required for phase separation. The larger the difference in molecular size between the two polymers, the more asymmetrical is the binodial. Qualitatively, the diagram in Figure 2.11 is in good agreement with what might be expected from theory.

In general, the higher the molecular weight the lower the concentration of polymers needed for phase separation, and the larger the difference in molecular weight between two polymers the more asymmetrical is the binodial.

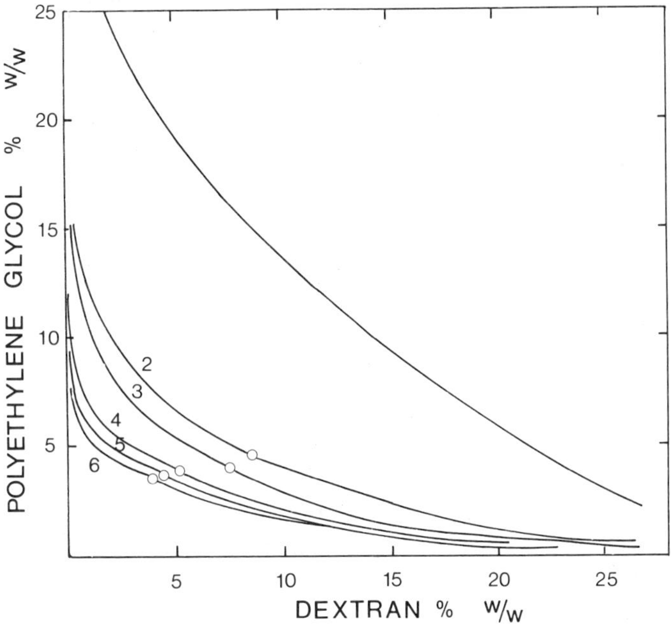

FIGURE 2.11. Binodials and critical points (○) of dextran–polyethylene glycol systems with the same polyethylene glycol (PEG 6000) and with the following fractions of dextran:
1 D5 (M_n = 2,300; M_w = 3,400)
2 D17 (M_n = 23,000; M_w = 30,000)
3 D24 (M_n = 40,500)
4 D37 (M_n = 83,000; M_w = 179,000)
5 D48 (M_n = 180,000; M_w = 460,000)
6 D68 (M_n = 280,000; M_w = 2,200,000)

Hydrophobicity of the Polymer

The binodials of different systems with dextran–hydroxypropyldextran–H_2O is shown in Figure 2.12. The degree of substitution of hydroxylpropyldextran, that is, the number of hydroxypropyl groups per glucose residue is different, but the molecular weight is of about the same order of magnitude for the different polymers. The more the dextran is loaded with hydroxypropyl groups, the less compatible it is with unsubstituted dextran. Also, any pair of the different hydroxypropyldextrans give phase separation. The more different they are with respect to hydroxypropyl content the less compatible they are, and the less is the binodial displaced from the original.

Polyethylene glycol forms two phases with dextran or hydroxypropyldextran

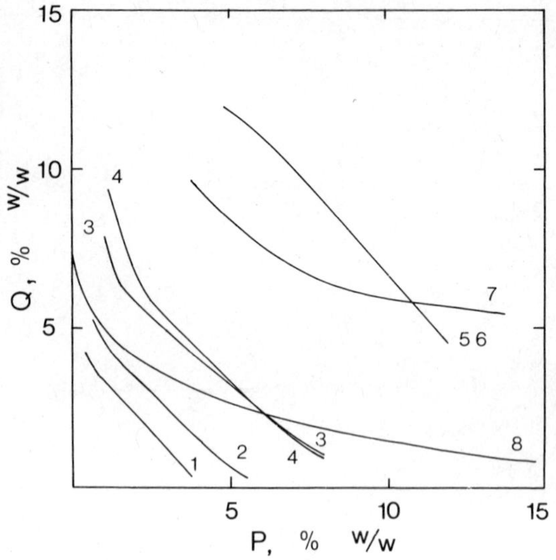

FIGURE 2.12. Binodials of pairs of polymers showing the effect of difference in hydrophobicity on compatibility. P is Dextran 500; Q is hydroxypropyl dextran made from Dextran 500; DS = degree of substitution of HP Dextran in number of hydroxypropyl groups per glucose unit.

1 Dextran – HP Dextran (DS = 1.5)
2 Dextran – HP Dextran (DS = 0.76)
3 Dextran – HP Dextran (DS = 0.39)
4 HP Dextran (DS = 0.39) – HP Dextran (DS = 1.50)
5 HP Dextran (DS = 0.76) – HP Dextran (DS = 1.50)
6 HP Dextran (DS = 0.39) – HP Dextran (DS = 0.76)
7 HP Dextran (DS = 0.39) – PEG 6000
8 Dextran-PEG 6000

A, which has a low hydroxypropyl content, but not with hydroxypropyldextran B or C, which have higher hydroxypropyl contents, Figure 2.12. Thus, the more the number of hydrophobic groups on the dextran backbone the more compatible the polymer is with polyethylene glycol. Usually one can correlate a certain difference in hydrophobicity between two polymers with their tendency toward phase separation.

The Viscosity

Generally it might be expected that, the higher the molecular weight of the polymers used, the higher the viscosity of the phases. This is true in many cases. This increase in viscosity due to greater molecular weight is, however, partly offset by the fact that lower concentrations of polymers with larger molecular weights are required for phase separation. Since the viscosity of a polymer solution is greatly dependent on concentration, this opposing effect sometimes may be considerable. The viscosity of one of the phases thus may be reduced by using a higher molecular weight fraction of the polymer which collects in the other phase. This is illustrated in Table 2.3, where the viscosity of the phases is given for two dextran–PEG systems, one with PEG 8000 and the other with PEG 40000. Using PEG 40000 reduces the viscosity of the lower phase, and the viscosity of the upper phase is increased only marginally compared to what one would expect from the presence of the larger PEG 40000 molecules (Johansson, 1978).

The dextran–polyethylene glycol, dextran sulfate–polyethylene glycol, and salt–polyethylene glycol systems have convenient viscosities. The most viscous systems are those containing methylcellulose and polyvinyl alcohol. The molecules of the latter substances have a linear chain structure and, for a given molecular weight and concentration, such molecules give rise to solutions with higher viscosities than solutions of branched molecules such as dextran. More or less compact molecules are the most suitable as far as viscosity is concerned. A sphere would be the ideal form of molecule since the viscosity of a suspension of spherical particles does not depend on the size of the particles but only on their volume fraction (Einstein, 1909). By increasing the molecular weight, a lower volume fraction would be necessary for phase separation, and, therefore, lower viscosity could be obtained.

TABLE 2.3 Viscosities of the Phases for Different Phase Systems[a]

Phase System Composition, % w/w			Viscosity of Phase Relative to Water	
Dextran 500	PEG 35000	PEG 6000	Upper	Lower
3.9	1.8		5.6	12.8
4.0	2.0		5.7	16.2
5.0	2.5		6.9	26.7
6.0	3.5		9.9	51.1
7.0	4.5		14.6	89.2
5.0		3.5	4.9	15.7
5.2		3.8	3.7	27.9
6.2		4.4	4.0	50.6
7.0		5.0	4.4	95.7

[a]One series with Dextran 500–PEG 35000 and the other with Dextran 500–PEG 6000 (Johansson, 1978).

Temperature

The phase diagram of the phase systems depends on the temperature. However, the effect of temperature is very different for different phase systems depending on the type of polymer used. For example, a system of polyethylene glycol–potassium phosphate–water will form two phases at higher temperatures more easily, that is, a smaller concentration of polymer or salt is needed for phase separation. By contrast, the system dextran–polyethylene glycol–water forms phases more easily at lower temperatures, while the system dextran–methylcellulose is not much affected by temperature.

The dextran–polyethylene glycol system gives two-phase systems from 0°C up to at least 100°C, and this system can, therefore, be used over a wide temperature range. For example, it has been used to study the partition of DNA in its melting range and also for biotechnical conversion of cellulose by cellulases at 50°C (see Chapter 9). Phase diagrams at different temperatures are shown in Figure 2.13.

The Time of Phase Separation

The time required for the phases to separate varies considerably for the different systems. It depends not only on the difference in density between the two phases and their viscosities, but also on the time needed for the small droplets, formed

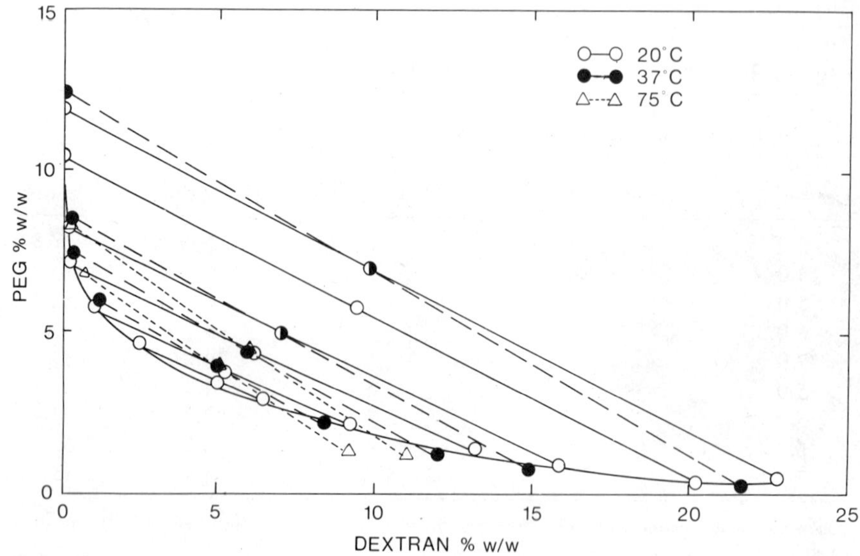

FIGURE 2.13. Phase diagram of the dextran–polyethylene glycol–water system (Dextran 500–PEG 6000) at different temperatures.

during shaking, to coalesce into larger drops. Near the critical point the density difference is small and the settling time, therefore, long. Far away from the critical point, the polymer concentration and, therefore, the viscosity are high, so that the settling time also becomes long for such systems. Thus at intermediate compositions the settling time is the shortest.

In a given phase system the settling time also depends on the volume ratio of the two phases if these have different viscosities. If the more viscous phase is larger in volume than the other phase, the settling time is longer than if the more viscous phase has a volume about equal to or smaller than the other phase. This holds, for example, for the dextran–polyethylene glycol system where the top phase is less viscous than the bottom phase. When the bottom phase is the larger in volume these systems separate more slowly than when the phases have equal volumes or the top phase is larger. Figure 2.14 shows examples of the effect of phase volume ratio on the settling time.

The salt–polyethylene glycol and dextran–polyethylene glycol systems have the shortest settling time (5–30 min) while the dextran–Ficoll or dextran–methylcellulose systems usually have longer settling time (1–6 hr). The settling of a dextran–Ficoll system can, however, be speeded up by addition of polyethylene

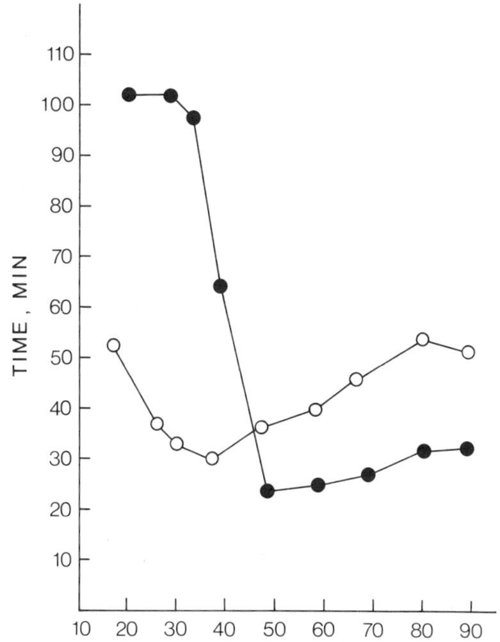

FIGURE 2.14a. Settling time as a function of phase volume ratio. Phase system: Dextran 500–PEG 4000. ○, 6% w/w; ●, 7% w/w of each polymer.

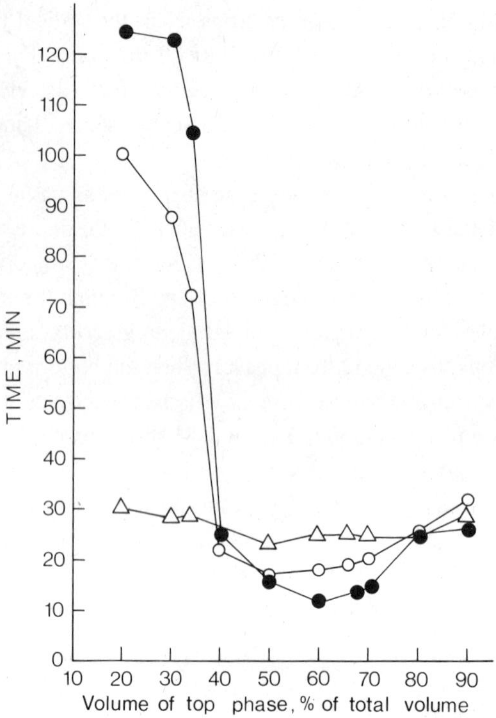

FIGURE 2.14*b*. Phase system: ○, 7% w/w Dextran 500–7% w/w PEG 4000 and 0.01 *M* Na phosphate, pH 6.8; ●, the same plus 0.1 *M* NaCl. △, 9% w/w Dextran 40–5% w/w PEG 6000, 0.01 *M* Na phosphate, pH 6.8. Total volume of phase system, 10 mL. Height of phase system, about 10 cm. The time recorded is that when the interface was 0.2 mL or 0.2 cm from its final position. (Experiment by Tjerneld, unpublished).

glycol. Thus, a system of 6% dextran 500, 5% Ficoll and 2% PEG 6000 settles in 2 to 5 min.

The times given above are those required for the main bulk of the phases to separate, that is, until a horizontal interface has been formed. Usually small drops of one phase remain in the other for a long time after the horizontal interface is apparent. The amount of this emulsification varies considerably for different phase systems and depends also on the volume ratio. For liquid-phase systems in general it is a common experience that there is less emulsification in the smallest phase; in fact, a transparent phase may be obtained if it is very small compared with the other phase (Figure 2.14*c*). Thus, the bottom phase of the system dextran–polyethylene glycol, with a volume less than one-tenth of the total volume, is almost optically clear. The large top phase then contains an

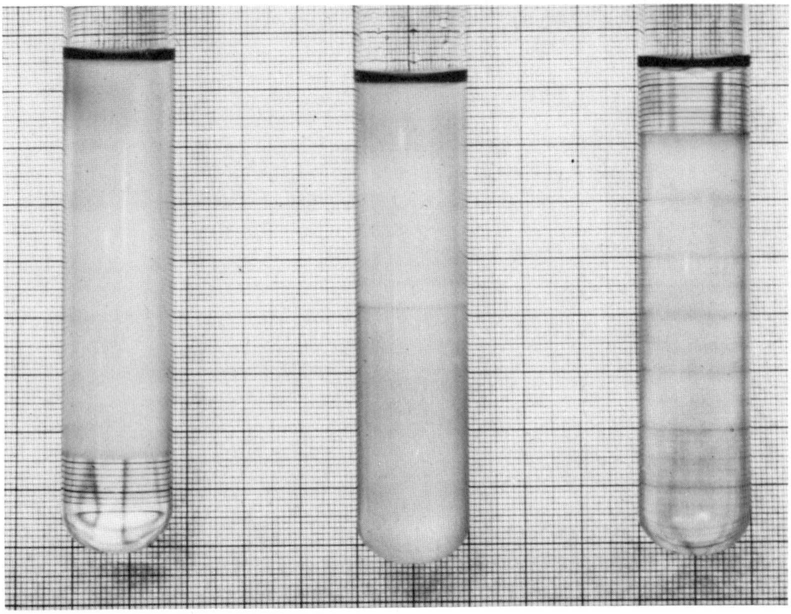

FIGURE 2.14c. Influence of the volume ratio on emulsions in the phases of a dextran–polyethylene glycol system. (Left) The volume of the bottom phase is about one-tenth that of the top phase. (Middle) The two phases have about equal volumes. (Right) The volume of the top phase is about one-tenth that of the bottom phase. The bottom phase (left) and the top phase (right) are clear, while the other phases are turbid.

emulsion. In the same way, the top phase can be made clear by reducing its volume. This effect is at least partly due to the fact that only the smallest phase which, for geometrical reasons, can form drops in the other phase. Emulsification is, however, a rather complicated phenomenon which depends on many factors.

Density of the Phases

Because of the high content of water the densities of the phases are close to 1. Densities of dextran and PEG solutions and the phases of some dextran–PEG systems are given in Tables 2.4–2.6.

In most cases the polymer which concentrates in the lower, denser phase does so at all compositions of the phase diagram. In the dextran–polyethylene glycol systems, for example, the dextran-rich phase is always denser than the polyethylene glycol-rich phase. In the dextran–Ficoll system, however, the dextran-rich phase is the lighter or the heavier phase depending on the composition of the

TABLE 2.4 Density of Aqueous Solutions of Dextran[a]

Polymer Concentration, % w/w	d^{20}_{4}
1.0	1.001
2.0	1.005
3.0	1.009
4.0	1.013
5.0	1.017
6.0	1.022
8.0	1.029
10.0	1.039
15.0	1.057
20.0	1.079

[a]D 48 at 20°C.

system. Close to the critical point the Ficoll-rich phase is denser than the dextran-rich phase while the reverse holds for a system far removed from the critical point. At intermediate compositions the densities of the two phases are equal, and one obtains isopycnic phases. If such a system is allowed to stand after shaking, a large drop of the one phase in the other is formed; it neither sinks nor floats.

The Interfacial Tension

The interfacial tension between the two phases is very small. In a test tube held in vertical position the liquid–liquid interface forms a 90° angle with the tube walls, and no tendency for a curvature close to the glass wall can be seen. The

TABLE 2.5 Density of Aqueous Solutions of Polyethylene Glycol[a]

Polymer Concentration, % w/w	d^{20}_{4}
2.0	1.001
5.0	1.007
8.0	1.012
10.0	1.015
15.0	1.024
20.0	1.033
25.0	1.043

[a]PEG 6000 at 20°C.

TABLE 2.6 Densities of the Phases of Some Dextran–Polyethylene Glycol–H_2O Two-Phase Systems[a]

Phase System, % w/w		Density at 20°C	
Dextran	PEG	Upper Phase	Lower Phase
8	6	1.0127	1.0779
7	4.4	1.0116	1.0594
5	4	1.0114	1.0416
5	3.5	1.0114	1.0326

[a]Rydén and Albertsson (1971).

interfacial tension of the dextran–polyethylene glycol–water system has been determined and is given in Table 2.7 and Figure 2.15. The lower the polymer concentration, that is, the closer the composition is to the critical point, the lower is the interfacial tension. A plot of the logarithm of the interfacial tension against the length of the tie line gives an almost straight line (Rydén and Albertsson, 1971; Schurch et al., 1981). Also, a plot of the logarithm of the interfacial tension against the logarithm of the length of the tie line gives a straight line as shown by Bamberger et al. (1984), who also studied the effect of the ionic composition on the interfacial tension.

The Osmotic Pressure

The osmotic pressure or the tonicity of the phases is rather small due to the high molecular weight of the phase-forming polymers. For the dextran–polyethylene glycol system the tonicity of each phase is of the same order of magnitude

TABLE 2.7 Interfacial Tension Between the Two Phases in Some Dextran–Polyethylene Glycol Systems[a]

Composition, % w/w		Interfacial Tension
Dextran 500	PEG 6000	(dyne/cm)
5	3.5	0.00046
5	4	0.0031
5.2	3.8	0.0021
6	4	0.007
7	4.4	0.020
8	6	0.066

[a]Rydén and Albertsson (1971).

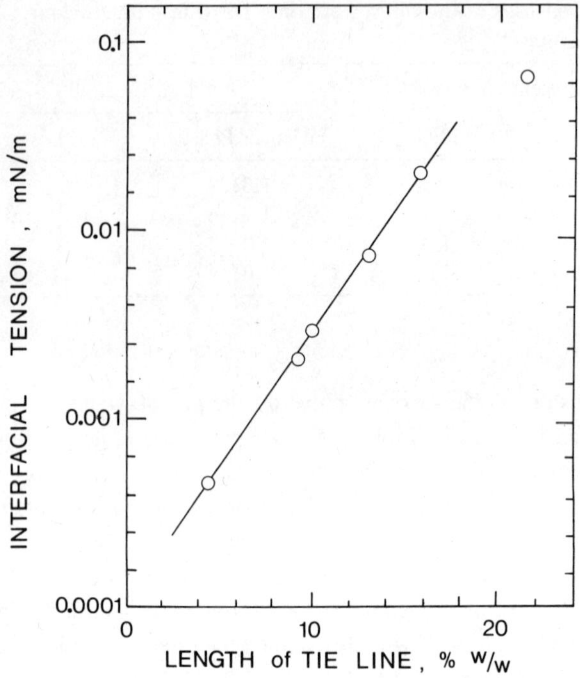

FIGURE 2.15. Interfacial tension of the dextran–polyethylene glycol–water system plotted against the length of the tie line. The longer the tie line the more the system is removed from the critical point (Rydén and Albertsson, 1971).

as 30 m*M* sucrose. To obtain suitable tonicity for cell organelles or cells one can include sucrose or salts in the phase systems. These usually partition fairly equally between the phases. When the phases are in equilibrium the osmotic pressure is the same for both phases.

At high polymer concentrations, the osmotic pressure is independent of the molecular weight. For example, a 10% solution of polyethylene glycol is in equilibrium with a 19% solution of dextran, independent of the dextran molecular weight, Figure 2.16.

Influence of Low-Molecular-Weight Substances on the Phase Systems

Polymer phase systems may be completed by low-molecular-weight substances such as sucrose, sorbitol, or salts in order to obtain a suitable medium for cells and cell organelles. The systems containing nonionic polymers only are hardly affected by the addition of, for example, 0.1–1 *M* sucrose or NaCl. Thus the

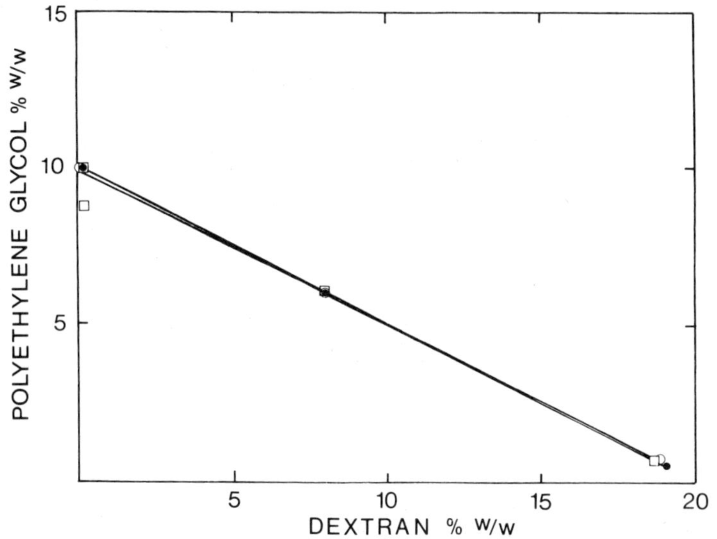

FIGURE 2.16. Three tie lines connecting the two phases of three systems of dextran and polyethylene glycol with different dextran fractions. ○, D68; ●, D48; ×, D37. The polyethylene glycol was PEG 6000 in all systems. Composition of the total system 8% w/w dextran and 6% polyethylene glycol at 20°C.

critical composition of the system D37–PEG 6000–H_2O at 0°C is 5.2% w/w D37 and 3.1% w/w PEG 6000; and the critical composition of the system D37–PEG 6000–10% w/w sucrose is 5.2% w/w dextran, 3.3% w/w PEG 6000, and 10% w/w sucrose. Similarly, the binodial of the system dextran–polyethylene glycol is not much changed by the addition of 0.1 *M* NaCl.

It is only at higher salt concentrations that an effect on the phase system may be observed. Thus, the volume ratio of the phases of the dextran–polyethylene glycol system is constant at salt concentrations up to about 0.8 *M* NaCl and 0.3 *M* phosphate buffer, pH = 7. As may be seen in Figure 12.20, NaCl distributes almost equally between the phases of the dextran–polyethylene glycol system even at rather high salt concentrations. Systems containing a polyelectrolyte are, however, highly dependent on the ionic composition (see below).

Polyelectrolyte Systems—Liquid Ion Exchangers

Phase separation in mixtures containing a polyelectrolyte depends highly on both the ionic strength and the kind of ions present. Thus a mixture of sodium dextran sulfate and polyethylene glycol gives a homogeneous solution if no salt is added, at least below polymer concentrations of about 20%. If NaCl is added up to a concentration of 0.15 *M*, the mixture immediately becomes turbid and a bottom phase containing most of the dextran sulfate separates out while most of the polyethylene glycol stays in the upper phase. As the NaCl concentration is increased, the lower concentration of the polymers is needed for phase separation; see Figure 12.36. Thus, in a mixture containing 5% NaDS68 in 0.15 *M* NaCl at 20°C, at least 8% PEG 6000 is necessary for phase separation, while in the presence of 1 *M* NaCl only 1.5% PEG 6000 is necessary for phase separation.

The concentration of polyethylene glycol which is necessary for phase separation with a certain amount of dextran sulfate depends also on the kind of salt added. The tendency to favor phase separation in a dextran sulfate–polyethylene glycol mixture thus increases with the following series of cations: lithium, sodium, cesium, and potassium, when these are added as chlorides. The phase separation also depends on the kind of anions: the tendency to favor phase separation decreases with the following series of anions: nitrate, chloride, and sulfate as sodium salts.

If enough salt is added, polyethylene glycol may even be omitted; a system of dextran sulfate–salt–water is then obtained. Thus, a two-phase system is obtained with 20% dextran sulfate (NaDS70) in 1 *M* NaCl in the cold. The bottom phase, which is clear and liquid, contains most of the dextran sulfate. At room temperature no phase separation is obtained even in 5 *M* NaCl. However, KCl precipitates dextran sulfate at concentrations above about 0.3 *M* at room temperature; the dextran sulfate phase is in this case turbid and highly viscous. The same consistency of the bottom phase is obtained when CsCl is used. No phase separation is obtained with dextran sulfate and LiCl or NH_4Cl, either in the cold or at room temperature. Dextran sulfate is precipitated by $BaCl_2$ but not by $CaCl_2$.

When polyelectrolytes with other acid groups are mixed with various salts, entirely different results are obtained. Thus, the sodium salt of carboxymethyldextran does not phase separate with any of the alkali halides but it is precipitated by $BaCl_2$. In systems with polyethylene glycol and Na carboxymethyldextran the polymer concentration necessary for phase separation depends on the kind of salt present.

An interesting observation is that a mixture of Na dextran sulfate and Na carboxymethyldextran forms a phase system even without the addition of electrolytes. In this case the bottom phase contains most of the Na dextran sulfate and the top phase most of the Na carboxymethyldextran. Similar systems are Na dextran sulfate–Na carboxymethylcellulose and Na carboxymethyldextran–Na carboxymethylcellulose. In these cases we obtain two liquid ion exchangers in equilibrium. Although these systems are formed without the addition of salt, their phase separation still depends highly on various salts added. Phase separation is, for example, favored at higher concentrations of salts added, as with the Na dextran sulfate–polyethylene glycol systems.

The results obtained on mixing solutions of Na dextran sulfate and the chloride form of diethylaminoethyldextran at various NaCl concentrations are particularly interesting. Without, or at low concentrations of, added NaCl a more-or-less solid precipitate or a highly viscous bottom phase, which contains both polymers, is obtained. When more salt is added the mixture becomes clear. These results are to be expected from previous experimental results on phase separation between gelatin and gum arabic when these carry opposite net charges (Bungenberg de Jong, 1949) and also with the theory for such systems. The fact that the mixture becomes homogeneous when NaCl is added is thus explained as a reduction of the attractive coloumbic forces between the two oppositely charged macromolecules.

However, in the case of Na dextran sulfate–diethylaminoethyldextran, when the addition of NaCl is increased, the mixture becomes turbid again. Two phases are formed, and the two polymers go to different phases. Two liquid ion exchangers in equilibrium, one cationic and one anionic, are formed. This phase separation, which occurs at high NaCl concentrations, is probably partly due to the presence of the diethylaminoethyl groups on the one polymer. At high salt concentrations, when the attraction between the two oppositely charged macromolecules is considerably depressed, the presence of the nonpolar ethyl groups on the diethylaminoethyldextran becomes of more importance and favors the incompatibility of the two polymers. The solubility of dextran sulfate, which is reduced at high salt concentrations, also favors this phase separation.

Apparently polyelectrolytes allow the construction of a large number of liquid-phase systems with highly diversified properties, which in turn increases the freedom of choosing a suitable phase system for fractionation purpose. It must be stressed, however, that polyelectrolytes usually interact more strongly with biological materials than nonionic polymers. Thus, there is a greater chance of obtaining irreversible complex formation between a protein or a cell particle and the polymer when a polyelectrolyte is used instead of a nonionic polymer. In general, though, these interactions should be depressed at high salt concentrations.

Influence of the Polydispersity of the Polymers

As mentioned before, the fact that the polymers are polydisperse means that there is not always a sharp change from a one-phase system to a two-phase system, such as indicated by a binodial. Figure 2.17 shows the phase diagram of the system dextran–methylcellulose. The curve represents the binodial obtained by

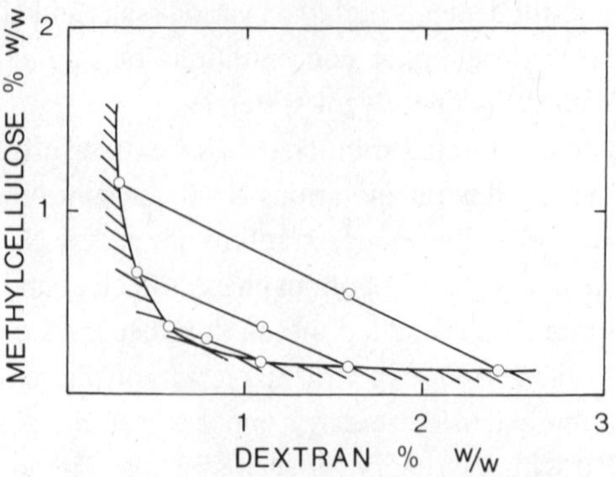

FIGURE 2.17. Influence of polydispersity of the polymers in the dextran–methylcellulose system D68–MC4000 at 20°C. The binodial of this diagram is drawn through the experimental points (○) representing the composition of the phases. The shaded area represents a region where a turbid mixture is obtained (see text).

drawing a line through the points representing the phase compositions. The shaded areas indicate regions where the mixtures are more or less turbid, although they are represented by points outside the binodial. As may be seen in Figure 2.17, the shaded regions are broader near the coordinate axes but absent near the critical point. At this point there is a sharp change from a one-phase to a two-phase system, also in systems with polydisperse polymers. This is to be expected, since at the critical composition when the two phases theoretically become equal, no polymer fractionation takes place. Figure 2.18 shows the system dextran–polyethylene glycol. In this case the polymers are not so polydisperse as in the dextran–methylcellulose system depicted in Figure 2.17. Therefore, the shaded region is almost absent except in the range when dextran has a low concentration. This phase system may therefore, practically, be regarded as a ternary system.

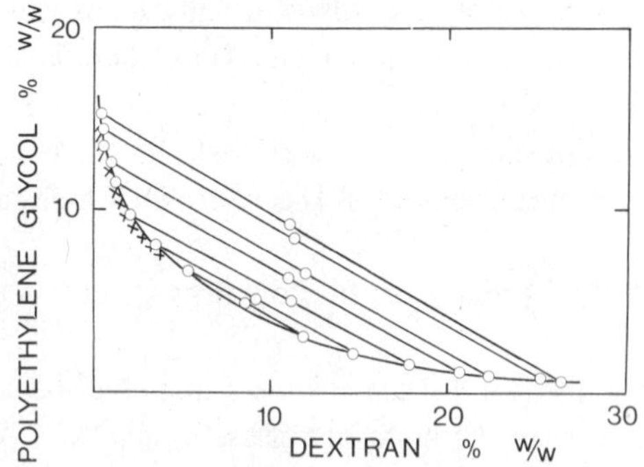

FIGURE 2.18. The same as Figure 2.17 for the dextran–polyethylene glycol system D17–PEG 6000.

REFERENCES

Albertsson, P.-Å. (1958). *Biochim. Biophys. Acta,* **27,** 378.

Albertsson, P.-Å. (1971). *Partition of Cell Particles and Macromolecules,* Wiley, New York.

Bamberger, S., Seaman, G.V.F., Sharp, K.A., and Brooks, D.E. (1984). *J. Colloid Interf. Sci.,* **99,** 194–200.

Beijerinck, M.W. (1896). *Zbl. Bakt.,* **2,** 627, 698.

Beijerinck, M.W. (1910). *Kolloid-Z.,* **7,** 16.

Bungenberg de Jong, H.G. (1949). In H.R. Kruyt, Ed., *Colloid Science,* Vol. II, Elsevier, Amsterdam.

Brooks, D.E., Sharp, K.A., Bamberger, S., Tamblyn, C.H., Seaman, G.V.F., and Walter, H. (1984). *J. Colloid Interf. Sci.,* **102,** 1–13.

Dobry, A. (1938). *J. Chim. Phys.,* **35,** 387.

Dobry, A. (1939). *J. Chim. Phys.,* **36,** 102.

Dobry, A. (1948). *Bull. Soc. Chim. Belg.,* **57,** 280.

Dobry, A., and Boyer-Kawenoki, F. (1947). *J. Polymer Sci.,* **2,** 90.

Edmond, E., and Ogston, A.G. (1968). *Biochem. J.,* **109,** 569.

Einstein, A. (1909). *Ann. Physik,* **19,** 289.

Flory, P.J. (1953). *Principles of Polymer Chemistry,* Cornell Univ. Press, Ithaca, NY.

Huggins, M.L. (1958). *Physical Chemistry of High Polymers,* Wiley, New York.

Johansson, G. (1978). *J. Chromatogr.,* **150,** 63–71.

Koningsveld, R. (1968). *Adv. Colloid. Interface Sc.,* **2,** 151.

Morawetz, H. (1975). *Macromolecules in Solution,* Wiley-Interscience, New York.

Rydén, J., and Albertsson, P.-Å. (1971). *J. Colloid Interf. Sci.,* **37**(1), 219–222.

Schürch, S., Gerson, D.F., and McIver, D.J.L. (1981). *Biochim. Biophys. Acta,* **640,** 557–571.

Scott, R.L. (1949). *J. Chem. Phys.,* **17,** 279.

Tompa, H. (1956). *Polymer Solutions,* Butterworths, London.

3 THEORY OF PARTITION

THE DISTRIBUTION OF PARTICLES DUE TO BROWNIAN MOTION AND THE INTERFACIAL FORCES

Two tendencies oppose each other in determining the distribution of particles in a two-phase system. One is the thermal motion of the particles—the so-called Brownian motion—which tends to distribute the particles randomly throughout the entire space of the phase system. The other is the forces acting upon the particles at the interface tending to distribute them unevenly so that they collect in the phase in which they have their lowest energy. If the work needed to move a particle from phase 1 to phase 2 against these forces is ΔE, then according to the theory of Brownian motion the following relation should hold:

$$\frac{C_2}{C_1} = e^{-\Delta E/kT} \tag{1}$$

where C_1 and C_2 are the concentrations of particles in phases 1 and 2, k is the Boltzmann constant, and T is the temperature.

In the following, ΔE is calculated for a spherical particle of radius R assumed to have a perfectly uniform interface separating it from the surrounding medium, and gravitational forces are neglected. Interfaces are characterized by possessing a certain amount of free surface energy per unit surface area. This is called the interfacial tension and may be expressed in ergs/cm^2; an interface with an interfacial tension γ ergs/cm^2 and an area of A cm^2 therefore has a free energy of $A\gamma$ ergs. This energy is stored in the interface. It has to be supplied when the interface is formed and is released when the interface disappears. The interfacial free energy of the particle in a liquid two-phase system can be considered as a function of the position of the particle in the phase system (Fig. 3.1). There are three different kinds of interfaces:

1. The liquid–liquid interface between the two phases of the system. Let this interfacial tension be γ_{12} ergs/cm^2.
2. The particle–liquid interface between the particle and the top phase. Let this interfacial tension be γ_{P1} ergs/cm^2.
3. The particle–liquid interface between the particle and the bottom phase. Let this interfacial tension be γ_{P2} ergs/cm^2.

When the particle is entirely in the lower phase, as at position *a* in Figure 3.1, its interfacial free energy G^s is

$$G^s{}_2 = 4\pi R^2 \gamma_{P2} \tag{2}$$

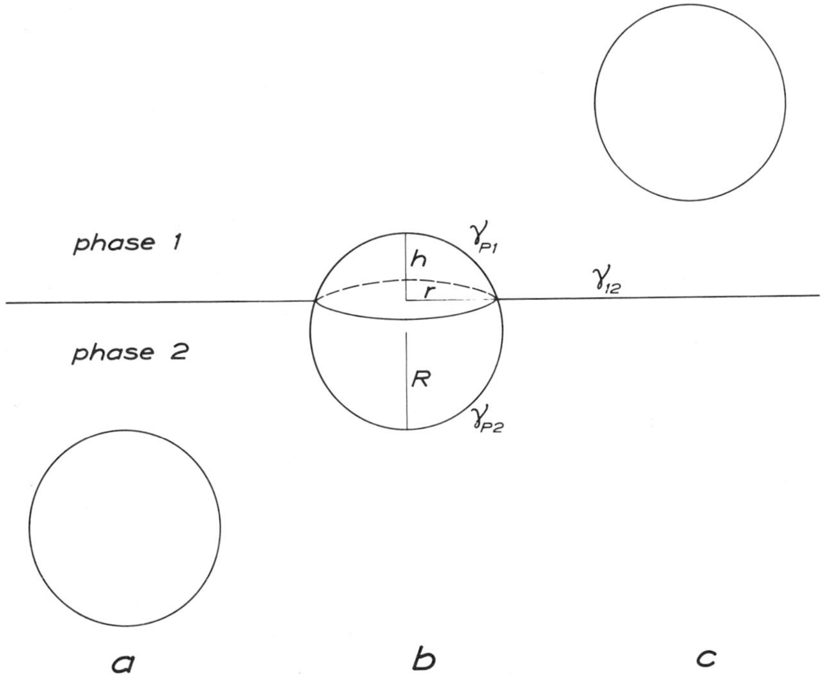

FIGURE 3.1. A spherical particle at three different positions in a liquid two-phase system. γ_{12} is the interfacial tension between the two phases; γ_{P1} between the particle and phase 1; and γ_{P2} between the particle and phase 2.

Let us suppose that the particle is shifted perpendicularly upward. When the particle has moved through the distance h partly into the top phase, as at position b in Figure 3.1, then a new interface between the particle and the top phase of area $2\pi Rh$ has been formed. This increases the free surface energy of the particle by $2\pi Rh\gamma_{P1}$ ergs. Simultaneously, the area of the interface between the particle and the bottom phase has decreased by the same extent resulting in a loss of $2\pi Rh\gamma_{P2}$ ergs in free surface energy of the particle. In addition, an area of πr^2 of the liquid–liquid interface between the top phase and the bottom phase has disappeared (r is the radius of the circle of contact between the particle and the liquid–liquid interface). Hence, there is a further decrease in free surface energy of $\pi r^2\gamma_{12}$ ergs. The potential energy of the particle caused by interfacial forces at position b will therefore be

$$G^s{}_h = 4\pi R^2\gamma_{P2} + 2\pi Rh\gamma_{P1} - 2\pi Rh\gamma_{P2} - \pi r^2\gamma_{12} \tag{3}$$

but

$$r^2 = h(2R - h)$$

hence

$$G^s{}_h = \pi[2Rh(\gamma_{P1} - \gamma_{P2} - \gamma_{12}) + h^2\gamma_{12}] + 4\pi R^2\gamma_{P2} \tag{4}$$

When h exceeds $2R$, as at position c in Figure 3.1, that is, when the particle is entirely in the top phase, the interfacial free energy of the particle is

$$G^s{}_1 = 4\pi R^2\gamma_{P1} \tag{5}$$

Differentiation of Eq. (4) gives

$$\frac{dG^s}{dh} = \pi[2R(\gamma_{P1} - \gamma_{P2} - \gamma_{12}) + 2h\gamma_{12}] \tag{6}$$

$$\frac{d^2G^s}{dh^2} = 2\pi\gamma_{12} \tag{7}$$

Analysis of Eqs. (6) and (7) shows that G^s will have a minimum at the h value

$$h_{\min} = R\left(1 - \frac{\gamma_{P1} - \gamma_{P2}}{\gamma_{12}}\right) \tag{8}$$

Since $0 < h_{\min} < 2R$ it follows that this minimum value of G^s only occurs when the value of the term on the right hand side of Eq. (8) lies between 0 and $2R$. G^s will therefore only have a minimum value when the following condition is fulfilled:

$$\left| \frac{\gamma_{P1} - \gamma_{P2}}{\gamma_{12}} \right| < 1 \tag{9}$$

The particle will then lie at its most stable position at the interface.

However, if the following relation holds

$$\left| \frac{\gamma_{P1} - \gamma_{P2}}{\gamma_{12}} \right| \geqq 1 \tag{10}$$

then G^s will have no minimum and the particle will have its most stable position in one of the phases. These two latter relations were first deduced by Des Coudres (1898).

Further analysis of Eq. (4) shows that the graph relating G^s to h will have different shapes depending on the relative values of the different surface tensions. The five main cases are shown in Figure 3.2. In the first and second cases (Figs. 3.2*a* and *b*) the relation (10) is assumed to hold and G^s has no minimum for a unique value of h between $h = 0$ and $h = 2R$. If $\gamma_{P1} > \gamma_{P2}$ (Fig. 3.2*a*), the particle will have its lowest energy in the bottom phase. The free surface energy difference resulting when the particle is moved from phase 2 to phase 1 is obtained by subtracting Eq. (2) from (5) and is

$$G^s_1 - G^s_2 = 4\pi R^2(\gamma_{P1} - \gamma_{P2}) \tag{11}$$

Similarly, if $\gamma_{P1} < \gamma_{P2}$ (Fig. 3.2*b*) the particle will have its lowest energy in the top phase. The free surface energy difference between the two phases is then obtained by subtracting Eq. (5) from (2) and is

$$G^s_2 - G^s_1 = 4\pi R^2(\gamma_{P2} - \gamma_{P1}) \tag{12}$$

In the remaining cases (Figs. 3.2*c*, *d*, and *e*), condition (9) is assumed to be fulfilled, and the particle will have its lowest energy at the interface. Depending on whether γ_{P1} is larger, smaller, or equal to γ_{P2} the three cases *c*, *d*, and *e*, respectively, are obtained. The energy difference of the particle between the two phases will be that expressed by Eqs. (11) and (12). The energy difference of the particle between the top phase and the G^s minimum position at the interface

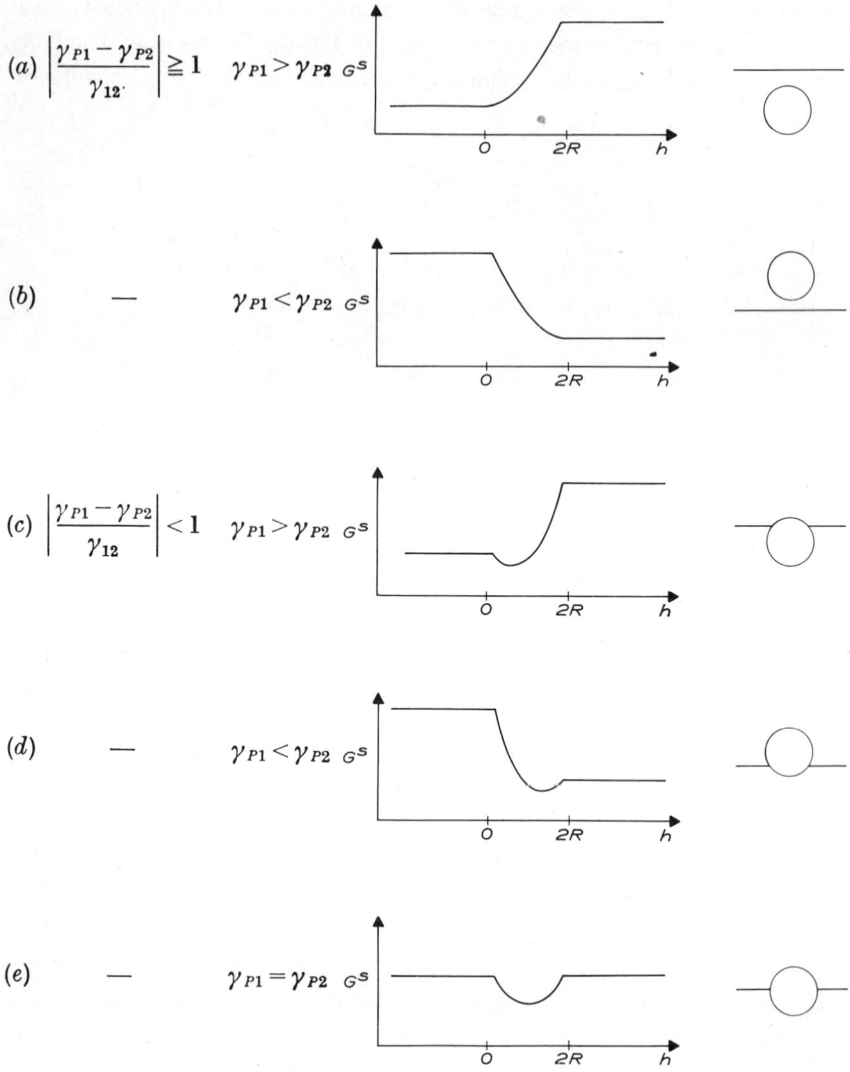

FIGURE 3.2. The potential energy G^s of a spherical particle as a function of h (see Fig. 3.1) at different mutual relations between γ_{12}, γ_{P1}, and γ_{P2}. When $h < 0$ the particle is entirely in the bottom phase; when $0 < h < 2R$ the particle passes through the interface; and when $h > 2R$ the particle is entirely in the top phase. All the curved lines are paraboloid. To the right, the G^s minimum position of the particle is shown.

is obtained by substituting the h_{min} value of Eq. (8) into Eq. (4) and subtracting (4) from (5) yielding

$$G^s{}_1 - G^s{}_{min} = \frac{\pi R^2(\gamma_{P2} - \gamma_{P1} - \gamma_{12})^2}{\gamma_{12}} \tag{13}$$

Similarly the energy difference between the bottom phase and the G^s minimum position at the interface is

$$G^s{}_2 - G^s{}_{min} = \frac{\pi R^2(\gamma_{P1} - \gamma_{P2} - \gamma_{12})^2}{\gamma_{12}} \tag{14}$$

When $\gamma_{P1} = \gamma_{P2}$ (Fig. 3*e*), the energy difference between the top (or the bottom phase) and the G^s minimum position at the interface is

$$G^s{}_{1 \text{ or } 2} - G^s{}_{min} = \pi R^s \gamma_{12} \tag{15}$$

The distribution of a number of similar particles between the two phases is now obtained by substituting the difference in energy of one particle, as calculated above, into Eq. (1). Thus, in the case of Figure 3.2*a* the distribution will be

$$\frac{C_1}{C_2} = K = \exp \frac{-4\pi R^2(\gamma_{P1} - \gamma_{P2})}{kT} \tag{16}$$

where K is the partition coefficient; more generally if a particle is not spherical,

$$K = \exp \frac{-A(\gamma_{P1} - \gamma_{P2})}{kT} \tag{17}$$

where A is the surface area. From this important equation we may infer that the partition coefficient in a given phase system is highly dependent both on the surface area of the particle and its surface properties. For particles with identical surface properties but with different surface areas the partition coefficient depends only on the surface area, $\gamma_{P1} - \gamma_{P2}$ being a constant; and

$$K = \exp \frac{A\lambda}{kT} \tag{18}$$

where $\lambda = -(\gamma_{P1} - \gamma_{P2})$. This is of a similar form as the expression derived by Brønsted (1931)

$$K = \exp \frac{M\lambda}{kT} \tag{19}$$

(where M = molecular weight). It was deduced for so-called isochemical substances, that is substances which differ in molecular weight only. As pointed out by Brønsted, M should be replaced by the surface area for large and more or less spherical molecules.

As may be seen from Eq. (17), the partition coefficient tends to become larger ($\gamma_{P1} < \gamma_{P2}$) or smaller ($\gamma_{P1} > \gamma_{P2}$) as the surface area of the particles increases. A more unilateral distribution is therefore obtained and two kinds of particles will be more completely separated the larger their particle size, if for one species $\gamma_{P1} > \gamma_{P2}$ and for the other $\gamma_{P1} < \gamma_{P2}$. From a theoretical point of view two-phase systems, therefore, seem to be very suitable for the fractionation of large particle-weight substances.

The distribution of the particles between one of the phases and the G^s minimum position at the interface is obtained by putting the energy difference of the particle as expressed by Eq. (13) into Eq. (1). Thus

$$\frac{C_{\min}}{C_1} = \exp \frac{\pi R^2(\gamma_{P2} - \gamma_{P1} - \gamma_{12})^2}{\gamma_{12}kT} \tag{20}$$

where $C_{\min}$ is the number of particles per cubic centimeter in a thin layer at the G^s minimum position and C_1 is the number of particles per milliliter of top phase.

In the special case when $\gamma_{P1} = \gamma_{P2}$ the adsorption will follow the equation

$$\frac{a_{\min}}{C} = \exp \frac{\pi R^2 \gamma_{12}}{kT} \tag{21}$$

where C is the concentration of particles in the top and bottom phases. All particles located between $h = 0$ and $h = 2R$ (Fig. 3.2) should be considered as adsorbed at the interface and their number per square centimeter (a) is obtained by summing up all particles present at different positions between $h = 0$ and $h = 2R$.

The following important conclusions can be drawn from Eqs. (20) and (21).

1. For different particles with similar interfacial tensions in a given phase system, that is when γ_{P1}, γ_{P2} and γ_{12} are constant, the tendency for adsorption at the interface will be greater the larger the radius of the particles.

2. For any particle having the same partition coefficient in different phase systems (that is, when R, γ_{P1} and γ_{P2} are constant, see Eq. (16)), the tendency for adsorption at the interface will be greater the larger the value of γ_{12}.

The special case of Eq. (21) when $\gamma_{P1} = \gamma_{P2}$ (Fig. 3.2*e*) is particularly interesting, since the distribution between the phases and the interface is determined only by the radius of the particle and the liquid–liquid interfacial tension, γ_{12} of the phase system. Thus, any particle, irrespective of its nature, having the same affinity for both phases of a phase system (that is, its partition coefficient = 1) must be more or less concentrated at the interface. This is probably one of the reasons why it is often so difficult to distribute large molecules such as proteins with a partition coefficient around unity in low molecular weight phase systems. Both the size of the protein molecules and the γ_{12} value of these systems are so large that the protein is almost entirely adsorbed at the interface when the system is such that the protein has about the same affinity for both phases. As an example, we may calculate the quotient a_{min}/c in Eq. (21) for a particle of radius 3 mμ and a phase system with an interfacial tension of 1 erg/cm^2 at 20°C. These values correspond in order of magnitude to a protein molecule with a molecular weight of about 100,000 and the interfacial tension of a water–butanol system. The chosen values will show that a_{min}/c is about 10^3; for larger particles it will be even higher.

Another interesting result of Eq. (21) is that it would be possible to estimate the interfacial tension of a phase system by studying the adsorption of a number of particles all having a partition coefficient of unity but differing in radius.

THE INFLUENCE OF GRAVITY

When the particle is denser than the phases, the particle is pulled down by a force F (see Figure 3.3) which depends upon the size and density of the particle and the density of the phases. This force is counteracted by an upward force depending on the vertical component of γ_{12} and the circumference of the circle of contact between the particle and the liquid–liquid interface. When the latter force can overcome the gravitational force the particle remains at the interface. Equations that describe the relationship fulfilling this condition have been deduced for particles of different shapes in connection with flotation studies; see for example the books of Gaudin (1957) and Taggart (1951).

We shall now consider the magnitude of the forces involved. Suppose that a spherical particle of radius one micron and a density of 1.3 is attached at the

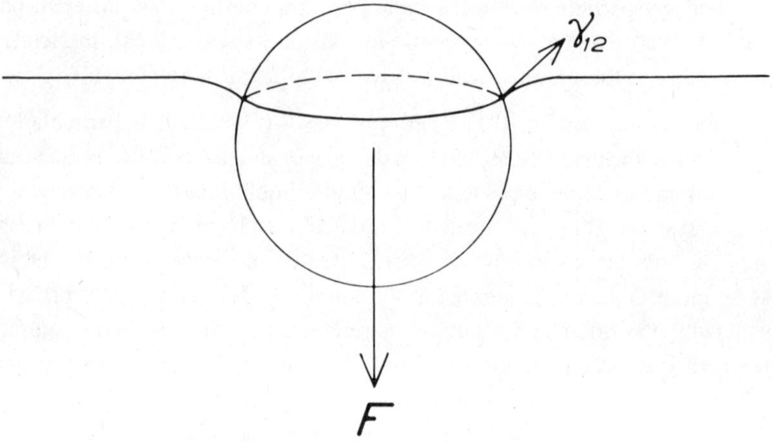

FIGURE 3.3. A particle at an interface under the gravitational force F, which is counteracted by an upward force equal to the vertical component γ_{12} times the circumference of the circle of contact between the particle and the liquid–liquid interface.

interface of a system having an interfacial tension of 0.001 dyne/cm and that the circle of contact between the particle and the interface equals the diameter of the particle. For simplicity, the densities of the phases are considered to be unity. The gravitational force F acting downwards on the particle will then be

$$F = g \cdot \tfrac{4}{3}\pi \times 10^{-12} \times 0.3 \text{ dynes}$$

where $g(= 981 \text{ cm/sec}^2)$ is the gravitational constant. The upward force on the particle is

$$0.001 \times 2\pi \times 10^{-4} \text{ dynes}$$

F will be of the order of 10^{-9} dynes while the upward force will be of the order of 10^{-6} dynes.

For smaller particles the interfacial force will be still larger compared with the gravitational force and we may therefore conclude, that for the kind of particles studied here, gravitation has in most cases only a small influence on the distribution of a single particle. If, however, an interface becomes overloaded with particles, so that several layers of particles are collected on it, then clumps of particles may fall from the interface due to gravitation.

THE INFLUENCE OF SHAKING

Theoretically, equilibrium may be obtained by adding the particles to a phase system and then allowing them to distribute by their own thermal motion. In practice this would take too long since the diffusion of the particles is very slow. To achieve equilibrium quickly one shakes the phase system containing the particles, so that a close contact between the two phases and the particles is obtained within a short time. During the shaking the particles may be subjected to various mechanical forces, for example, centrifugal and frictional forces supplying the particles with kinetic energy in addition to that of the thermal motion. These effects are probably proportionally much greater for large particles such as whole cells.

It should also be pointed out that, while the distribution of the particles takes place during shaking, the experimental determination of the particle distribution is done after the two phases have separated and formed the final horizontal interface. The partition coefficient of a substance has probably the same value during shaking and after separation. However, the magnitude of adsorption at the large interface formed during shaking may be different from that occurring at the final horizontal interface. At present there seems to be no method available for determining the total interfacial area (and hence the value of a in Eqs. (20) and (21)) formed during shaking.

Nevertheless as a rough guide we may use Eqs. (20) and (21) if we wish to study the factors determining the adsorption at the final horizontal interface when condition (9) is fulfilled.

THE RELATION BETWEEN THE PARTITION COEFFICIENT AND THE ACTIVITY

When two phases are in equilibrium the chemical potential of the ith component is the same in both phases

$$\mu_{i,1} = \mu_{i,2} \tag{22}$$

where 1 and 2 denote the top and bottom phases, respectively. Choosing the same standard state $\mu^0{}_i$ for both phases we may write

$$\mu^0{}_i + RT \ln f_{i,1}C_{i,1} = \mu^0{}_i + RT \ln f_{i,2}C_{i,2} \tag{23}$$

where C is the molar concentration and f is the activity coefficient. Thus

$$\frac{C_{i,1}}{C_{i,2}} = \frac{f_{i,2}}{f_{i,1}} = K \tag{24}$$

where K is the partition coefficient. With the standard state chosen here the partition coefficient is thus inversely proportional to the ratio between the activity coefficients of the substance in the top and bottom phases.

THE DONNAN EFFECT—THE PARTITION POTENTIAL

When partitioned macromolecules or particles carry a net electrical charge and they distribute in a manner different from other ions, this can create an electrical potential between the phases. This is the well known Donnan effect, which, for example, plays a part when a solution of charged macromolecules is separated from a salt solution by a membrane permeable to the small ions but not to the macromolecules. The Donnan effect has thus to be considered in measurements of the osmotic pressure of proteins (Donnan, 1911). It also enters in ultracentrifugation (Svedberg and Pedersen, 1940; Tiselius, 1932) where charged macromolecules and small ions move with different velocities.

The condition for equilibrium between two phases involving charged molecules is that the so-called electrochemical potential (φ) of thc ith component is the same in both phases.

This potential may be written as

$$\varphi_i = \mu_i + FzU \tag{25}$$

where F is the Faraday constant, z the net electronic charge of the molecules and V the electrical potential of the phase. Thus, at equilibrium

$$\mu^0{}_i + RT \ln f_{i,1}C_{i,1} + Fz_iU_1 = \mu^0{}_i + RT \ln f_{i,2}C_{i,2} + Fz_iU_2 \tag{26}$$

By solving Eq. (26) we obtain

$$\ln K^*{}_i = \ln \frac{C_{i,1}}{C_{i,2}} = \ln \frac{f_{i,2}}{f_{i,1}} + \frac{Fz_i(U_2 - U_1)}{RT} \tag{27}$$

or

$$\ln K^*_i = \ln K_i + \frac{Fz_i(U_2 - U_1)}{RT} \tag{28}$$

where K is the partition coefficient in the absence and K^* the partition coefficient in the presence of an electrical potential difference.

Let us now consider the distribution of a salt $A_{Z^-} \cdot B_{Z^+}$ which dissociates into the ion A^{Z+} carrying Z^+ positive charges and the ion B^{Z-} carrying Z^- negative charges. (Z^+ and Z^- are both positive numbers). For the positive ion, according to Eq. (28),

$$\ln K^*_{A^{Z+}} = \ln K_{A^{Z+}} + \frac{FZ^+(U_2 - U_1)}{RT} \tag{29}$$

and for the negative ion

$$\ln K^*_{B^{Z-}} = \ln K_{B^{Z-}} - \frac{FZ^-}{RT}(U_2 - U_1) \tag{30}$$

Electroneutrality in the two phases requires that

$$Z^+C_{A^{Z+},1} = Z^-C_{B^{Z-},1} \tag{31}$$

and

$$Z^+C_{A^{Z+},2} = Z^-C_{B^{Z-},2} \tag{32}$$

which means that

$$\frac{C_{A^{Z+},1}}{C_{A^{Z+},2}} = \frac{C_{B^{Z-},1}}{C_{B^{Z-},2}}$$

that is,

$$K^*_{A^{Z+}} = K^*_{B^{Z-}} = K^*_{A^{Z-}B^{Z+}} \tag{33}$$

Combination of Eqs. (29), (30), and (33) gives

$$\ln K^*_{A^{Z-}B^{Z+}} = \frac{\ln [(K_{A^{Z+}})^{Z^-}(K_{B^{Z-}})^{Z^+}]}{Z^+ + Z^-} \tag{34}$$

or the partition coefficient of the salt

$$K^*_{A^{Z-}B^{Z+}} = [(K_{A^{Z+}})^{Z^-} \cdot (K_{B^{Z-}})^{Z^+}]^{1/(Z^+ + Z^-)} \tag{35}$$

For example, for NaCl,

$$K^*_{NaCl} = [K_{Na^+} \cdot K_{Cl^-}]^{1/2} \tag{36}$$

for Na_2SO_4,

$$K^*_{Na_2SO_4} = [K^2_{Na^+} \cdot K_{SO_4^{2-}}]^{1/3} \tag{37}$$

and for a polyelectrolyte Na_ZP,

$$K^*_{Na_ZP} = [K^{Z^-}{}_{Na^+} \cdot K_{P^{Z-}}]^{1/(Z^- + 1)} \tag{38}$$

The electrical potential difference $U_2 - U_1$ between the two phases is

$$U_2 - U_1 = \frac{RT}{Z^+F} \ln \frac{K^*}{K_{A^{Z+}}} \tag{39}$$

or

$$U_1 - U_2 = \frac{RT}{Z^-F} \ln \frac{K^*}{K_{B^{Z-}}} \tag{40}$$

Combination of Eqs. (39) and (40) gives

$$U_2 - U_1 = \frac{RT}{(Z^+ + Z^-)F} \ln \frac{K_{B^{Z-}}}{K_{A^{Z+}}} \tag{41}$$

An electrical potential difference will therefore be created between the phases if the two ions of a salt have different affinities for the two phases, that is, $K_{A^{Z+}}$ is different from $K_{B^{Z-}}$, or the ratio between the activity coefficients in the two phases f_1/f_2 is different for two ions (Eq. 24). Also the electrical potential difference becomes smaller the larger the sum of the number of charges on the ions of the salt (Eq. (41)). This electrical potential difference we may call a partition potential. It is a kind of a Donnan potential.

We will now consider some special cases:

1. A polyelectrolyte, for example a protein or a nucleic acid, together with an excess of a salt. We may denote the polyelectrolyte A_ZP and assume that it dissociates into ZA^+ ions and the P^{Z-} ion. The salt is AB and dissociates into the A^+ and the B^- ions.

Equilibrium between the phases requires for the polyelectrolyte (compare Eq. 30)

$$\ln K^*_{P^{Z-}} = \ln K_{P^{Z-}} - \frac{FZ^-}{RT}(U_2 - U_1) \tag{42}$$

for the A^+ ion

$$\ln K^*_{A^+} = \ln K_{A^+} + \frac{F}{RT}(U_2 - U_1) \tag{43}$$

and for the B^- ion

$$\ln K^*_{B^-} = \ln K_{B^-} - \frac{F}{RT}(U_2 - U_1) \tag{44}$$

Electroneutrality in each phase requires

$$C_{A^+,1} = C_{B^-,1} + Z^-C_{P^{Z-},1} \tag{45}$$

$$C_{A^+,2} = C_{B^-,2} + Z^-C_{P^{Z-},2} \tag{46}$$

However, if the salt AB is in such an excess that $C_{AB} \gg Z^-C_{P^{Z-}}$ we may neglect the $Z^-C_{P^{Z-}}$ terms and put $C_{A^+,1} = C_{B^-,1}$ and $C_{A^+,2} = C_{B^-,2}$. The potential between the phases will then be determined by the differential distribution of the A and B ions such that, according to Eq. (41),

$$U_2 - U_1 = \frac{RT}{2F}\ln\frac{K_{B^-}}{K_{A^+}} \tag{47}$$

By putting this into Eq. (42) we obtain

$$\ln K^*_{p^{z-}} = \ln K_{p^{z-}} + \frac{Z^-}{2} \ln \frac{K_{A^+}}{K_{B^-}} \tag{48}$$

Thus, the partition coefficient of the polyelectrolyte K^* will depend strongly on the partition coefficient of the small ions. If these have different affinities for the phases so that the ratio K_{B^-}/K_{A^+} is different from 1, the potential thus created affects the partition coefficient of the polyelectrolyte. Since the latter has many charges (Z^- is high), the effect can be large even if the potential difference is small. In the special case when the two ions A and B have equal values, $U_1 - U_2$ becomes zero and the $K^*_{p^{z-}}$ of the polyelectrolyte is the same as $K_{p^{z-}}$.

2. A polyelectrolyte plus a salt, as in 1, but the polyelectrolyte is in excess over the salt so that $Z^- C_{p^{z-}} \gg C_{A^+B^-}$. Equations (31) and (32) then become, respectively,

$$C_{A^+,1} = Z^- \cdot C_{p^{z-},1}$$

$$C_{A^+,2} = Z^- C_{p^{z-},2}$$

We may then use Eq. (38) so that

$$\ln K^*_{p^{z-}} = \frac{1}{1 + Z^-} \ln K_{p^{z-}} \cdot \frac{Z^-}{1 + Z^-} \ln K_{A^+} \tag{49}$$

If K_{A^+} is close to 1, as for example in the case of Na^+, one obtains

$$\ln K^*_{p^{z-}} = \frac{\ln K_{p^{z-}}}{1 + Z^-} \tag{50}$$

As a result the numerical value of ln K^* is reduced or the K^* value approaches unity the larger the charge on the protein. The Donnan effect thus tends to distribute a charged molecule more equally between the two phases.

The considerations presented above should apply generally to equilibria between two phases, not only liquid–liquid systems. For example, the partition between a gel and a liquid could be treated in a similar way. Thus an unequal distribution of ions between a gel and a liquid should create a potential which would affect the distribution of polyelectrolytes, for example, in gel chromatography.

SUMMARY

The theoretical considerations of this chapter may be summarized as follows.

If a suspension of particles with identical properties is shaken in a liquid–liquid two-phase system, their tendency to collect in one of the phases will be greater the larger the particle size, provided there is no adsorption [relation (10)] at the interface.

If the same particles are shaken in a system where the conditions for adsorption at the interface [relation (9)] are fulfilled, then the attachment of the particles to the interface will be more favored the larger their size and the larger the interfacial tension of the system. The unequal distribution of the ions of a salt creates a potential difference between the phases which, even if it is small, has a strong influence on the partition of polyelectrolytes.

These phenomena will be discussed in Chapter 4 in relation to experimental results.

REFERENCES

Brønsted, J.N. (1931). *Z. Phys. Chem., A* (Bodenstein-Festband), 257.

Brønsted, J.N., and Warming, E. (1931). *Z. Phys. Chem., A* **155,** 343.

Des Coudres, T. (1898). in L. Rhumbler, *Wilhelm Roux' Arch. Entwicklungsmech. Organ.*, Vol. 7, p. 225.

Donnan, F.G. (1911). *Z. Elektrochem.*, **17,** 572.

Einstein, A. (1906). *Ann. Physik,* **19,** 371.

—— (1956). *Investigations on the Theory of the Brownian Movement,* Dover, New York.

Gaudin, A.M. (1957). *Flotation,* McGraw-Hill, New York.

Svedberg, T., and Pedersen, K.O. (1940). *The Ultracentrifuge,* Clarendon Press, Oxford.

Taggart, A.F. (1951). *Elements of Ore Dressing,* Wiley, New York, 246.

Tiselius, A. (1932). *Kolloid-Z.*, **59,** 306.

4 FACTORS DETERMINING PARTITION

Partition between two phases depends on many factors. This is to be expected since interaction between the partitioned substance and the components of each phase is a complex phenomenon involving hydrogen bonds, charge interaction, van der Waals forces, hydrophobic interactions, and steric effects. Thus, the partition depends on the molecular weight and chemical properties of the phase-forming polymers and the size and chemical properties of the partitioned molecules or particles. For particles, it is mainly the exposed groups of the surface which come in contact with the phase components, and partition of particles is, therefore, a surface-dependent phenomenon. Since the ions of a salt have different affinity for the two phases, an electrical potential difference is established between the phases, and this has a strong influence on the partition of charged molecules or particles.

Since partition depends on so many factors it seems to be a complication making it difficult to plan experiments rationally for separation or to interpret results. It also means, however, that partition can be used for a multitude of separations. The different factors which determine partition can be explored separately or in combination to achieve an effective separation. We can also enlarge some of the factors so that they will dominate the partition behavior. With respect to the partitioned substance the following types of partition can be distinguished:

1. Size-dependent partition: molecular size or the surface area of the particles is the dominating factor.

2. Electrochemical: electrical potential between the phases is used to separate molecules or particles according to their charge.

3. Hydrophobic affinity: hydrophobic properties of a phase system is used for separation according to the hydrophobicity of molecules or particles.

4. Biospecific affinity: the affinity between sites on the molecules or particles and ligands attached to the phase polymers is used for separation.

5. Conformation-dependent: conformation of the molecules and particles is the determining factor.

6. Chiral: enantiomeric forms are separated.

Formally, we can split the logarithm of the partition coefficient into several terms:

$$\ln K = \ln K^0 + \ln K_{\text{el}} + \ln K_{\text{hfob}} + \ln K_{\text{biosp}} + \ln K_{\text{size}} + \ln K_{\text{conf}} \quad (1)$$

where el, hfob, biosp, size, and conf stand for electrochemical, hydrophobic, biospecific, size, and conformational contributions to the partition coefficient and $\ln K^0$ includes other factors.

The different factors are more or less independent of each other though none is probably completely independent of the others. For example, when we introduce a hydrophobic group on one of the phase-forming polymers we may slightly influence both the distribution of ions and the electrical potential. When the molecular weight of a macromolecule is increased its net charge may also increase. When the conformation of a macromolecule is changed new groups of atoms are exposed to the surrounding phase.

DIFFERENCE PARTITION

The factors determining partition can be studied by difference partition. By this method we can also learn about the properties of the surface of the particles or molecules.

Partition is compared for two phase systems which differ in one parameter only (Fig. 4.1). The first phase system, the reference system, will give a certain partition coefficient K_1. The second phase system is identical to the reference system except for one parameter, for example, the interfacial electrical potential, number of hydrophobic groups, or biospecific ligands, and so forth. The partition coefficient in this phase system is K_2. The difference $\Delta \log K = \log K_2 - \log K_1$ will reflect the effect of the parameter on partition. Thus, if the ionic composition is changed, and hence the electrical potential, the $\Delta \log K$ will give

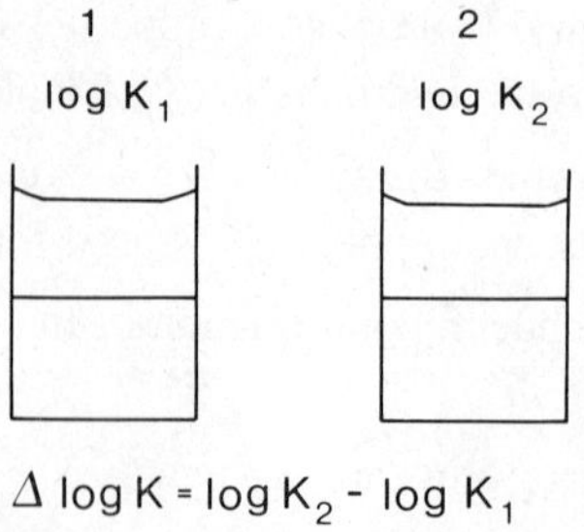

FIGURE 4.1. The effect of different factors on the partition can be studied by difference partition. System 1 is the reference phase system, for example a dextran–PEG phase system with a certain composition. System 2 has an identical composition except for one component, for example, a salt or a polymer with hydrophobic or biospecific groups. The two phase systems may differ in interfacial potential, hydrophobicity, and so forth, and the effect of these parameters can be obtained by determining the difference in log *K*.

information on the surface charge if the change in potential is known. Conversely, if the net surface charge of the particle is known we can determine the potential. If we attach hydrophobic groups on one of the polymers we can, by difference partition, get information on the hydrophobicity of the partitioned particles or molecules. By binding a biospecific ligand to one of the polymers we can study interaction with the partitioned particles or molecules.

Below some of the main factors determining partition in the dextran–polyethylene glycol system are described.

POLYMER CONCENTRATION—INTERFACIAL TENSION

Close to the critical point of the phase system, molecules such as proteins partition almost equally between the phases. If the polymer concentration is increased, that is, the composition of the phase system deviates more from the critical point, the partition of proteins will be more one-sided, that is, K will either increase above 1 or decrease below 1. There are exceptions, however, to this general rule, for example, the partition coefficient may first increase upon the addition of polymers, pass through a maximum, and then decrease upon further addition of polymers. This means that the effect of polymer concentration in each phase on the thermodynamic activity coefficient of the protein is different for the two polymers.

In the case of cell particles, these will favor the upper or the lower phase

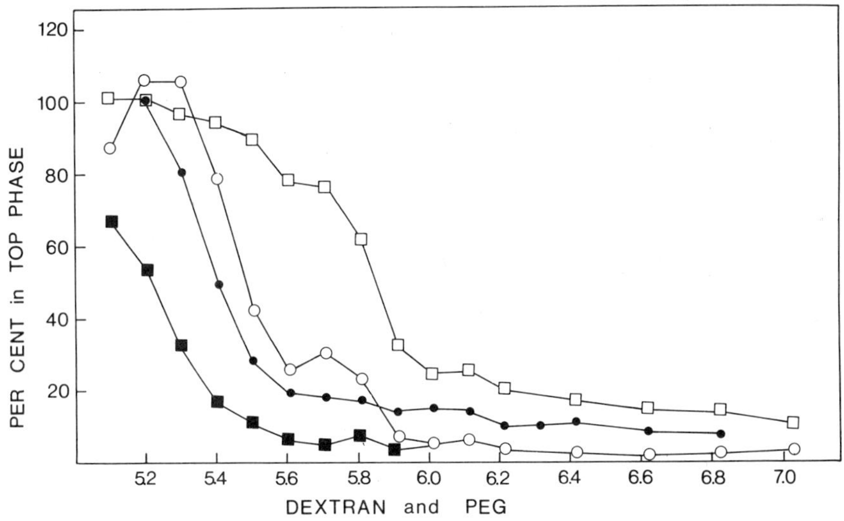

FIGURE 4.2*a*. Effect of polymer concentration on the partition of cell particles. When the concentration of the two polymers dextran and polyethylene glycol (PEG) are each increased the phase system is more and more removed from the critical point and the tie line of the phase diagram is increased. The particles favor more the interface and the bottom phase and the concentration in the top phase decreases. Symbols: ○, chloroplasts (class II); □, press disintegrated chloroplast vesicles; ●, small stroma thylakoid vesicles (100 *K* fraction); ■, inside-out thylakoid vesicles.

completely when the system is close to the critical point, that is, there is no adsorption at the interface. As the polymer concentration is increased the particles will have a tendency to be more absorbed to the interface. Thus, if the particles are in the upper phase close to the critical point they will be removed from this phase into the interface and sometimes to the bottom phase when the polymer concentration is increased. Depending on the surface properties of the particles they will be removed selectively from the upper phase. See Figure 4.2. An experiment such as that in Figure 4.2 is commonly used to find a suitable system for fractionation of a mixture of particles.

When we increase the polymer concentration of a phase system its composition is removed from the critical point and the interfacial tension is increased (see Fig. 2.15). A plot of the logarithm of the partition coefficient of membrane vesicles against the interfacial tension gives an almost straight line. See Figure 4.2*b*, demonstrating an exponential relationship between partition and interfacial tension.

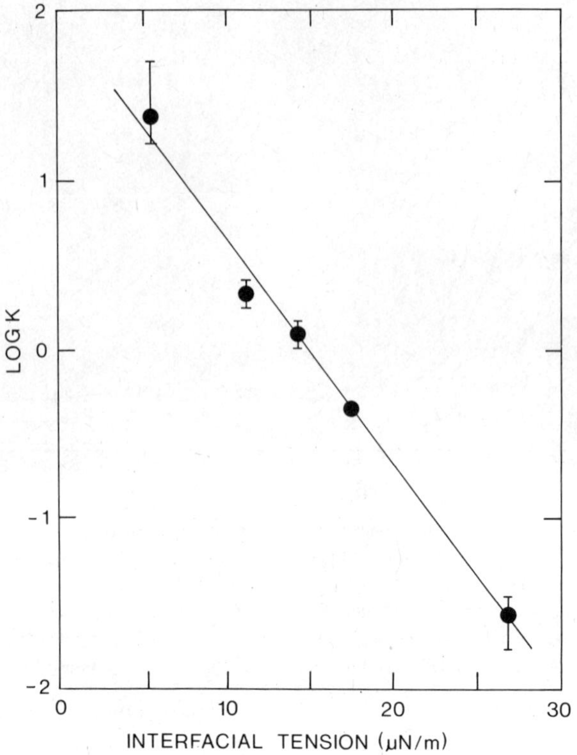

FIGURE 4.2*b*. Relation between partition coefficient of material in upper phase divided by sum of material in lower phase and interface of membrane vesicles and the interfacial tension of the phase system (Brooks et al., 1984).

MOLECULAR WEIGHT OF POLYMER

The following general rule holds for the effect of the molecular weight of polymers on partition: If, for a given phase composition, one polymer is replaced by the same type of polymer having a smaller molecular weight, partitioned molecules such as proteins and nucleic acids or particles such as cells and cell organelles will favor more the phase containing the polymer with decreased molecular weight. For example, the partition coefficient of a protein in the dextran–polyethylene glycol system will increase if the molecular weight of polyethylene glycol is decreased or the molecular weight of dextran is increased. Conversely the partition of the protein will decrease if the molecular weight of the polyethylene glycol is increased or the molecular weight of the dextran is decreased. That is, the partitioned protein behaves as if it is more attracted by smaller polymer molecules and more repelled by larger polymer molecules, provided all other factors, such as, polymer concentration, salt composition, temperature, and so forth, are kept constant. No exception to this rule has been found. It holds for different types of polymer-phase systems and for different types of partitioned substances, such as, proteins, nucleic acids, cell organelles, and cells.

The Polymer Molecular Weight Effect Depends on the Molecular Weight of the Partitioned Substance

The above rule states only the direction of the partition change. The magnitude of change depends on the molecular weight of the partitioned substance. Small molecules, such as, amino acids or small proteins are not so much affected as larger protein molecules (see Tables 4.1 and 4.2). Compare, for example, the shift in partition coefficient when replacing Dextran 40 (MW = 40000) with Dextran 500 (MW 500,000) for cytochrome *c* or ovalbumin with the same shift for catalase or phycoerythrin. For the larger protein molecules the partition coefficient is about 6 to 7 times higher in the Dextran 500 system compared to the Dextran 40 system. For cytochrome *c* the partition coefficient is about the same, and for ovalbumin it is only 1.3 times higher in the Dextran 500 system compared to the Dextran 40 system.

To replace one polymer with the same type of polymer with another molecular weight is very useful in order to select a phase system to give a desired partition of a substance, particularly when it is essential to have a constant ionic composition. Since proteins with different molecular weights are influenced differently (Tables 4.1 and 4.2) it can be used also to increase the separation between two proteins having different molecular weights. Table 4.3 shows the ratio between the partition coefficients (separation factor) of some proteins in two phase systems differing only in the molecular weight of the dextran.

MOLECULAR WEIGHT OF THE PARTITIONED SUBSTANCE

The larger the size of a molecule, the larger its exposed surface which can interact with the surrounding phase components. One would expect, therefore, an effect of the molecular weight of a substance on its partition. This also has been demonstrated experimentally. In the dextran–methylcellulose system a linear relation was found between the logarithm of the partition coefficient and the surface area for a number of proteins and viruses [Fig. 4.3 (Albertsson, 1958;

TABLE 4.1 Effect of Molecular Weight of Dextran on Partition Coefficient of Proteins with Different Molecular Weights

Protein	Molecular Weight	Dextran Component[a]				
		Dextran 40	Dextran 70	Dextran 220	Dextran 500	Dextran 2000
Cytochrome	12.384	0.18	0.14	0.15	0.17	0.21
Ovalbumin	45.000	0.58	0.69	0.74	0.78	0.86
Bovine serum albumin	69.000	0.18	0.23	0.31	0.34	0.41
Lactic dehydrogenase	140.000	0.06	0.05	0.09	0.16	0.10
Catalase	250.000	0.11	0.23	0.40	0.79	1.15
Phycoerythrin	290.000	1.9	2.9	—	12	42
β-Galactosidase	540.000	0.24	0.38	1.38	1.59	1.61
Phosphofructokinase	800.000	<0.01	0.01	0.01	0.02	0.03
Ribulose diphosphate carboxylase	800.000	0.05	0.06	0.15	0.28	0.50

[a]In phase systems with 6% PEG 6000 and 8% dextran having different molecular weights; 10mM Na phosphate, pH 6.8.

TABLE 4.2 Effect of Molecular Weight of Polyethylene Glycol on the Partition Coefficient of Proteins with Different Molecular Weights

Protein	Molecular Weight	Components of System[a]			
		$Dex^{500}{}_{9}$–$PEG^{4000}{}_{7.1}$	$Dex^{500}{}_{8}$–$PEG^{6000}{}_{6}$	$Dex^{500}{}_{8}$–$PEG^{20000}{}_{6}$	$Dex^{500}{}_{8}$–$PEG^{40000}{}_{6}$
Cytochrome *c*	*K*	0.17	0.17	0.13	0.12
Ovalbumin	*K*	1.25	0.85	0.50	0.50
Bovine serum albumin	*K*	0.52	0.34	0.14	0.11
Lactic dehydrogenase	*K*	0.13	0.08	0.05	0.03
Catalase	*K*	0.82	0.38	0.16	0.10

[a] In systems with 8% Dextran 500 and 6% PEG 6000, 20000, or 40000, and 9% Dextran 500 and 7.1% PEG 4000; phosphate, see Table 4.1.

TABLE 4.3 Ratio Between the Partition Coefficients of Different Proteins and the Partition Coefficient of Cytochrome c[a]

	Phase System	
Protein	Dextran 40	Dextran 500
Ovalbumin	3.2	4.6
Bovine serum albumin	1	2
Catalase	0.6	4.6
Phycoerythrin	11	71
β-Galactosidase	1.3	9.4
RudP Carboxylase	0.3	1.6

[a] Separation factor = *K* of protein/*K* of cytochrome *c*.
[b] The two phase systems differ in the molecular weight of dextran. Polymer composition as in Table 4.1.

Albertsson and Frick, 1969)]. For nonheme-proteins a relation between partition coefficient and molecular weight was found (Sasakawa and Walter, 1972). Surprisingly heme proteins partitioned almost independent of molecular weight. For DNA the log *K* was proportional to the molecular weight in the dextran–methylcellulose or dextran sulfate–methylcellulose system (Lif et al., 1961). In the dextran–PEG system a similar relation was found (Müller et al., 1979). For DNA the number of charged groups is proportional to the molecular weight, so the relation found might be due to a linear relation between log *K* and the number of charges as the formula in Chapter 3 would predict.

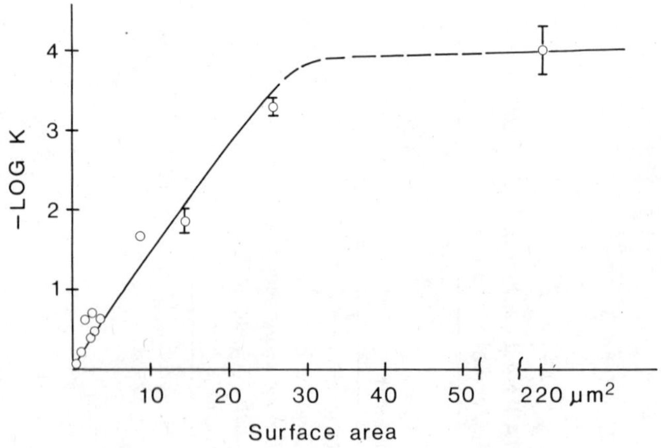

FIGURE 4.3. Partition coefficient depends exponentially on the surface area of the partitioned particles (Albertsson, 1962).

ELECTROCHEMICAL PARTITION

The effects of salts on the partition of charged macromolecules is most dramatic. Minor changes in the ionic composition can transfer DNA, for example, almost completely from one phase to the other. The partition of biopolymers is determined mainly by the kind of ions present and the ratio between different ions. The ionic strength is not so important. This can be explained by an electrical potential between the two phases, created by an unequal distribution of the ions, which have different affinities for the phases, as described in Chapter 3. Careful studies on the partition of inorganic salts in a dextran–polyethylene glycol–water system have shown that different salts have small yet significant differences in their partition coefficient (Johansson, 1970). See Table 5.1.

Such partition differences between salts mean that the different ions have different affinities for the two phases. Consequently, an electrical potential difference between the phases is created. For a salt, the ions of which have the charges Z^+ and Z^-, the interfacial potential $U_2 - U_1$ is given by

$$U_2 - U_1 = \frac{RT}{(Z^+ + Z^-)F} \ln \frac{K_-}{K_+} \tag{2}$$

where R is the gas constant, F is the Faraday constant, T is the absolute temperature, and K_- and K_+ are hypothetical partition coefficients of the ions in the absence of a potential. The interfacial potential will be larger the larger the K_-/K_+ ratio. A salt with two ions which have very different affinities for the two phases will generate a larger potential difference than a salt with ions which have similar affinities for the two phases.

Furthermore, it can be shown [Chapter 3, Eq. (42)] that in the presence of excess of salt a protein will partition according to

$$\ln K_P = \ln K^0{}_P + \frac{FZ}{RT}(U_2 - U_1) \tag{3}$$

where K_P is the partition coefficient of the protein, $K^0{}_P$ is the value of this coefficient when the interfacial potential (generated by the excess salt) is zero, or when the protein net charge Z is zero. Thus, the difference in partition of the ions of the salt generates an electrical potential difference according to Eq. (2), which in turn affects the partition coefficient of the protein according, to Eq. (3). Even if $U_2 - U_1$ is small, it will strongly influence K_P because Z is a large number for most proteins. K_P changes exponentially with Z. Figure 4.4 shows

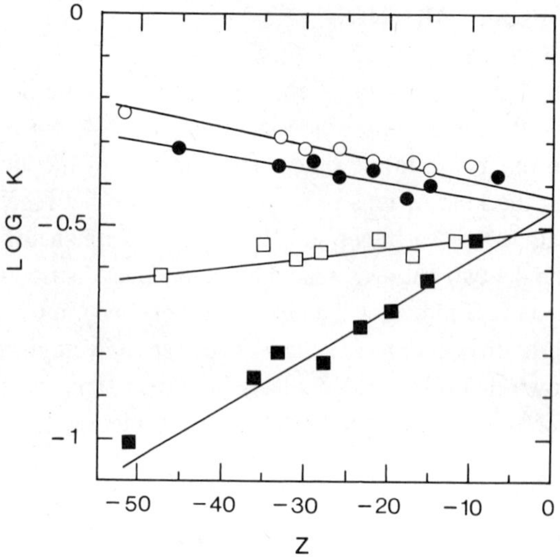

FIGURE 4.4. Partition coefficient depends exponentially on the net charge of proteins. The partition coefficient for serum albumin was determined at different pH in four phase systems having different interfacial potentials, created by the presence of different salts. The net charge (Z) was calculated from titration data (Johansson and Hartman, 1974; Johansson, 1974*b*).

experimental results which agree with Eq. (3). The strong effect of the ionic composition holds for all charged macromolecules and also for cell particles which each carry a vast number of charges.

What is the value of the interfacial potential? It is a central dogma in electrochemistry that an absolute electrical potential difference between two chemically different phases cannot be measured. However, if one is not too puritan, one may calculate it under certain assumptions. For example, if $K^0{}_P$ in Eq. (2) is assumed to be independent of pH, the $U_2 - U_1$ value is obtained from the slope of the lines in Figure 4.4. The interfacial potentials obtained are in the range 0–20 mV. Other measurements, using electrodes, give similar values (Johansson, 1974; Reitherman et al., 1972). If the interfacial potential of a phase system is known, we can use partition in this system for the determination of the net charge per molecule of a protein without knowing its molecular weight. If we determine the partition coefficient of a protein by a specific method (enzyme activity or immunologically) the protein does not even have to be pure. Figure 4.5 illustrates the use of partition for determination of the net charge of different enolase forms.

It is striking that, with a given salt, the interfacial potential varies very little with ionic strength (Albertsson and Nyns, 1961). Further, partition of proteins

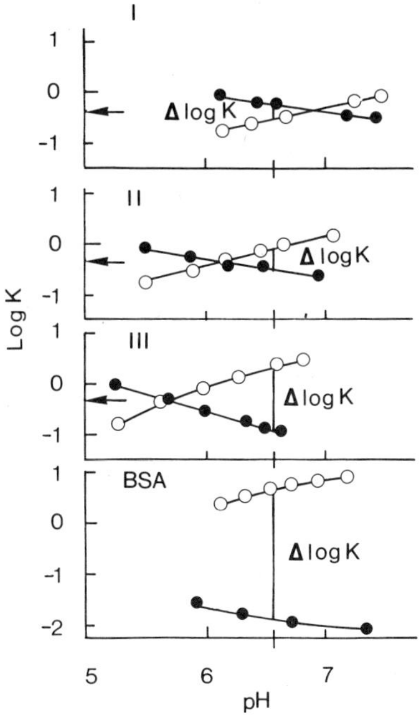

FIGURE 4.5. Cross partition of isoenzymes (I, II, and III) of enolase. The partition coefficient at the cross point (= isoelectric point) is the same for the tree forms. The isoelectric points are different. The net charge at a given pH, for example 6.57, can be determined by comparing the Δ log *K* at this pH with the Δ log *K* of a protein with known net charge at the same pH, for example BSA, lower figure (Blomquist, 1976).

in a given salt does not significantly depend upon the ionic strength in the range 5–100 m*M*. That is, for a given salt the effective net charge *Z* of the protein is independent of ionic strength. This behavior is in contrast to other phenomena where charge is involved, such as electrophoresis and ion exchange chromatography. Electrophoretic mobility and adsorption on an ion-exchange column strongly depend on ionic strength.

HYDROPHOBIC INTERACTIONS

The electrical effects can be nullified by choosing a suitable salt composition of the phase system so that the electrical potential difference between the phases becomes zero. Other factors determining the partition then will come to the fore such as the hydrophobic–hydrophilic properties of the particle surface.

The hydrophobic effect can be increased by binding covalently hydrophobic groups to one of the polymers (Shanbhag and Johansson, 1974; Shanbhag and Axelsson, 1975). If a protein, for example, contains surfaces or pockets which bind hydrophobic groups, its partition coefficient will be changed. This so-called hydrophobic affinity partition can be used both for characterizing the hydrophobic properties of proteins or cell particles and also for separating molecules or particles differing in hydrophobicity.

Examples of hydrophobic affinity partition of proteins are given in Chapter 5 in Figures 5.11 and 5.12.

Hydrophobic affinity partition has also been studied on cells (Eriksson et al., 1976) and is described in Chapter 8, Figure 8.23. Cell organelles have been separated by hydrophobic affinity partition. By using PEG esterified with deoxycholate or fatty acids different chloroplasts should be separated (Westrin et al., 1976; Johansson and Westrin, 1978).

BIOSPECIFIC AFFINITY PARTITION

Biospecific ligands coupled to one of the polymers will strongly influence the partition of molecules or particles which have affinity for the ligand. Several applications of such biospecific affinity partition are reported in the literature (see Chapter 5, Table 5.6).

Theoretically, one should expect very large effects on the partition of a protein if it binds strongly to a ligand coupled, for example, to polyethylene glycol of a dextran–polyethylene glycol phase system as calculated by Flanagan and Barondes (1975).

If the partition coefficient of the ligand–polyethylene glycol is $K_{\text{L–PEG}}$, that of the protein is K_{P}, and that of the complex is $K_{\text{P–L–PEG}}$, and if the dissociation constant is K_1 in the upper phase and K_2 in the lower phase then the following relation holds:

$$K_{\text{P–L–PEG}} = K_{\text{P}} \times K_{\text{L–PEG}} \times \frac{K_2}{K_1} \tag{4}$$

If the protein has N independent binding sites:

$$K_{\text{P–L–PEG}} = K_{\text{P}} \times \left(K_{\text{L–PEG}} \times \frac{K_2}{K_1}\right)^N \tag{5}$$

If we assume the same dissociation constant in the top and bottom phase:

$$K_{\text{P-L-PEG}} = K_{\text{P}} \times (K_{\text{L-PEG}})^N \qquad (6)$$

or

$$\log K_{\text{P-L-PEG}} = \log K_{\text{P}} + N \times \log K_{\text{L-PEG}} \qquad (7)$$

Since the ligand–PEG may have a K value in the range of 10 to 100 we should expect a 10 to 100-fold increase in partition for every binding site, provided ligand–PEG is added in excess so that all binding sites are filled. If a protein has several binding sites, one would, therefore, expect changes in K value of several orders of magnitude. In practice, the observed increase in K value lies between 10 to 10000-fold upon the addition of a PEG–ligand. Figure 4.6 shows the effect of Cibachrome-PEG on the partition coefficient of phosphofructokinase

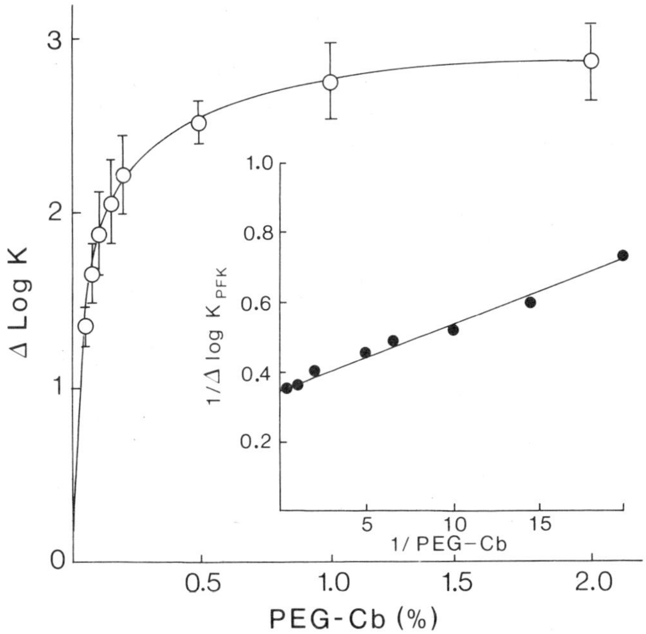

FIGURE 4.6. Affinity partition of phosphofructokinase. The change in partition coefficient of the enzyme as a function of added Cibachrome blue-PEG. The log K value for the enzyme was -1.25. When Cibachrome blue-PEG is added the K value increases. Δ log K is the difference in log K between a system with and without Cibachrome blue. B is an inverse plot of data in A (Johansson et al., 1983).

(Johansson et al., 1983). This enzyme has 16 binding sites, and according to theory, the log K for the enzyme should increase by at least 16 upon addition of saturating amounts of Cibachrome-PEG. The fact that it reaches a value of only 3 may have several explanations. For example, binding of one or two ligand–PEG molecules to the enzyme may hinder binding of more ligand–PEG molecules, that is, the binding sites are not independent as supposed in the theory. The dissociation constant may also be different in the two phases. Furthermore, the fact that part of the surface of the protein molecule is withdrawn from direct contact with the phases by the binding of the ligand–polymer must be taken in consideration (see Chapter 10). This is discussed in detail by Johansson et al. (1983).

Self-aggregation of the polymer-bound dye molecules might also reduce the affinity partition effect.

Affinity partition offers an exceptionally effective procedure for rapid, selective separation which can easily be scaled up, as described in Chapter 9.

CONFORMATION

Phase systems with zero interfacial potential can be constructed by choosing a suitable salt or salt mixture. Partition of proteins in such a system should be independent of the net charge of the protein, that is, independent of pH. Some proteins also show a constant partition coefficient over a wide pH range in zero-potential phase system (Johansson and Hartman, 1974). Other proteins show changes at certain pH intervals. In some cases this occurs when the protein undergoes a conformational change, such as in the case of serum albumin at low pH, or when protein molecules form dimers, as with lysozyme. For such proteins, K^0 is not independent of pH, and partition in zero-potential systems could, therefore, be used to detect conformational changes, when, for example, previously hidden groups are exposed on the protein surface, or to study association–dissociation phenomena among proteins. The effect of conformation of nucleic acids is described in Chapter 5.

CHIRAL PARTITION

Enantiomeric forms of molecules should partition differently in a phase system containing optically active components. Dextran, Ficoll, starch, and cellulose polymers are all optically active and should discriminate D and L forms of molecules. Also, a binding protein which can select D or L forms and be enriched in one of the phases might be used for chiral partition. D and L tryptophane have been separated by a phase system containing serum albumin (Ekberg et al. 1985).

TEMPERATURE

Temperature is an important factor in partition. However, the effect of temperature is mainly indirect by changing the polymer composition of the phases. As shown in Chapter 2, the phase diagram depends on the temperature and, particularly close to the critical point, a small change in temperature will have considerable influence on the polymer composition and thereby also the partition of a substance in the phase system. The temperature has much less influence on the partition when it is far away from the critical point.

REFERENCES

Albertsson, P.-Å. (1958). *Nature,* **182,** 709–711.

Albertsson, P.-Å. (1962). *Meth. Biochem. Anal.,* **10,** 229–262.

Albertsson, P.-Å, *Partition of Cell Particles and Macromolecules,* 2nd ed. (1971). Almquist & Wiksell, Stockholm, and Wiley, New York.

Albertsson, P.-Å, and Frick, G. (1960). *Biochim. Biophys. Acta,* **37,** 230–237.

Albertsson, P.-Å, and Nyns, Ed. J. (1961). *Arkiv Kemi,* **17,** 197–206.

Albertsson, P.-Å, Andersson, B., Larsson, C., and Åkerlund, H.-E. (1982). *Meth. Biochem. Anal.,* **28,** 115–150.

Blomquist, G. (1976). *Biochim. Biophys. Acta,* **420,** 81–86.

Borgström, B., and Erlanson, C. (1978). *Gastroenterology,* **75,** 382–386.

Brooks, D.E., Bamberger, S.B., Harris, J.M., and Van Alstine, J., (1984). Proc. 5th Europ. Symp. Material Sc. Microgravity (Europ. Space Agency, SP-222).

Ekberg, B. Sellergren, B. and Albertsson, P-Å. (1985). J. Chromatogr. 333, 211–214.

Eriksson, E., Albertsson, P.-Å., and Johansson, G. (1976). *Mol. Cell. Biochemistry,* **10,** 123–128.

Flanagan, S.D., and Barondes, S.H. (1975). *J. Biol. Chem.,* **250,** 1484.

Frick, G., and Lif, T. (1962). *Arch. Biophys. Suppl.,* **1,** 271.

Johansson, G. (1970). *Biochim. Biophys. Acta,* **221,** 387–390.

Johansson, G. (1971). *Proc. International Solvent Extraction Conference, The Hague, Soc. of Chem. Industries,* Vol. 2, p. 928.

Johansson, G. (1974a). *Acta Chem. Scand.,* **B28,** 873–882.

Johansson, G. (1974b). *Mol. Cell. Biochem.,* **4,** 169–180.

Johansson, G., and Hartman, A. (1974). *Proc. International Solvent Extraction Conference, Lyon, Soc. Chem. Industries,* Vol. 1, pp. 927–942.

Johansson, G., and Westrin, H. (1978). *Plant Science Letters,* **13,** 201–212.

Johansson, G., Kopperschläger, G., and Albertsson, P.-Å. (1983). *Eur. J. Biochem.,* **131,** 589–594.

Lif, T., Frick, G., and Albertsson, P.-Å. (1961). *J. Mol. Biol.,* **3,** 727.

Müller, W., Schuetz, H.J., Guerrier-Takada, C., Cole, P.E., and Potts, R. (1979). *Nucleic Acid Res.,* **7,** 2483–2500.

Reitherman, R., Flanagan, S.D., and Barondes, S.H. (1972). *Biochim. Biophys. Acta,* **297,** 193.

Sasakawa, S., and Walter, H. (1972). *Biochemistry,* **11,** 2760–2765.

Shanbhag, V.P., and Axelsson, C.G. (1975). *Eur. J. Biochem.,* **60,** 17–22.

Shanbhag, V.P., and Johansson, G. (1974). *Biochem. Biophys. Res. Comm.,* **61,** 1141–1146.

Westrin, H., Albertsson, P.-Å., and Johansson, G. (1976). *Biochim. Biophys. Acta,* **436,** 696–706.

5 PARTITION OF SOLUBLE COMPOUNDS

LOW-MOLECULAR-WEIGHT SUBSTANCES

Generally, low-molecular-weight substances, such as inorganic salts, sugars, amino acids, and nucleotides, partition almost equally between the two phases, that is, the K values are mostly around 1. However, the K values of some of these substances deviate considerably from 1, as shown by the studies of Johansson (1970a). Table 5.1 shows the K values of different salts, acids, and aromatic compounds. It is seen that various halide salts have K values close to 1. There is a significant tendency, however, for the K to increase in the order $K^+ < Na^+ < Li^+$ and $Cl^- < Br^- < I^-$. The same order is observed for the cations when the corresponding sulfates are partitioned.

Sulfates and phosphates give K values significantly lower than 1. Polybasic acids, such as, phosphoric, citric and oxalic, have K larger than 1, while their salts have K lower than 1. In the case of phosphate, K decreases with increasing charge of the phosphate ion.

Among the aromatic compounds, pyridine has a K of about 1, while phenol and naphthol have K much higher than 1.

It is clear that although most low-molecular-weight substances distribute fairly equally between the phases there are many exceptions which deviate from this rule. Particularly, the unequal partition of ions is of interest in connection with their effect on the electrical potential between the phases (see Chapter 3) and the partition of macromolecules and particles (see below).

TABLE 5.1 Partition Coefficient *K* of Various Salts, Acids, and Aromatic Compounds[a]

Compound	Concentration, *M*	*K*	Compound	Concentration, *M*	*K*
LiCl	0.1	1.05	NaH_2PO_4,	Mixture of	0.96
LiBr	0.1	1.07	Na_2HPO_4	0.03 each	0.74
Li I	0.1	1.11	Na_3PO_4	0.06	0.72
NaCl	0.1	0.99	Citric acid	0.1	1.44
NaBr	0.1	1.01	Na_3 citrate	0.1	0.81
Na I	0.1	1.05	Oxalic acid	0.1	1.13
KCl	0.1	0.98	K_2 oxalate	0.1	0.85
KBr	0.1	1.00			
K I	0.1	1.04	Pyridine[b]		0.92
			Phenol[b]		1.34
Li_2SO_4	0.05	0.95	Naphthol-1[b]		1.76
Na_2SO_4	0.05	0.88			
K_2SO_4	0.05	0.84			
H_3PO_4	0.06	1.10			

[a] Dextran–polyethylene glycol two-phase system (7% w/w Dextran 500 and 7% w/w PEG 4000). From Johansson (1970a).
[b] In 0.025 *M* Na phosphate buffer, pH 6.9.

TABLE 5.2 Partition of Two Charged Polymers, One Positive, DEAE–Dextran, and One Negative, CM–Dextran[a]

Salt, at 0.01 M	*K* of DEAE–Dextran	*K* of CM–Dextran
LiCl	3.1	0.96
NaCl	5.2	0.18
KCl	5.5	0.15
CsCl	5.1	0.18
NaF	0.6	0.59
NaCl	5.2	0.18
NaBr	6.7	0.13
NaI	13.6	0.12
Li_2SO_4	0.20	4.9
Na_2SO_4	0.22	0.8

[a] Two-phase system of dextran–polyethylene glycol with different salts present. Phase system: 5% w/w Dextran 500, 4% w/w PEG 6000, and 0.1% w/w of either DEAE–dextran or CM–dextran, at 20–23°C. The polyelectrolytes were determined by acid–base titration. (Experiments by S. Nilsson in the author's laboratory.)

POLYELECTROLYTES

Partition of DEAE-dextran and carboxymethyldextran in the dextran–polyethylene glycol system in the presence of different salts is shown in Table 5.2. The effect of different ions is very different for the two substances. Thus, for example, the positively charged DEAE-dextran has a lower K value in the presence of Li^+ compared to Na^+, while the reverse is true for the negatively charged carboxymethyldextran. In fact, replacement of one salt by any of the others always results in opposite changes of the K value of these two charged polymers. This is analogous to the salt effect on the partition of a protein at pH's on the two sides of the isoelectric point. (See Figs. 5.2 and 5.4.)

PROTEINS

Dextran–Polyethylene Glycol System

Partition of proteins in this phase system depends on the ionic composition, pH, size of the protein molecule, and concentration and molecular weight of the two polymers (Albertsson, 1958, 1971; Albertsson and Nyns, 1959, 1961).

Ionic Composition. Figure 5.1 shows a typical pattern of influence of different ions on the partition of a protein. Phosphate buffer is present at a concentration of 10 m*M*. The other salts are added at higher concentration and they therefore dominate in creating the interfacial potential.

The cations decrease the partition coefficient in the order $Li^+ < NH^+{}_4 < Na^+ = Cs^+ < K^+$ and the anions in the order $F^- < Cl^- < Br < I^-$. The divalent anions monohydrogenphosphate, sulfate, and citrate increase the partition coefficient relative to the monovalent ions. The effect of ions is algebraically additive. For example, the difference in effect between the ammonium and sodium ion is the same whether the anion is chloride, nitrate, or sulfate.

It has been demonstrated (Albertsson and Nyns, 1961) that it is mainly the ratio between the ions and not so much the total ionic strength which determines the partition at least in the range 10–100 m*M*. Figure 5.1b shows how the ratio of chloride to phosphate determines the partition of proteins.

The order of ions in their effect upon the partition given above is typical for a negatively charged protein. The order is reversed if a positively charged protein

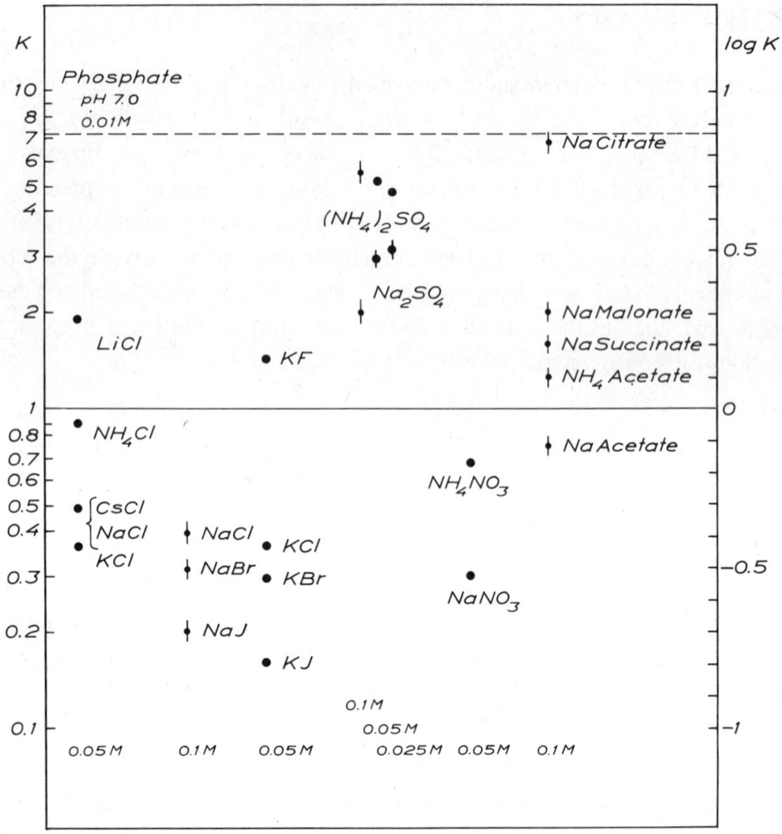

FIGURE 5.1a. Partition coefficients K of phycoerythrin in a dextran–polyethylene glycol system with 0.005 M KH_2PO_4 and 0.005 M K_2HPO_4, and different electrolytes; the concentrations of the latter are indicated in the bottom row. Polymer composition = 7% w/w Dextran 500; 4.4% w/w PEG 6000. Temperature: 20°C.

is used, as shown in Figure 5.2. (See also Johansson, 1971.) The partition coefficient of the negatively charged ovalbumin decreases while that of the positively charged lysozyme increases in the series F, Cl, Br, I or Li, Na, K.

Figure 5.2*b* shows the effect of the concentration of NaCl in the absence of other salts. At zero concentration, the partition coefficient of the proteins is close to one. This can be predicted from Eq. (50), Chapter 3. In this case, the ionic concentrations of the proteins are in excess over the salts' concentrations. The Donnan effect tends to distribute the proteins equally between the phases. As the NaCl concentration is increased, the salts' influence will predominate over that

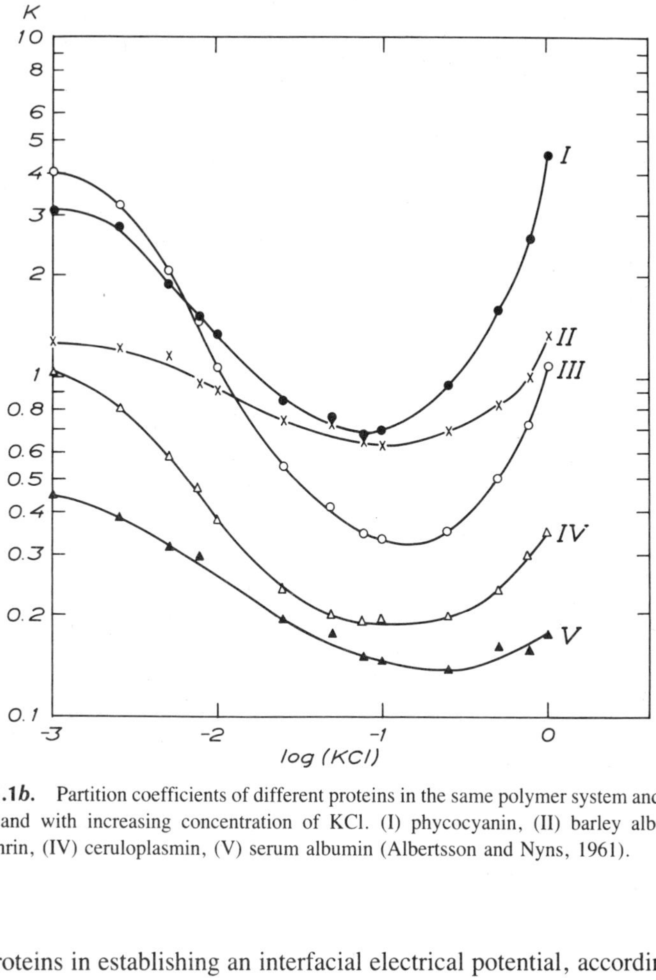

FIGURE 5.1*b*. Partition coefficients of different proteins in the same polymer system and phosphate as in (*a*) and with increasing concentration of KCl. (I) phycocyanin, (II) barley albumin, (III) phycoerythrin, (IV) ceruloplasmin, (V) serum albumin (Albertsson and Nyns, 1961).

of the proteins in establishing an interfacial electrical potential, according to Eq. (47), Chapter 3, and the protein will partition according to Eq. (48), Chapter 3. Since the two proteins have charges of opposite signs their partition curves will diverge as NaCl is increased, approaching a constant value when the NaCl concentration is in excess over that of the proteins (above 25 m*M*).

Obviously, therefore, the separation of two oppositely charged proteins can be considerably improved by selecting suitable ionic conditions.

High NaCl Concentrations. It seems to be a general phenomenon that proteins are partitioned more into the upper phase when NaCl is increased in the range 1–5 *M* (Fig. 5.3). The log *K* increases almost linearly with NaCl concentration

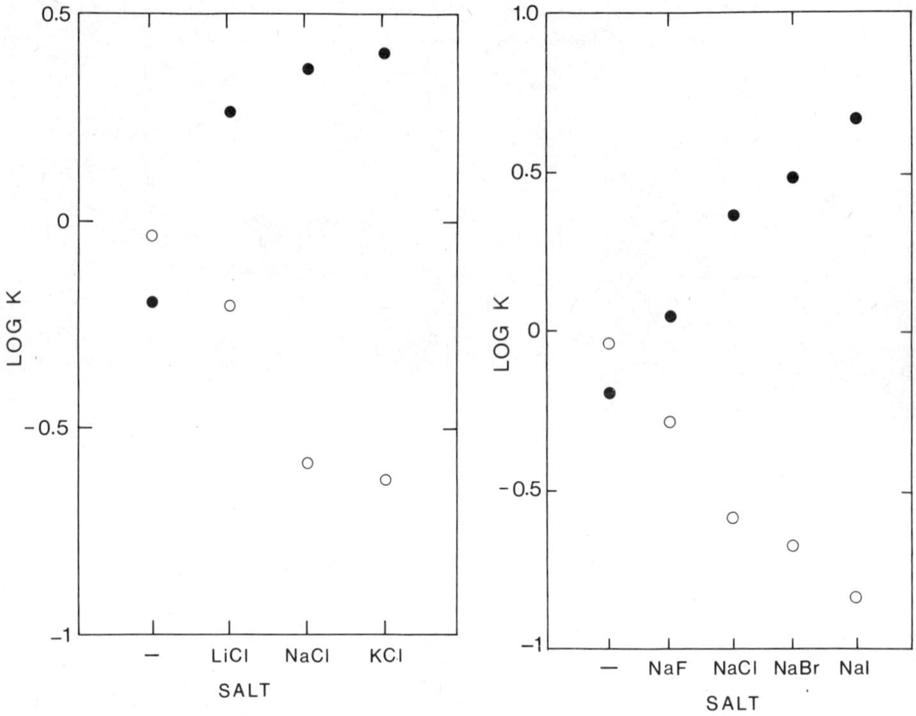

FIGURE 5.2a. Effect of ionic composition on the partition coefficient of a positively and negatively charged protein. Lysozyme (●) and ovalbumin (○) were partitioned in a dextran–PEG phase system with 0.5 m*M* Na phosphate at pH 6.9 and 25 m*M* of the indicated salts. At this pH lysozyme is positively charged and ovalbumin negatively charged. Polymer composition 8% w/w Dextran 500; 8% w/w PEG 4000.

in this range. The increase is different for different proteins; the log K values become increasingly divergent as the salt concentration is raised. For fractionation purpose the conditions, therefore, improve as more salt is added. The three serum proteins, albumin, caeruloplasmin, and immunoglobulin, have not very different K values at 0.1 M NaCl (K = 0.15, 0.23, and 0.38, respectively) while at 3 M NaCl the K values are 0.4, 1.5, and 7, respectively. The separation effect is still greater if the polymer composition is further removed from the critical point.

Cross Partition. The influence of the net charge of a protein on its partition can be studied by measuring the partition coefficient at different pH. In order to suppress the influence of different buffer ions on the electrical potential between

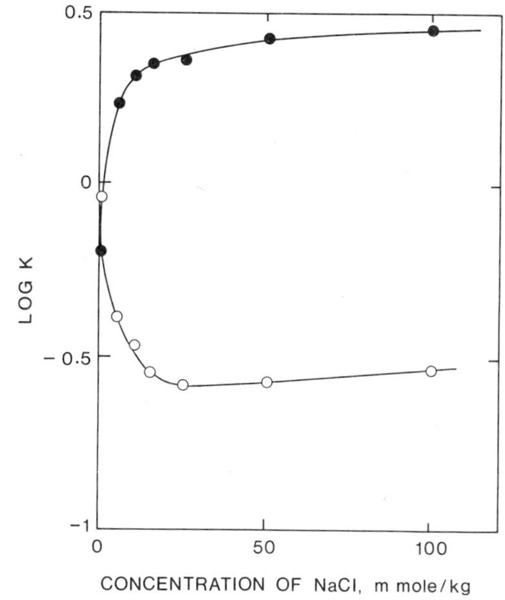

FIGURE 5.2*b*. Effect of NaCl concentration on partition of ovalbumin and lysozyme at pH 6.9. Polymer composition as in Figure 5.2*a* (Johansson, 1971).

the phases a neutral salt such as NaCl or Na_2SO_4 is added in excess (at ten times higher concentration than the buffer). If the partition coefficient K is plotted against pH, one curve is obtained for NaCl and a different one for Na_2SO_4 since these salts give rise to different electrical interfacial potentials. The two curves will cross at the isoelectric point (Albertsson, 1970; Albertsson et al., 1970). Figure 5.4 shows examples of such cross-partition curves for different proteins. Agreement between the cross point and the isoelectric point has been demonstrated for a large number of proteins (Albertsson et al., 1970) and cross partition can, therefore, be used for the determination of the isoelectric point of proteins (Sasakawa and Walter, 1972; Walter and Sasakawa, 1971; Walter et al., 1972) and also other macromolecules or particles. It has also been used for the determination of the isoelectric point of chloroplasts, mitochondrial membrane vesicles, and bacteria (see Chapter 7). Even small molecules such as amino acids and peptides show a cross point (Sasakawa and Walter, 1974).

The curve representing the partition coefficient as a function of pH may have different shape for different neutral salts used to create the potential or at different concentration of these salts (Fig. 5.4). But the two curves for two different salts

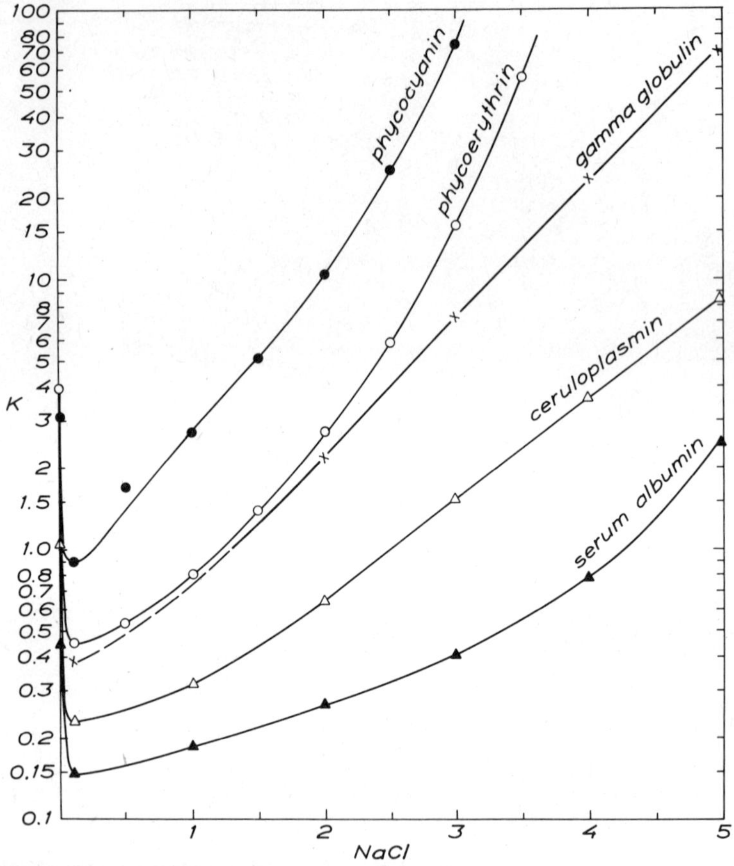

FIGURE 5.3. Partition coefficients K of a number of proteins in the dextran–polyethylene glycol system in 0.005 M KH_2PO_4, 0.005 M K_2HPO_4, and with increasing concentration of NaCl; the latter is expressed as moles NaCl added per kilogram standard phase system. For the polymer and salt compositions of the phases, see Figures 12.19 and 12.20. Similar curves are obtained with Tris-buffer instead of phosphate buffer. Tris-buffer has the advantage in this system that the pH does not drop so much as with phosphate when high concentration of NaCl is added.

always cross at the isoelectric point (Albertsson et al., 1970). This is to be expected from Eq. (3) of Chapter 4.

For any U_2–U_1 value (which is determined by the salt composition) the value of ln K_P will be equal to ln $K^0{}_P$ when Z is zero, that is, at the isoelectric point.

Comments on the Effect of Ionic Composition. The behavior of the proteins in the dextran–polyethylene glycol system with different ionic compositions can be explained by the theory presented in Chapter 3. According to Eq. (47), the

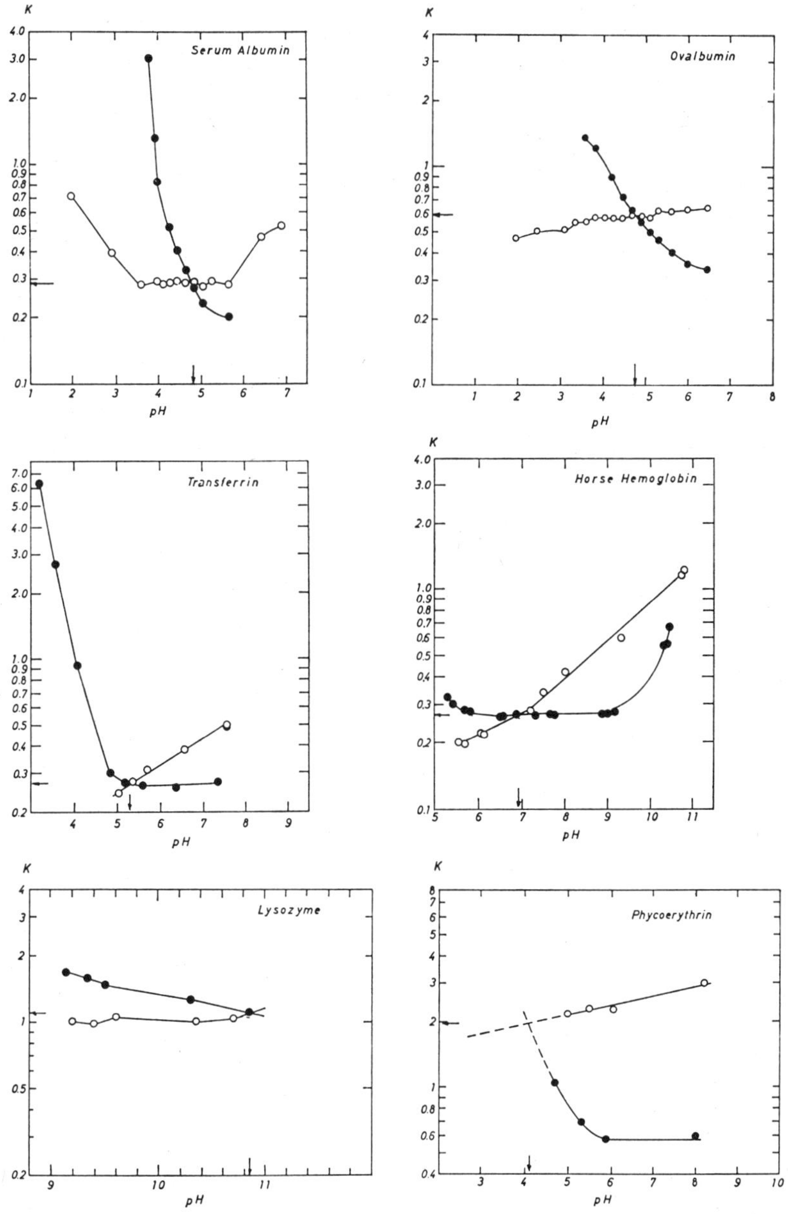

FIGURE 5.4. Cross partition of proteins. For each protein the K values in two different salt media are plotted against pH. ●, 0.1 M NaCl; ○, 0.05 M Na_2SO_4. The two curves cross at the isoelectric pH (Albertsson et al., 1970).

presence of salts will create an electrical potential between the two phases. The sign and magnitude depends on the affinity of the ions for the two phases. If the salts are mixed, the potential will be governed by the salt in excess (Fig. 5.1*a* and *b*). Proteins with different net charges will react differently with a change in the salt composition according to Eq. (48) of Chapter 3. It is of interest that at a given ionic ratio, the partition does not depend much on the ionic strength in the range 25–100 m*M*. This is predicted by theory and also found empirically. See, for example, Figure 5.2*b* and Albertsson and Nyns (1961). In this respect the partition differs from other separation methods where charge is involved, such as ion exchange chromatography and electrophoresis, where ionic strength is a very critical factor.

An effect of high NaCl concentration (Fig. 5.3) on the partition of proteins is very common. At NaCl concentrations above 1 *M* most proteins tend to favor the upper, polyethylene glycol rich phase. This is probably a result either of hydrophobic interactions with the polyethylene glycol or a phenomenon related to salting out. Although polyethylene glycol is weakly hydrophobic, its interactions with hydrophobic domains on proteins may increase at high salt concentrations. However, this effect is mainly observed for NaCl in the dextran–polyethylene glycol system. If phosphate is increased to high concentrations, many proteins favor the lower phase.

Protein Concentration. The *K* value as a function of the protein concentration has not been studied for a large number of proteins. Countercurrent experiments indicate, however, that the *K* value is constant at different protein concentrations, even at fairly high concentrations, and independent of the presence of other proteins (see Chapter 6). An experiment with serum albumin at high concentrations is shown in Figure 5.5. The partition isotherm is linear up to about 5% serum albumin in the bottom phase and deviates only slightly from linearity at higher concentrations.

Dextran-Charged PEG Systems

As described in Chapter 3, the electrical effect on partition can be magnified by introducing charged groups on the polymer. If, for example, one charged group is covalently attached to PEG this ion will have a high affinity for the PEG phase of a dextran–PEG phase system. If its counter ion has a more even affinity for the two phases a relatively large electrical potential is created provided no other salts are present. We can predict this from Eq. (2) in Chapter 4 describing the

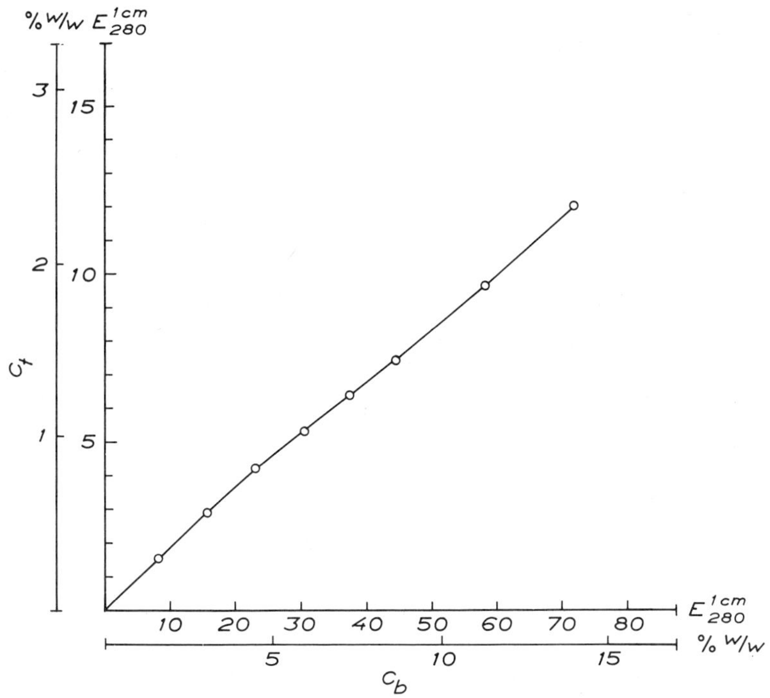

FIGURE 5.5. Partition isotherm of human serum albumin in the dextran–polyethylene glycol system, 7% w/w D48, 4.4% w/w PEG 6000, 0.01 *M* Na phosphate, pH = 6.8, and 0.1 *M* NaCl, at 20°C, C_b and C_t = concentration of serum albumin in bottom and top phase.

relation between the electrical potential difference

$$U_2 - U_1 = \frac{RT}{2F} \ln \frac{K_-}{K_+} \tag{1}$$

and the partition coefficients of the two ions when the potential difference is zero. Suppose that the phase system contains $(\text{TMA-PEG})^+$ and Cl^- as the only ions and that the partition coefficient of PEG^+ at a zero potential difference is 10 and that of Cl^- is 1. Then a potential difference of

$$\frac{RT}{2F} \ln 10 = 29.6 \quad \text{mV} \tag{2}$$

is created, which is much larger than the few millivolts created by inorganic salts

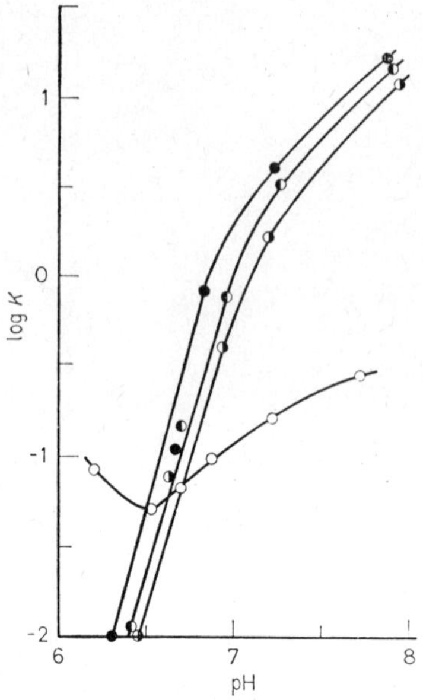

FIGURE 5.6. Effect of charged PEG on partition of CO-hemoglobin. Phase system: 8% Dextran 500 and 8% PEG 6000 (I) plus trimethylamino-PEG 6000 (II) ○, I only; ◐, 50% I and 50% II; ◑, 25% I and 75% II; ●, II only (Johansson et al., 1973).

(Chapter 4). From Eq. (1) we also learn that even larger potential difference would be obtained if we put one charged group of opposite sign on the dextran molecules. Thus, if PEG^+ $dextran^-$ was the only salt present, and the partition coefficients of these two ions at zero potential are assumed to be 10 and 0.1, respectively, the potential difference would be

$$\frac{RT}{2F} \ln 100 = 59 \quad \text{mV} \tag{3}$$

Furthermore, we can learn from Eq. (2) in Chapter 4 that the number of charges on each polymer should not be large because the sum is in the denominator and therefore the larger it is the smaller the potential difference. For application one should, therefore, strive to have a small number of charges per polymer molecule, and avoid the presence of other inorganic salts. Highly charged polymers such as dextran sulfate or DEAE-dextran with a large number of charges per molecule would, therefore, not be very effective in creating a large interfacial potential. Polyethylene glycol, however, is very suitable in this respect. It has only two hydroxyl groups which can be modified or coupled to charged molecules in order

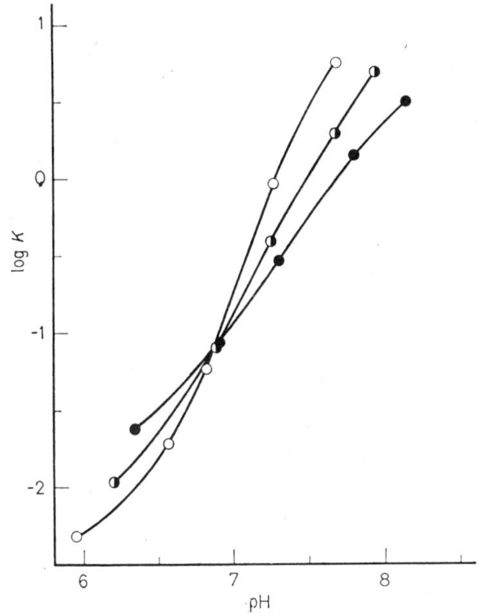

FIGURE 5.7. The effect of charged PEG is diminished by addition of inorganic salts. Partition of CO-hemoglobin in 8% Dextran 500, 8% TMA-PEG with 2 m*M* (○); 5 m*M* (◑) and 10 m*M* (●) potassium phosphate (Johansson et al., 1973).

to give a positively or negatively monovalent or divalent PEG ion. The following derivatives have been synthesized (Johansson, 1970b; Johansson et al., 1973).

Trimethylamino-PEG (TMA-PEG)
PEG-sulfonate (S-PEG)
Amino-PEG (PEG-NH_2)
Carboxyl-PEG (PEG-COOH)

Figure 5.6 shows the effect of charged PEG (TMA-PEG) on the partition profile of CO-hemoglobin. At low pH, when hemoglobin is positively charged (isoelectric point 7), the protein is pressed down to the lower phase by the positively charged upper phase, while at higher pH, when the protein is negatively charged, it is attracted by the upper phase. Thus the *K* value of hemoglobin shifts from 0.01 to over 10 in the range between pH 6 and 8.

The system of Figure 5.6 did not contain any extra salt or buffer. If an inorganic salt is added it will suppress the effect of the charged PEG (Fig. 5.7) because

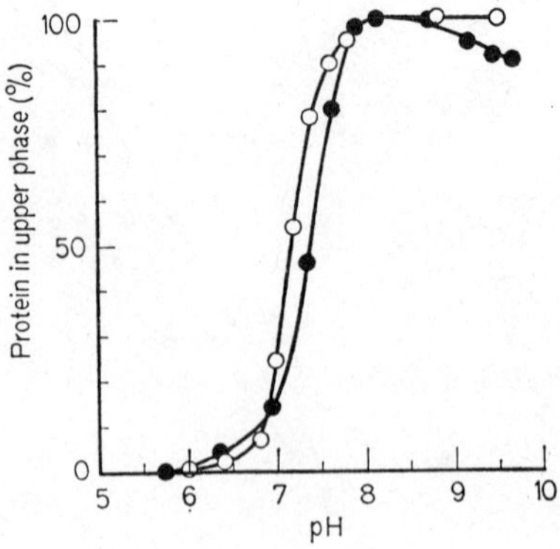

FIGURE 5.8. Instead of salts, a zwitterion can be added without much change in the effect of charged PEG. Phase system 8% Dextran 500, 8% TMA-PEG, with (●) and without (○) 50 m*M* glycine (Johansson et al., 1973).

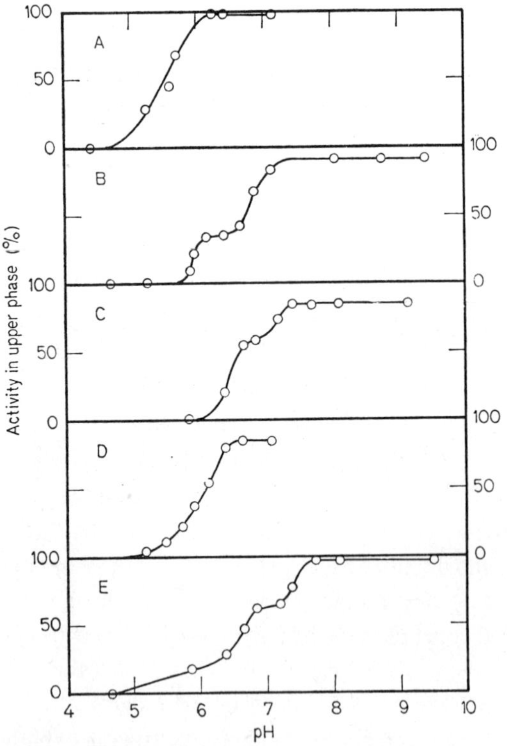

FIGURE 5.9. Extraction of glycolytic enzymes from baker's yeast with charged PEG. Phase system 6.6% Dextran 500 and 6.4% TMA-PEG. (A) Hexokinase; (B) pyruvate kinase; (C) 3-phosphoglycerate kinase; (D) triosephosphate isomerase; (E) enolase (Johansson et al., 1973).

the electrical potential difference will be diminished and will approach that of a dextran–PEG system with salt only. This puts certain limitations to the practical use of charged PEG since some proteins may need a certain amount of salt present to keep them in solution. The salt may be replaced however by a zwitterion, such as glycine or betaine, which may facilitate solubility of the protein and, because they have no net charge, do not affect the electrical potential difference (Fig. 5.8).

Figure 5.9 shows partition profiles of different glycolytic enzymes in a system with TMA-PEG. Each enzyme is extracted into the upper phase at higher pH and the shape of the profile depends on the isoelectric point of the protein, its net charge at different pH, and also the presence of several species of each enzyme. Enzymes, therefore can be separated from each other more effectively by these systems than by the dextran–PEG system alone. Figure 5.10 shows a separation of glycolytic enzymes by countercurrent distribution. See also Chapter 6.

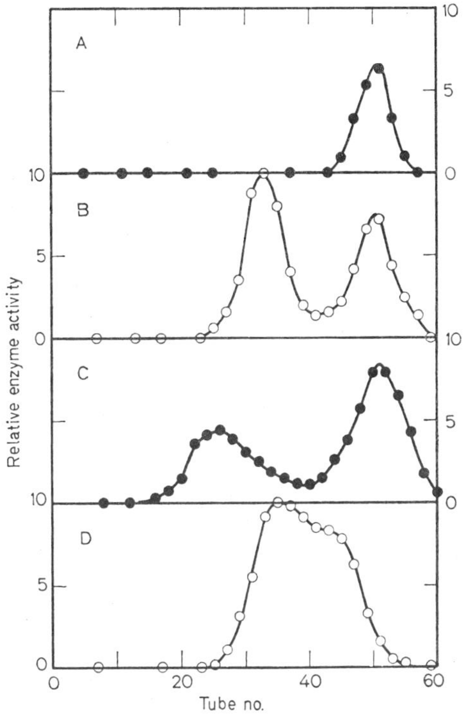

FIGURE 5.10. Countercurrent distribution of glycolytic enzymes from baker's yeast. Phase system 6.6% Dextran 500, 6.4% TMA-PEG, and 5 mM potassium phosphate, pH 6.5. (A) Hexokinase; (B) pyruvate kinase; (C) 3-phosphoglycerate kinase; (D) triosephosphate isomerase (Johansson et al., 1973).

Hydrophobic Affinity Partition

When a hydrophobic group is covalently bound to one of the polymers in a phase system, partition will depend to a high degree on the hydrophobic properties of the proteins. The hydrophobic effect depends on the size of the hydrophobic group bound to the polymer and on the number and binding strength of the hydrophobic areas of the protein molecule.

Johansson (1976) has prepared esters of PEG with fatty acids with different length of hydrocarbon tail. When their effect on the partition of protein was studied it was found that esters with short fatty acids (chain length less than 10) had no effect. An increasing effect was found the longer the fatty acid chain, up

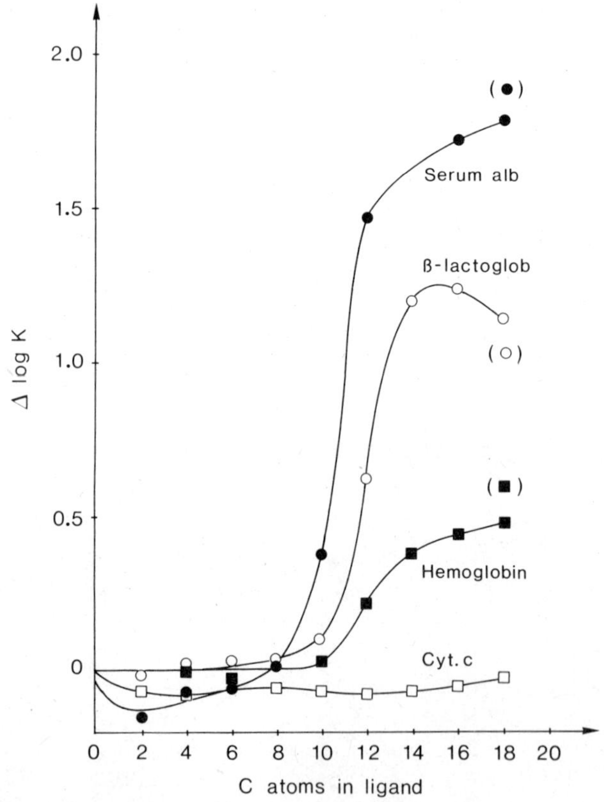

FIGURE 5.11. Effect of chain length of the aliphatic ligand bound to PEG on the partition of proteins. $\Delta \log K$ is the difference in $\log K$ of protein in a phase system of 7% Dextran 500 and 7% PEG 6000 with and without fatty acid esterified groups. When PEG was esterified, 10% of its weight contained PEG ester with a degree of substitution of 13–26% (Shanbhag and Axelsson, 1975).

to palmitate and stearate (Fig. 5.11). Upon increasing concentration of ester, saturated fatty acids esters gave a plateau in the log K, while log K seem to increase continuously with addition of unsaturated ester. Several other hydrophobic ligands, such as esters of deoxycholate and benzonate, have also been tested (Johansson, 1976).

The Hydrophobicity of Proteins Can Be Determined by Hydrophobic Partition. By comparing the partition coefficient of a protein in a dextran–PEG system with that in which part of the PEG is replaced by palmitoyl-PEG one can measure the hydrophobicity of a protein, as described by Shanbhag and Axelsson (1975). The difference in log K of the two systems will be related to the free energy of transfer of the protein from a PEG phase to a PEG phase with palmitoyl group (Fig. 5.12*a*) if all other components such as salts and the temperature of the two systems are the same. It is a very simple and elegant way of measuring the hydrophobicity of proteins and also other macromolecules and particles while still in their native conformation. Figure 5.12*b* shows examples where the hydrophobicity of different proteins are compared. Serum albumin and lactoglobulin, which are known for binding nonpolar compounds, are very effectively extracted into the upper phase even at relatively low concentrations of palmitoyl–polyethylene glycol. Also, hemoglobin and myoglobin display hydrophobicity, but chymotrypsinogen and ovalbumin do not. Interestingly, the hy-

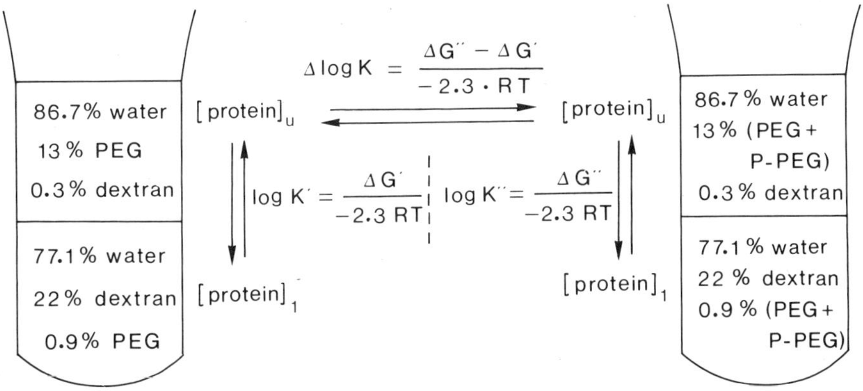

FIGURE 5.12*a*. Determination of hydrophobicity of proteins by difference partition. The total composition of system I is 8% Dextran T70, 8% PEG 6000, 84% H_2O. In system II a part of the PEG is replaced by PEG-palmitate, the rest of the system being identical. $\Delta G'$ and $\Delta G''$ are the free energies of transfer for the protein from the lower to the upper phase in systems I and II, respectively. Δ log K will be the free energy of transfer between the two upper phases and a measure of the interaction between PEG-palmitate and the protein (Shanbhag and Axelsson, 1975).

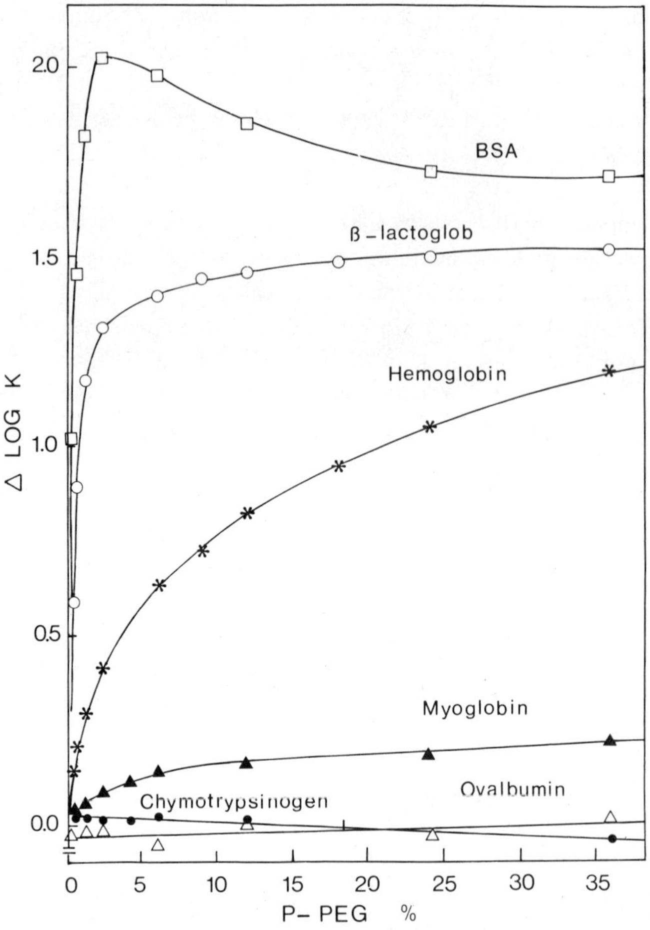

FIGURE 5.12*b*. Hydrophobic partition of proteins. Proteins were partitioned in a dextran–PEG phase system with and without addition of palmitoyl-PEG (P-PEG) Δ log *K* is the difference in log *K* between phase systems with and without P-PEG (see Figure 5.12*a*). The more hydrophobic surface of the protein molecules the stronger interaction with P-PEG and a larger Δ log *K*. All proteins except ovalbumin and chymotrypsinogen display hydrophobicity.

drophobicity of hemoglobin is about four times larger than that of myoglobin. Some proteins show negative hydrophobicity, that is, they are repelled by the palmitoyl-containing phase. These proteins behave as if they were lipophobic. Pancreatic lipase shows such a behavior (Fig. 5.12*c*). This behavior can, however, also be explained by a very different strength of binding between the protein and the hydrophobic ligand in the two phases. If the protein binds the ligand much stronger in the lower phase than in the upper phase, the complex will prefer the

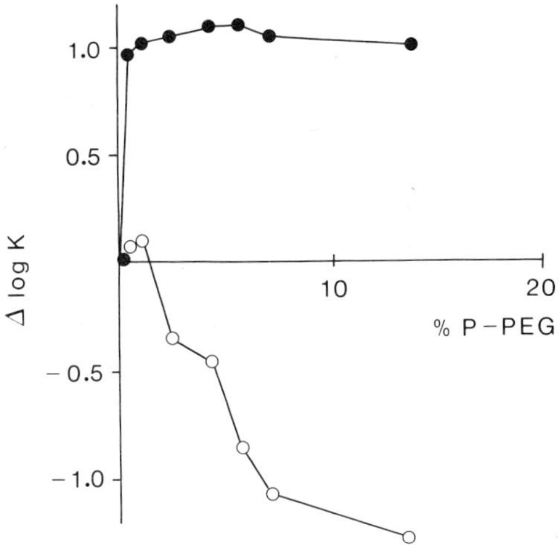

FIGURE 5.12c. Similar experiment as in Figure 5.12*b*. ○ Lipase in contrast to ● serum albumin is repelled by the upper phase. This may be the result of a stronger binding of lipase to P-PEG in the lower phase than the upper phase (Borgström and Erlanson-Albertsson, 1978).

lower phase [Eq. (12), Chapter 10]. Table 5.3 demonstrates the hydrophobicity of histones. By varying the salt content, the pH or the temperature in the two systems one can also study the hydrophobicity of the protein as a function of these parameters (Shanbhag and Axelsson, 1975; Shanbhag and Johansson, 1979).

Biospecific Affinity Partition

Biospecific affinity partition was introduced by Shanbhag and Johansson (1974), who demonstrated that human serum albumin could be selectively extracted from plasma by a PEG-palmitate. Takerkart et al. (1974) showed that the partition of

TABLE 5.3 Effect of Hydrophobic Ligands on PEG on the Partition of Histones[a]

Histone	Δ log[b] K in 10 mM NaCl	10 mM K_2SO_4
1	0.03	0.5
2a	1.5	2.4
2b	0.5	1.3
3	0.6	2.9
4	2.1	1.9

[a]Phase system Dextran 70–PEG 6000 with or without 20% of its weight replaced by PEG–palmitate; 2 mM potassium phosphate, pH 7.1, and two different salts (Axelsson and Shanbhag, 1976).
[b]Δ log K is the difference in log partition coefficient between a phase system with and without PEG–palmitate.

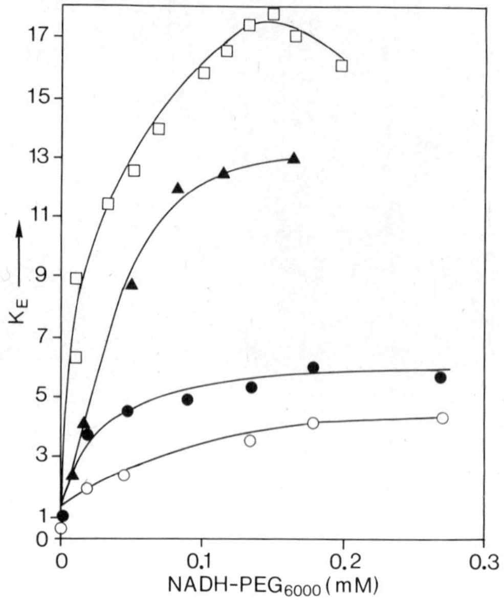

FIGURE 5.13. Affinity partition of dehydrogenases. The partition coefficient of the enzymes are plotted as a function of NADH ligand coupled to PEG. Phase system: 6% w/w Dextran 500, 7% w/w PEG 4000, 50 m*M K* phosphate, pH 7.5, and 0.1 m*M* mercaptoethanol, 20°C. □, Lactate dehydrogenase (rabbit muscle); ▲, alcohol dehydrogenase (yeast); ●, formate dehydrogenase (*Candida boidinii*); ○, formaldehyde dehydrogenase (*Candida boidinii*) (Kula et al., 1982).

trypsin was changed by introducing a competitive trypsin inhibitor, diamidinodiphenylcarbamyl, as a ligand coupled to PEG. Flanagan and Barondes (1975) used dinitrophenyl-PEG, which binds specifically to S-23 myeloma protein, to shift its partition in favor of the upper PEG phase of a dextran–PEG system. Hubert et al. (1976) used PEG–estradiol for purifying a 3-oxosteroid isomerase. Aminoethyl-NADH coupled to PEG was used for affinity partition of formate and formaldehyde dehydrogenases (Fig. 5.13) (Kula et al., 1978, 1982).

The triazine dye Cibachrome blue, which binds to nucleotide binding sites of enzymes, has been covalently coupled to PEG and used for affinity partition of several enzymes (Kopperschläger and Johansson, 1982; Kopperschläger et al., 1983; Johansson, 1984; Johansson and Andersson, 1984a, 1984b; Johansson et al., 1983, 1984).

Figure 5.14 shows countercurrent distribution of different enzymes in phase systems with and without a polymer-bound ligand.

Affinity partition can also be used for studying the binding of different ligands to a protein. For example, by studying the ability of free ligand to compete with

TABLE 5.4 Applications of Biospecific Affinity Partition on Proteins

Protein	Ligand	Reference
Trypsin	Diamino-α, ω-diphenyl carbamyl	Takerkart et al. (1974)
Serum albumin	Fatty acid	Shanbhag and Johansson (1974, 1979); Johansson (1976)
β-Lactoglobulin	Fatty acid	Shanbhag and Axelsson (1975)
Concanavalin	—	Flanagan and Barondes (1975)
S-23 myeloma protein	Dinitrophenyl	Flanagan and Barondes (1975)
Histones	Fatty acid	Axelsson and Shanbhag (1976)
3-Oxosteroid isomerase	Estradiol	Hubert et al. (1976)
Formaldehyde dehydrogenase	NADH	Kula et al. (1979)
Formate dehydrogenase		
Colipase	Lecithin	Erlanson-Albertsson (1980)
Myosin	Fatty acid	Pinaev et al. (1982)
Phosphofructokinase	Triazine dye	Kopperschläger and Johansson (1982); Johansson et al. (1983, 1984)
Interferon	Phosphate	Menge et al. (1983)
Pyruvate kinase	Triazine dye	Kopperschläger et al. (1983)
Glutamate dehydrogenase	Triazine dye	Kopperschläger et al. (1983)
Glycerol kinase	Triazine dye	Kopperschläger et al. (1983)
Hexokinase	Triazine dye	Kopperschläger et al. (1983)
Lactate dehydrogenase	Triazine dye	Kopperschläger et al. (1983)
Malate dehydrogenase	Triazine dye	Kopperschläger et al. (1983)
Transaminase	Triazine dye	Kopperschläger et al. (1983)
α-Fetoprotein	Triazine dye	Birkenmeier et al. (1984a)
Pre-albumin	Remazol yellow	Birkenmeier et al. (1984a, 1984b)
Glucose 6-phosphate dehydrogenase	Triazine dye	Johansson and Andersson (1984b); Johansson et al. (1984); Kroner et al. (1982)
Glyceraldehyde phosphate dehydrogenase	Triazine dye	Johansson and Andersson (1984b); Johansson et al. (1984)
3-Phosphoglycerate kinase	Triazine dye	Johansson and Andersson (1984b); Johansson et al. (1984)
Alcoholdehydrogenase	Triazine dye	Johansson and Andersson (1984b); Johansson et al. (1984)
Nitrate reductase	Triazine dye	Schiemann and Kopperschläger (1984)

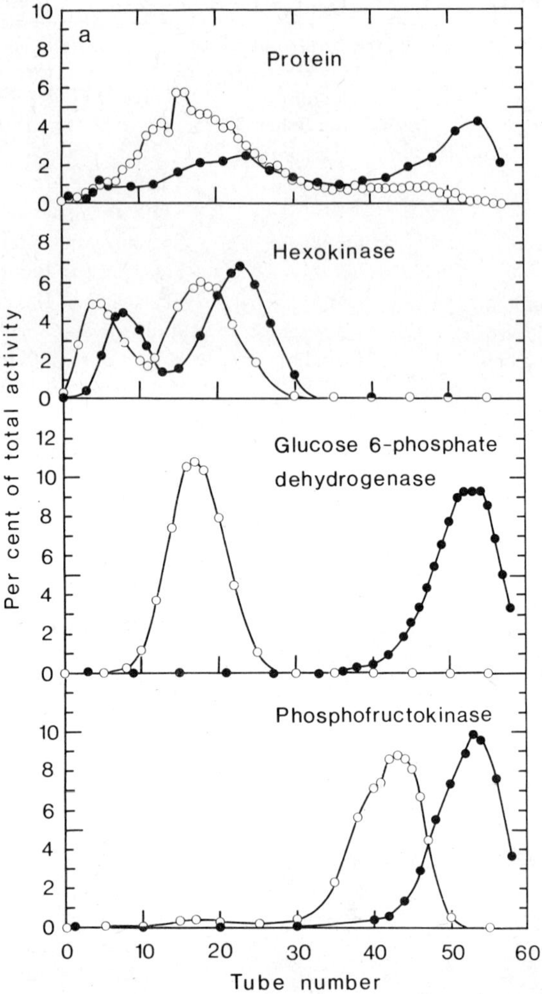

FIGURE 5.14. Countercurrent distribution of yeast extract in a phase system with Cibachrome Blue bound to PEG. Phase system: 7% w/w Dextran 500, 5% w/w PEG 6000, 50 m*M* Na phosphate, pH 7.0, 0.2 m*M* EDTA, and 5 m*M* 2-mercaptoethanol; with 0.5% (○) or 5% (●) of total PEG in the form of Cibachrome Blue-PEG (Johansson et al., 1984).

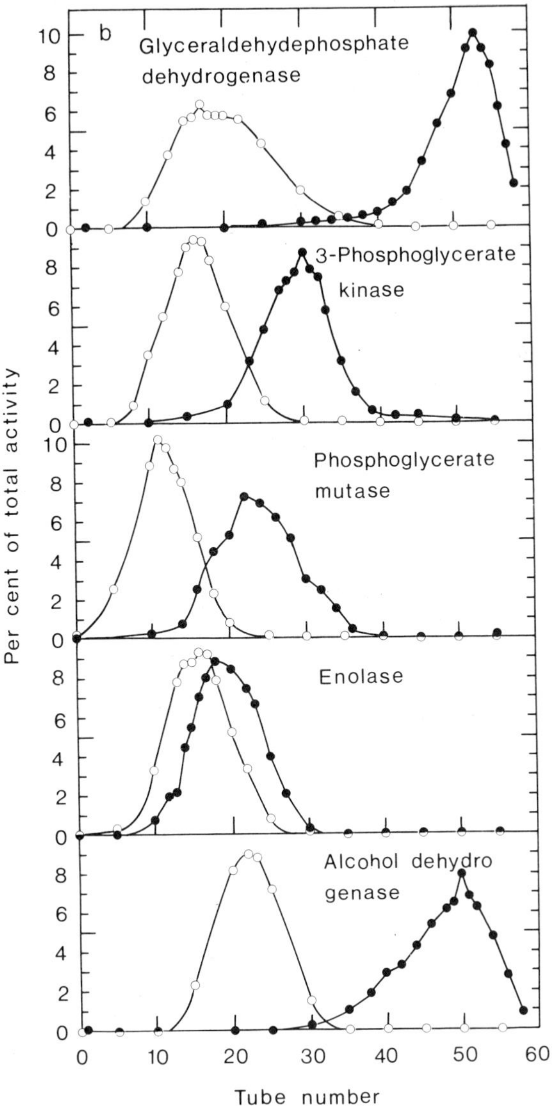

b
Glyceraldehydephosphate dehydrogenase
3-Phosphoglycerate kinase
Phosphoglycerate mutase
Enolase
Alcohol dehydro genase
Per cent of total activity
Tube number
0
10
20
30
40
50
60
2
4
6
8

Cibachrome Blue–PEG, information on the binding strength of the ligands to the protein was obtained (Kopperschläger et al., 1983).

Binding between colipase and lecithin–PEG was used to determine the role of tyrosine in the binding between this enzyme and lecithin (Erlanson-Albertsson, 1980). While native colipase showed binding to lecithin–PEG, modified colipase, where the tyrosine groups had been acetylated, did not.

Table 5.4 gives examples of biospecific partition of proteins.

Other Phase Systems

Other phase systems which have been used for partition of proteins include: dextran–methylcellulose, dextran–Ficoll, dextran–Ficoll–PEG, starch–PEG, and salt–PEG.

The dextran–methylcellulose system has comparatively high viscosity and long settling time. This can be reduced, however, by low-speed centrifugation. There is a fairly good relation between the molecular weight of the protein and its partition coefficient in this system (Table 5.5 and Fig. 4.3). Even very-high-molecular-weight proteins, such as the hemocyanins (MW 1–10 × 10^6) are soluble in the dextran–methylcellulose system.

The dextran–Ficoll system is also fairly viscous, and it has, therefore, a rather long settling time. This can be reduced by centrifugation and also by including

TABLE 5.5 Partition Coefficient *K* and Surface Area of a Number of Protein and Virus Particles[a]

Particle	Surface area $(\mu m)^2$ ($\times 10^3$)	K
Phycoerythrin	0.3	0.95
Hemocyanin, "eighth"	0.86	0.65
Hemocyanin, "whole"	3.5	0.25
ECHO virus	2.3	0.2
Polio virus	1.3	0.3
Southern bean Mosaic virus	2.5	0.41
Phage ΦX 174	2.8	0.34
Phage T3	8.7	2.1×10^{-2}
Tobacco mosaic virus	14.4	$1–2 \times 10^{-2}$
Phage T2	25.5	$6–10 \times 10^{-4}$
Phage T4	25.5	$3–5 \times 10^{-4}$
Vaccinia	220	$4–12 \times 10^{-5}$

[a]Dextran–methylcellulose system D68-MC4000 (Albertsson, 1962b).

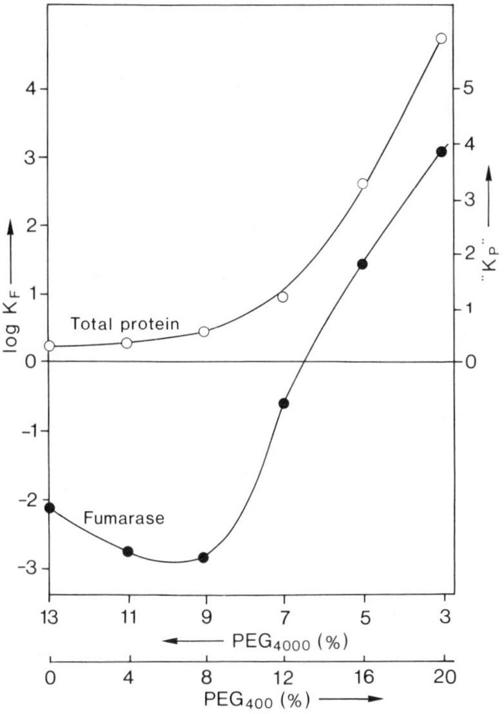

FIGURE 5.15. Partition of fumarase (K_F) and total protein (K_p) as a function of the PEG 400 + PEG 4000 content in a PEG phosphate system containing crude extract of Brevibacterium ammoniagenes (Kula et al., 1982).

a certain amount of PEG. For example, a system of 6% w/w Dextran 500, 5% w/w Ficoll, and 2% w/w PEG 6000 is useful for partition of proteins. Proteins which have a low solubility in the dextran–PEG system may have a better solubility in the dextran–Ficoll or dextran–Ficoll–PEG system since it is usually the presence of PEG which reduces the solubility of proteins in dextran–PEG systems.

The soluble starch–PEG phase system is attractive for technical applications because of its low price. The factors which determine the partition of proteins in this phase system are about the same as for the dextran–PEG system, that is, ionic composition, molecular weight, of polymer and so forth.

The salt–PEG system differs from the previous systems in that the ionic strength of the phases is very high, since salts such as $(NH_4)_2SO_4$, $MgSO_4$, or K_3PO_4 have to be used in the concentration range of about 0.5–2 *M*. These systems, however, can be used with advantage for purification of enzymes (Fig. 5.15) (Kula et al., 1982; Menge et al., 1983). (See also Chapter 9.)

TABLE 5.6 Approximate Concentrations of Polyethylene Glycol (PEG 6000) in Different Salt Media at which Proteins Start to Precipitate[a]

	Salt		
Electrolyte Composition	Serum Albumin	Ceruloplasmin	Phycoerythrin
0.005 M KH_2PO_4 + 0.005 M K_2HPO_4	40	50	>50
0.005 M KH_2PO_4 + 0.005 M K_2HPO_4 + 0.1 M KCl	22	12	>50
0.05 M KH_2PO_4 + 0.05 M K_2HPO_4	50	27	>50

[a]Concentration PEG 6000 in % w/w (Albertsson 1960).

Solubility of Proteins in PEG Solutions

As first demonstrated in the first edition of this book (Albertsson, 1960), different proteins are precipitated by different concentrations of polyethylene glycol in a solution. Precipitation of proteins by PEG is thus selective and may be used for purification purposes (Polson et al., 1964; Zeppezauer and Brishammar, 1965). It should be stressed, however, that the precipitated protein phase is not always a solid precipitate; frequently it is rather a highly concentrated protein solution of high viscosity. In fact, it may be considered as a phase of the two-phase system of protein, PEG, and solvent in a way similar to dextran, PEG, and solvent. A phase diagram of serum albumin–PEG–H_2O has also been described (see Chapter 2, Edmond and Ogston, 1968).

The concentration of PEG necessary for precipitation of a protein depends not only on the kind of protein but also on the ionic composition (see Table 5.6). Phycoerythrin is not precipitated in any salt media shown in Table 5.6, while ceruloplasmin and serum albumin are. The solubility of serum albumin and ceruloplasmin in PEG becomes less when KCl is added to the phosphate buffer. Qualitatively there is a correlation between high solubility in PEG, and a high K value in the dextran–PEG phase system. Phycoerythrin, for example, has the highest K value and also the highest solubility. Addition of KCl lowers the K value of serum albumin and ceruloplasmin and also lowers their solubility in PEG. In Chapter 2 it was mentioned that phase separation in mixtures of a polyelectrolyte (Na dextran sulfate or Na carboxymethylcellulose) and polyethylene glycol in water depends highly on the electrolyte content. Qualitatively, there is a striking similarity between the salt effect on these phase separations and the solubility behavior of proteins in polyethylene glycol solutions.

Protective Effect of Polymers

No sign of denaturation and loss in biological activity, such as enzyme activity, or virus infectivity, or change in color or solubility, has so far been encountered after a partition experiment with polymer two-phase systems when the partition is carried out at a suitable pH or salt concentration.

The polymer–polymer phase systems do not thus appear to affect the proteins irreversibly, which is probably due mainly to the high water content (80–99%) of both phases and the small liquid–liquid interfacial tension. However, the possibility that the presence of the polymers themselves stabilizes the proteins should also be considered. The polymers used are polysaccharides or polyols which are known to stabilize certain enzymes against inactivation. Thus, it is well known that many enzymes and proteins are stabilized against heat denaturation or freeze drying by such substances as glycerol, glucose, sucrose, starch, dextrins, amino acids, and proteins. Viruses are stabilized against gas–liquid interface inactivation by proteins. These also protect viruses against inactivation when subjected to sonic treatment.

The protective effect of some polymers used here may be easily demonstrated on egg albumin, which is rather sensitive to surface denaturation. If a test tube containing a solution of egg albumin is shaken for a while, the solution becomes turbid due to precipitated denatured egg albumin. If, however, methylcellulose or polyethylene glycol are previously added, the solution remains clear, even after prolonged, vigorous shaking (Albertsson, 1960, 1971). This protective effect of the polymers is particularly useful when a large number of shakings are necessary for an experiment with a countercurrent distribution apparatus.

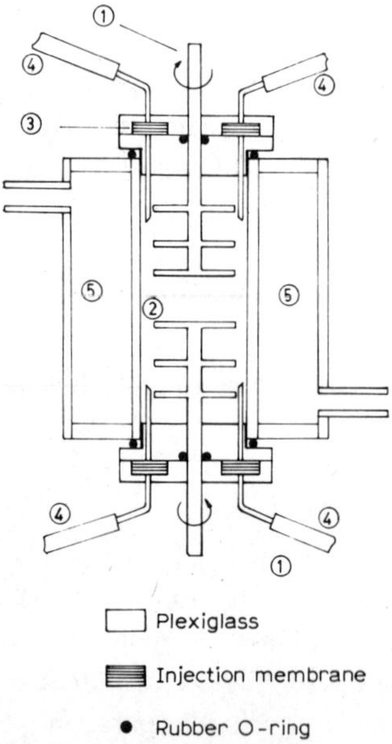

FIGURE 5.16*a*. Diffusion cell for measuring the transport of proteins across the interface of a liquid–liquid two-phase system. At zero time the protein is present in one of the phases only. Its concentration in the two phases is then measured until equilibrium is reached. (1) Stirrers; (2) (dashed line) interface; (3) injection membrane; (4) tubing to and from photometer; (5) water jacket (Shanbhag, 1973).

Diffusion of Proteins through the Interface

It is a general experience that equilibrium between the phases is reached quickly, in a matter of seconds, by shaking of an aqueous polymer two-phase system. This is probably due to the low interfacial tension which allows a very fine dispersion of one phase in the other, even on gentle mixing.

The rate of diffusion of various substances including proteins through a resting interface of the dextran–polyethylene glycol system has been studied by Shanbhag (1973). A top phase containing the protein was layered over the bottom phase without protein (Fig. 5.16*a*). Each phase was constantly stirred without disturbing the interface. By continuously monitoring the increase of protein in the bottom phase the rate of transport through the interface could be determined. In a similar experiment a top phase without protein was layered over a bottom phase with protein in order to study the transport from the bottom phase through the interface to the top phase. The rate depends on the molecular weight and the partition coefficient of the diffusing substance. When the protein is diffusing downhill, that is, from the phase with lesser affinity for the protein to the other phase, the rate of diffusion through the interface depends only on the diffusion coefficient of the protein (Fig. 5.16*b*). If, however, the protein moves uphill,

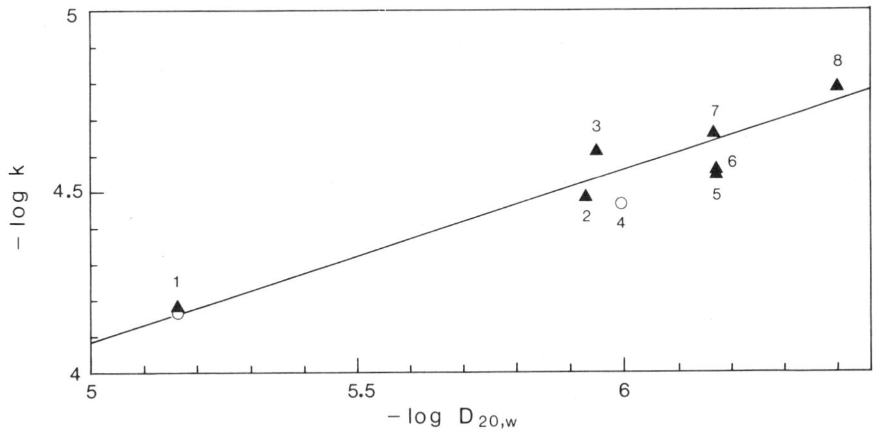

FIGURE 5.16*b*. Plot of diffusion of molecules downhill through the interface against the diffusion constant ($D_{20,w}$); k is the interfacial transfer coefficient for diffusion from the lower to the upper phase for substances with partition coefficient greater than one (○) or the interfacial transfer coefficient for diffusion from the upper to the lower phase for substances with partition coefficient less than one (▲). 1, phenylalanine; 2, ribonuclease; 3, myoglobin; 4, chymotrypsin; 5 and 6, hemoglobin (human); 7, hemoglobin (pig); 8, phycoerythrin.

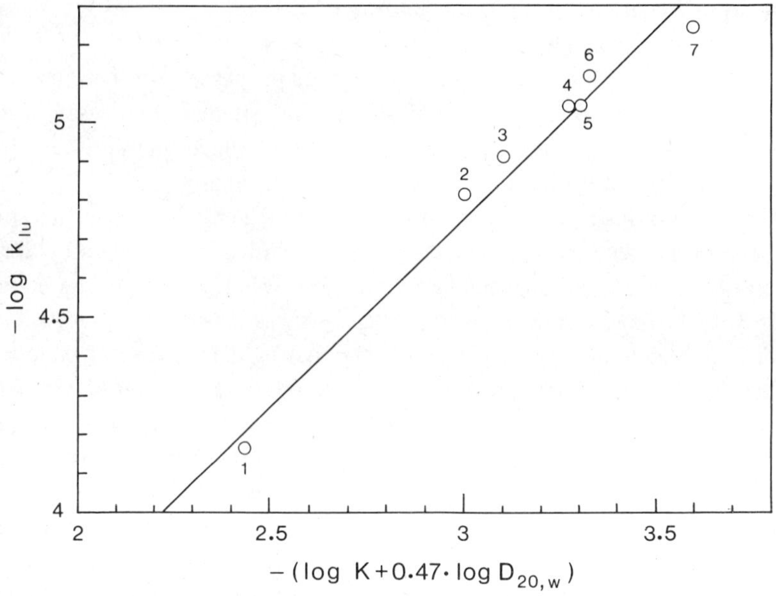

FIGURE 5.16c. Plot of diffusion of molecules uphill through the interface as a function of partition coefficient *K* and diffusion constant *D*. k_{lu} is the interfacial transfer coefficient for diffusion from the lower to the upper phase for substances with partition coefficients less than one. 1, phenylalanine; 2, ribonuclease; 3, myoglobin; 4 and 5, hemoglobin (human); 6, phycoerythrin; 7, hemoglobin (pig) (Shanbhag, 1973).

that is, into the phase with lesser affinity, then the rate of transport through the interface depends both on its diffusion constant and the partition coefficient (Fig. 5.16*c*). The interface offers no significant resistance to diffusion of the substances studied.

NUCLEIC ACIDS

The behavior of nucleic acids in the dextran–PEG system has been studied in detail (Albertsson, 1965). As with proteins, the partition coefficient depends on the molecular weights of the two polymers, the polymer composition, and the ionic composition. Partition of nucleic acids is extremely sensitive to the ionic composition, much more than the partition of proteins described previously in this chapter. The molecular weight and the conformation of the nucleic acids also influence their partition (Albertsson, 1962, 1965).

Influence of Salts

The partition coefficients of DNA in a dextran–PEG system in the presence of 10 m*M* Na phosphate, pH 6.8, and different salts is shown in Figure 5.17.

With the phosphate buffer alone, the *K* value of DNA in this system is 10. With different electrolytes, the *K* value obtained is lower than that with the buffer alone. The different ions follow the same series as for the proteins (Fig. 5.1).

Thus the chlorides lower the *K* value in the order Li < NH_4 < NaCl < Cs < K < Rb. This order is independent of the anion used.

The sodium halides lower the *K* values in the order $F^- < Cl^- < I^- < Br^-$. Experiments with other halides indicate that this order is independent of the kind of cation used.

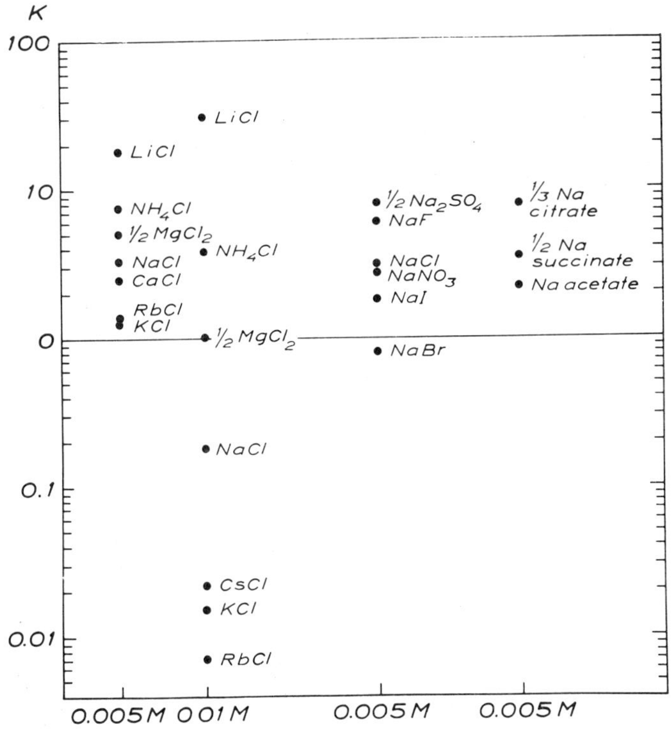

FIGURE 5.17. Partition coefficients of DNA in a dextran–PEG system with the following composition = 5% w/w Dextran 500, 4% w/w PEG 6000, 0.005 *M* NaH_2PO_4, 0.005 *M* Na_2HPO_4, and different electrolytes; the concentration of the latter are indicated in the bottom row (from Albertsson, 1965). Temperature: 4–5°C.

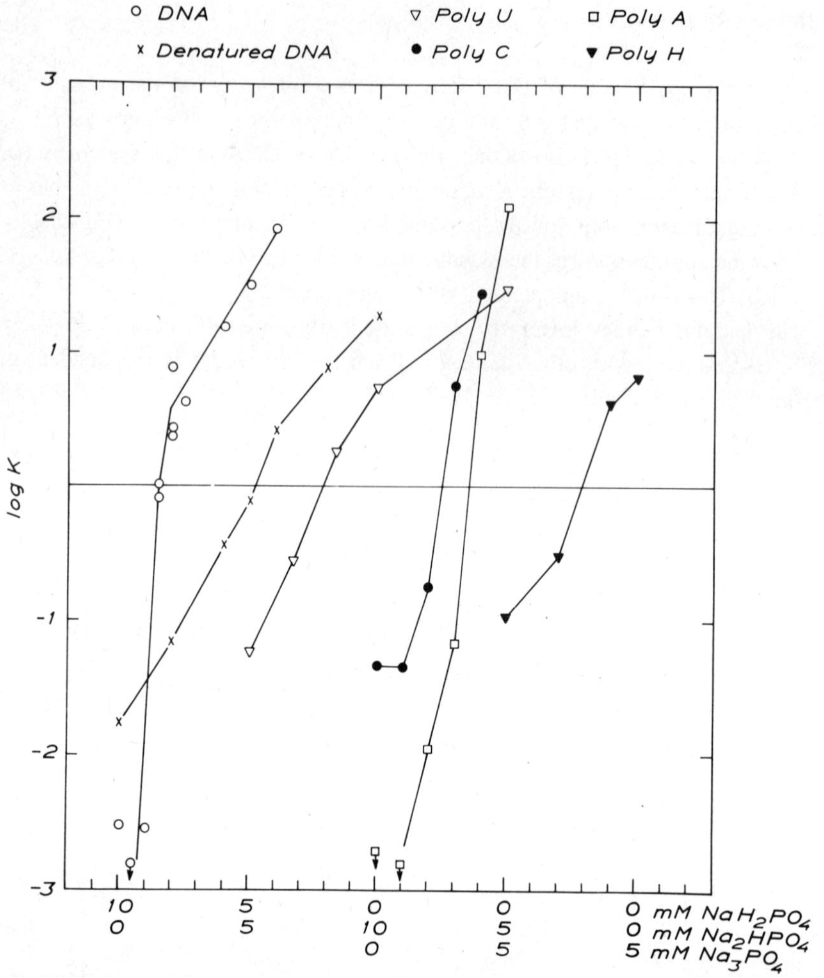

FIGURE 5.18. Partition coefficient of different nucleic acids in a dextran–PEG system as a function of different mixtures of sodium phosphates. Polymer composition: 5% w/w Dextran 500–4% w/w PEG 6000. Temperature: 4–5°C (Albertsson, 1965).

The divalent or trivalent anions monohydrogen phosphate, sulfate, and citrate give higher K values than the monovalent anions.

The effects of the different ions are additive. The highest K values are obtained with Li_2HPO_4, Li_3PO_4, or Li_2SO_4 and the lowest with KCl, KNO_3, KBr, or the corresponding Cs and Rb salts.

As with proteins, the ratio between the different ions has a very strong influence on the K value. This is illustrated in Table 5.7 and Figure 5.18. Table 5.7 gives

TABLE 5.7 Partition Coefficient of DNA in a Dextran–Polyethylene Glycol System[a]

NaCl (M)	LiCl (M)	Partition Coefficient
0.1	0	<0.02
0.075	0.025	<0.02
0.042	0.058	<0.02
0.032	0.068	0.04
0.024	0.076	6.5
0.018	0.082	<10

[a](5% w/w Dextran 500, 4% w/w PEG 6000) with 0.1 M Cl^- and different ratios between Na^+ and Li^+ (Albertsson, 1965).

K values for DNA in LiCl and NaCl mixtures at constant ionic strength but with different Li^+/Na^+ ratios. As can be seen, the K value shifts from 0.02 to 10 with a small change in the LI^+/Na^+ ratio. In Figure 5.18 the K values for some nucleic acids with different sodium phosphate mixtures are shown. Here we can study the effect of different ratios of the different phosphate ions. In 0.01 M

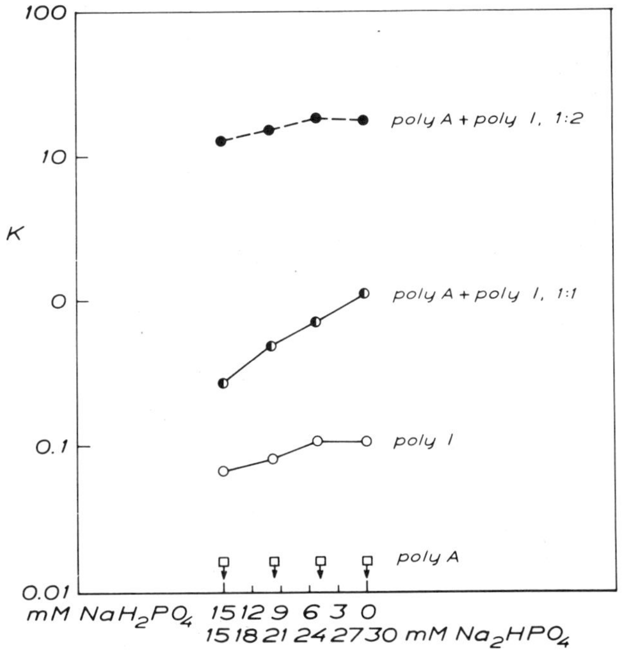

FIGURE 5.19. Partition of poly A and poly I separately or mixed in 1:1 or 1:2 ratios. Phase system as in Figure 5.18 (Albertsson, 1965).

NaH_2PO_4, for example, DNA is in the lower phase. If a fraction of the $H_2PO^-{}_4$ ions are replaced by $HPO^{2-}{}_4$ ions, the DNA is transferred to the upper phase, the transfer taking place over a small change in the ratio of the two phosphate ions.

The total ionic strength, with the ratio of the different ions being constant, also influences the *K* value, (Fig. 5.19). This is in contrast to the behavior of proteins, whose *K* values are influenced to a much smaller degree by the ionic strength.

Single-Stranded and Double-Stranded Nucleic Acids Can Be Separated by Partition

There is a great difference between the partition of native, double-stranded DNA and denatured, single-stranded DNA as may be seen in Figure 5.18. A strong influence of the conformation of a nucleic acid on the *K* value has also been demonstrated with mixtures of the homopolymers polyadenylic acid (Poly A), polyuridylic acid (Poly U), polycytidylic acid (Poly C), and polyinosinic acid (Poly I). It is known from a number of investigations that double- or triple-stranded structures are formed when Poly A and Poly U are mixed in certain proportions and in suitable salt media. Also, Poly A and Poly I or Poly I and Poly C can form such complexes. The partition coefficients of Poly A and Poly I, either alone or mixed are shown in Figure 5.19. In all cases, the *K* values of the mixtures Poly A + Poly U, Poly A + Poly I, and Poly I + Poly C are much higher than the corresponding single-stranded molecules. In contrast, the mixtures Poly A + Poly C, Poly U + Poly C or Poly I + Poly U in the same phase systems gave *K* values which one could expect when there was no interaction between the two polynucleotides.

In the case of the Poly A + Poly I mixture, the *K* value depends on the ratio between the concentrations of Poly A and Poly I (expressed as the concentration ratio between the two bases). Thus, the mixture 1 Poly A + 2 Poly I, which leads to a triple-stranded form, Poly(A + 2 I), has a much higher *K* value than the mixture 1 Poly A + 1 Poly I. The question then arises whether the 1:1 mixture consists of only double-stranded molecules, Poly(A + I), or of 50% triple-stranded Poly(A + 2I) molecules. In order to answer this question a countercurrent distribution of the 1:1 mixture was carried out (Albertsson, 1965). The result showed that the 1:1 mixture is composed of essentially two fractions, one having a high *K* value (>10) and moving with the upper phase and the other having a low *K* value (<0.1) and remaining in the lower phase. The 1:1 mixture thus cannot be composed of double-stranded molecules; it must be a mixture of single-stranded and triple-stranded molecules.

Pettijohn (1967) made the interesting observation that supercoiled DNA molecules differ in partition compared to DNA lacking this topology. The K value of supercoiled (form I) polyoma DNA is six- to twelve-fold smaller than the open circle and linear forms II and III. This difference may arise from the more compact conformation of supercoiled DNA relative to the open circle and linear forms or from a difference in exposure of partially uncoiled helical structures.

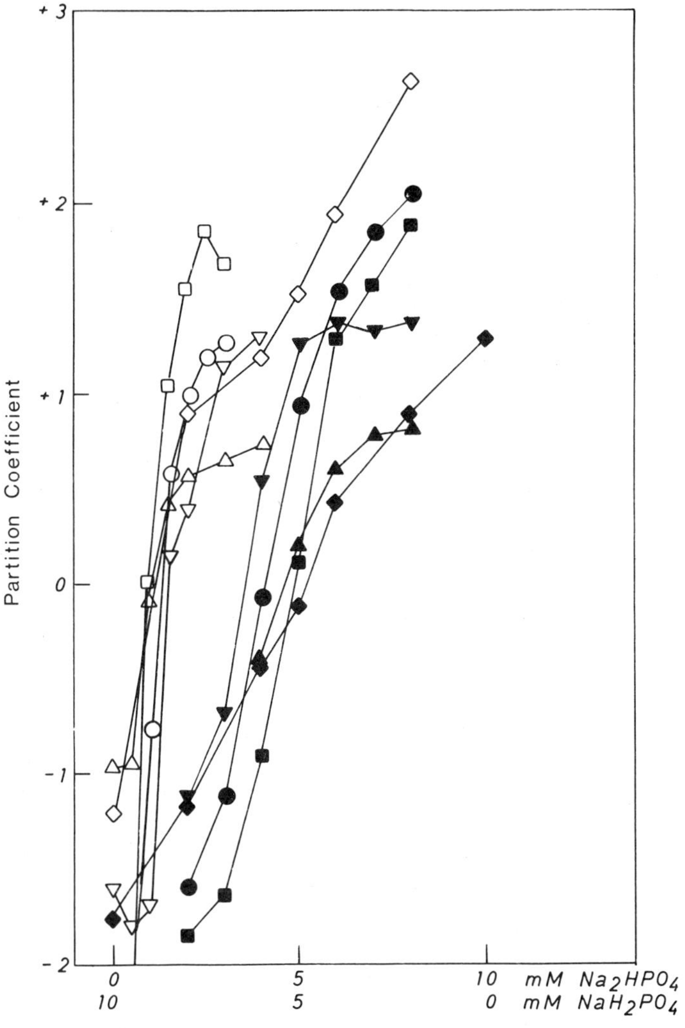

FIGURE 5.20. Partition of DNA of different origins in the same phase system as Figure 5.18. Open symbols represent native and filled denatured DNA. The different symbols represent: □, Phage T2; ○, *Eshcerichia coli*; ▽, *Micrococcus lysodeiticus*; ◇, Calf thymus; △, *Bacillus subtilis*.

Base Composition of DNA and Its Partition

The partition behavior of DNA from different species is shown in Figure 5.20. It may be seen that the curves for native DNA are very close to one another, whereas, denatured DNA from different species partition differently. The difference between native and denatured DNA is probably due to more exposure of the bases in the denatured state.

Except for denatured phage T2 DNA, there is a tendency for a higher GC content of the denatured DNA to give a higher *K* value. That phage DNA does not fit this pattern may be due to the glucosylated phases. The number of DNA molecules of diverse origin tested is too small, however, to make this tendency a general rule.

Affinity Partition Separates DNA According to Base Composition

The influence of base composition can be enhanced by using base-specific ligands coupled to PEG. Thus Müller and Eigel (1981) coupled a GC specific ligand (1,2-dimethyl-3-amino-5-(4′carboxylphenyl)-7-dimethylaminophenzinium) covalently to PEG and used this polymer in a dextran–PEG phase system for base-specific affinity partition of DNA. They could show that when the ligand was bound to PEG it was still base-specific in its interaction with DNA, although its binding strength decreased somewhat. Added to a phase system, one ligand–PEG per 50 base pairs of DNA could shift its partition coefficient about three orders of magnitude. Figure 5.21 shows the specific effect of the ligand–PEG on the partition of two different bacterial DNA. For GC contents around 50%,

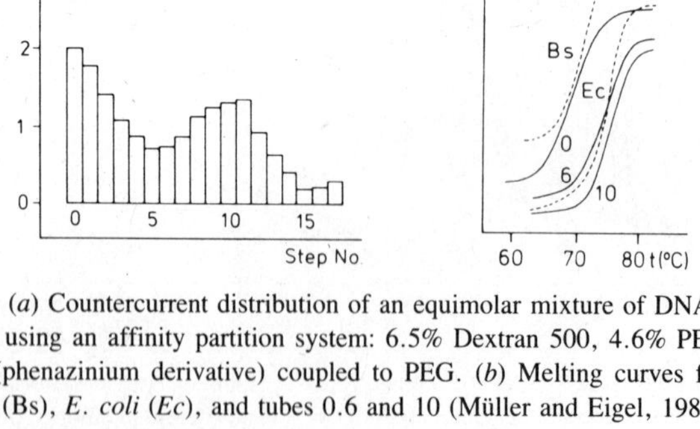

FIGURE 5.21. (*a*) Countercurrent distribution of an equimolar mixture of DNA from *E. coli* and *Bacillus subtilis* using an affinity partition system: 6.5% Dextran 500, 4.6% PEG 6000 with GC-specific ligand (phenazinium derivative) coupled to PEG. (*b*) Melting curves for the DNA from *Bacillus subtilis* (Bs), *E. coli* (*Ec*), and tubes 0.6 and 10 (Müller and Eigel, 1981).

1% difference in GC content yields a change in the partition coefficient by a factor of 1.25. From this one can calculate that about 15 countercurrent distribution steps would be enough to separate two DNAs differing by about 8% in GC. Satellites of calf thymus DNA have been separated by ten distribution steps. Müller and Eigel (1981) also describe procedures for removing DNA from the polymers.

DNA Can Be Separated According to Its Molecular Weight

The partition of DNA depends on its size. For a dextran–methylcellulose and a dextran sulfate–methylcellulose system the logarithm of the partition coefficient is linearly related to the sedimentation coefficient. Also, for the more convenient phase system dextran–PEG the partition of DNA depends markedly on its size (Lif et al., 1961; Frick and Lif, 1962). Müller et al. (1979) studied this effect in detail using restriction enzymes to obtain defined fragments of DNA. Based on these studies they could fractionate DNA fragments in the range 20–220,000 base pairs by liquid–liquid chromatography employing the dextran–PEG system, [Figs. 6.19–6.20 (Müller et al., 1979; Müller and Kutemeier, 1982)].

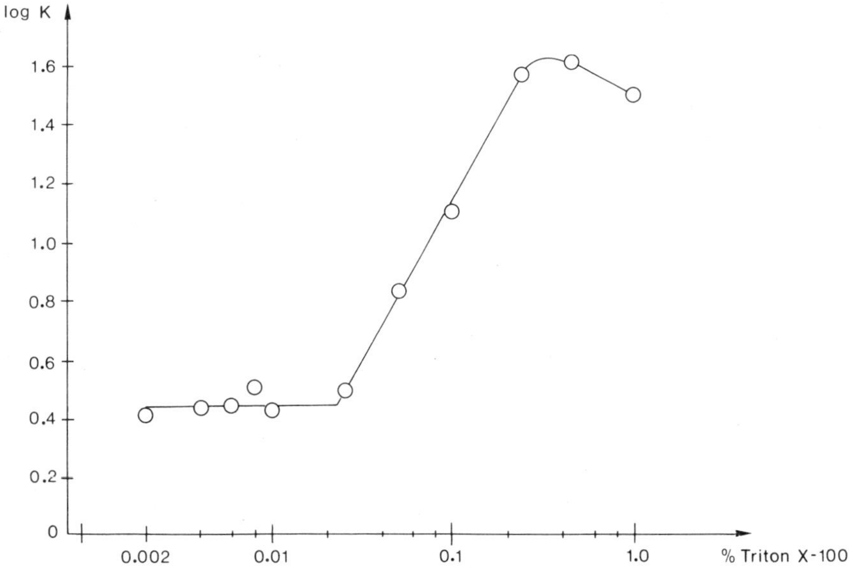

FIGURE 5.22. Partition of Triton X-100 in a phase system of 8% w/w Dextran 70 and 8% w/w PEG 6000 and 2 m*M* K phosphate, pH 6.0. The critical micelle concentration is 0.023% w/w Triton (Axelsson, 1978).

DETERGENTS

The partition of detergents depends on the type of hydrophobic and hydropholic parts of the detergent molecule and also on the critical micelle concentration.

Figure 5.22 shows the partition coefficient of Triton X-100 at different concentrations. Below the critical micelle concentration the partition coefficient is independent of detergent concentration. A sudden increase in partition coefficient indicates onset of micelle formation. In fact, such an experiment can be used to determine the critical micelle concentration. At sufficiently high Triton concentrations (more than 1.5% w/w) the detergent will separate out as a phase, and a three-phase system is obtained.

Octylglucoside partitions evenly between the phases, and no indication of micelle formation is formed (Svensson et al., 1985). Sodium dodecylsulphate also partitions evenly between the phases while digitonin has a partition coefficient less than one, that is, it prefers the bottom phase.

REFERENCES

Albertsson, P.-Å. (1958). *Nature,* **182,** 709.

Albertsson, P.-Å. (1960). *Partition of Cell Particles and Macromolecules, 1st ed.,* Almquist and Wiksell, Stockholm, and Wiley, New York.

Albertsson, P.-Å. (1962a). *Arch. Biochem. Biophys.,* **1,** 264.

Albertsson, P.-Å. (1962b). *Meth. Biochem. Anal.,* **10,** 229–262.

Albertsson, P.-Å. (1965). *Biochim. Biophys. Acta,* **103,** 1.

Albertsson, P.-Å. (1970). *Advances in Protein Chemistry,* **24,** 309.

Albertsson, P.-Å. (1971). *Partition of Cell Particles and Macromolecules, 2nd ed.,* Almquist and Wiksell, Stockholm, and Wiley, New York.

Albertsson, P.-Å., and Nyns, Ed. J. (1959). *Nature,* **184,** 1465.

Albertsson, P.-Å., and Nyns, Ed. J. (1961). *Arkiv kemi,* **17,** 197.

Albertsson, P.-Å., Sasakawa, S., and Walter, H. (1970). *Nature,* **228,** 1329.

Axelsson, C.G. (1978). *Thesis,* Umeå University, Umeå., Sweden.

Axelsson, C.G., and Shanbhag, V.P. (1976). *Eur. J. Biochem.,* **71,** 419–423.

Birkenmeier, G., Usbeck, E., and Kopperschläger, G. (1984a). *Anal. Biochem.,* **136,** 265–271.

Birkenmeier, G., Tschechonien, B., and Kopperschläger, G. (1984b). *FEBS Lett.,* **174,** 162–166.

Borgström, B., and Erlanson-Albertsson, C. (1978). *Gastroenterol.,* **75,** 382–386.

Erlanson-Albertsson, C. (1980). *FEBS Lett.,* **117,** 295.

Flanagan, S.D., and Barondes, S.H. (1975). *J. Biol. Chem.,* **250,** 1484.

Frick, G., and Lif, T. (1962). *Arch. Biochem. Biophys. Suppl.,* **1,** 271.

Hubert, P., Dellacherie, E., Neel, J., and Baulieu, E.-E. (1976). *FEBS Lett.,* **65,** 169.

Johansson, G. (1970a). *Biochim. Biophys. Acta,* **221,** 387.

Johansson, G. (1970b). *Biochim. Biophys. Acta,* **222,** 381.

Johansson, G. (1971). *Proceedings International Solvent Extraction Conference 1971, The Hague, Holland,* The Society of Chemical Industry, London.

Johansson, G. (1976). *Biochim. Biophys. Acta,* **451,** 517–529.

Johansson, G. (1984). *Meth. Enzymol.,* **104,** 356.

Johansson, G., and Andersson, M. (1984a). *J. Chromatogr.,* **291,** 175.

Johansson, G. and Andersson, M. (1984b). *J. Chromatogr.,* **303,** 39–51.

Johansson, G., Hartman, A., and Albertsson, P.-Å. (1973). *Eur. J. Biochem.,* **33,** 379–386.

Johansson, G., Kopperschläger, G., and Albertsson, P.-Å. (1983). *Eur. J. Biochem.* **131,** 589.

Johansson, G., Andersson, M., and Åkerlund, H.-E. (1984). *J. Chromatogr.,* **298,** 483–493.

Kopperschläger, G., and Johansson, G. (1982). *Anal. Biochem.,* **124,** 117.

Kopperschläger, G., Lorenz, G., and Usbeck, E. (1983). *J. Chromatogr.,* **259,** 97–105.

Kroner, H.K., Cordes, A., Schelper, A., Morr, M., Bückmann, A. F., and Kula, M.-R. (1982). In T.C.J. Gribnau et al., Eds., *Affinity Chromatography and Related Techniques,* Elsevier, Amsterdam, pp. 491–501.

Kula, M.-R., Johansson, G., and Buckman, A.F. (1979). *Biochem. Soc. Transact.,* **7,** 1–5.

Kula, M.-R., Kroner, K.H., and Hustedt, H. (1982). *Adv. Biochem. Engineer.,* **24,** 74–118.

Lif, T., Frick, G., and Albertsson, P.-Å. (1961). *J. Mol. Biol.,* **3,** 727.

Menge, U., Morr, M., Mayr, U., and Kula, M.-R. (1983). *J. Applied Biochem.,* **5,** 75–90.

Müller, W., and Eigel, A. (1981). *Anal. Chem.,* **118,** 269–277.

Müller, W., and Kütemeier, G. (1982). *Eur. J. Biochem.,* **128,** 231–238.

Müller, W., Schuetz, H.J., Guerrier-Takata, C., Cole, P.E., and Potts, R. (1979). *Nucleic Acid. Res.,* **7,** 2483–2499.

Pettijohn, D.E. (1967). *Eur. J. Biochem.,* **3,** 25.

Pinaev, G., Tartakovsky, A., Shanbhag, V.P., Johansson, G., and Backman, L. (1982). *Mol. Cell. Biochem.,* **48,** 65–69.

Polson, A., Potgieter, G.M., Largier, J.G., Mears, G.E.F., and Joubert, F.J. (1964). *Biochim. Biophys. Acta,* **82,** 463.

Rudin, L. (1967). *Biochim. Biophys. Acta,* **134,** 199.

Sasakawa, S., and Walter, H. (1972). *Biochemistry,* **11,** 2760.

Sasakawa, S., and Walter, H. (1974). *Biochemistry,* **13,** 29.

Shiemann, J., and Kopperschläger, G. (1984). *Plant. Sci. Lett.,* **36,** 205–211.

Shanbhag, V.P. (1973). *Biochim. Biophys. Acta,* **320,** 517–527.

Shanbhag, V.P., and Axelsson, C.G. (1975). *Eur. J. Biochem.,* **60,** 17–22.

Shanbhag, V.P., and Johansson, G. (1974). *Biochim. Biophys. Res. Comm.,* **61,** 1141.

Shanbhag, V.P., and Johansson, G. (1979). *Eur. J. Biochem.,* **93,** 363–367.

Svensson, P., Schröder, W., Åkerlund, H.-E., and Albertsson, P.-Å. (1985). *J. Chromatogr.,* **323,** 363–372.

Takerkart, G., Segard, E., and Minsigny, M. (1974). *FEBS Lett.,* **42,** 218.

Walter, H., and Sasakawa, S. (1971). *Biochemistry,* **10,** 108.

Walter, H., Sasakawa, S., and Albertsson, P.-Å. (1972). *Biochemistry,* **11,** 3880.

Zeppezauer, M., and Brishammar, S. (1965). *Biochim. Biophys. Acta,* **94,** 582.

6 STRATEGIES FOR SEPARATION—APPLICATIONS

Several approaches can be used in applying phase partition for fractionation and purification. The choice depends on the selectivity of the phase system, the amount of material to be purified, its stability, and so forth. If the partition coefficients of the substances are sufficiently different a desired separation may be obtained by one or few partition steps only. If several components are to be separated and their partition coefficients are not too different a multistage procedure, such as countercurrent distribution or partition chromatography, has to be used.

We can distinguish the following procedures:

1. Single-step partition.
2. Repeated batch extractions—gradient extraction.
3. Countercurrent distribution.
4. Liquid–liquid columns.
5. Partition chromatography.

SINGLE-STEP PARTITION

If the partition coefficients for two substances are sufficiently different an efficient separation can be obtained by one-partition step. The separation will be more efficient the larger the separation factor, that is, the ratio K_1/K_2 between the two partition coefficients. It is also important, however, that the partition ratios have suitable values. The partition ratio G is the ratio between the amount of substance in the top phase and the amount of substance in the bottom phase:

$$G = \frac{C_t V_t}{C_b V_b} \tag{1}$$

(C_t and C_b are concentrations in the top and bottom phases and V_t and V_b are the volumes of the top and bottom phases or

$$G = K \frac{V_t}{V_b} \tag{2}$$

that is, the partition coefficient times the volume ratio. Thus it is not enough that the partition coefficients be different. The phase-volume ratio also should be such that an efficient separation is obtained. Two examples may illustrate this point. In the first example the partition coefficients are 100 and 0.01. An efficient separation will be accomplished if the two volumes are equal. In the second example the partition coefficients are 1 and 0.0001; the ratio between the partition coefficients is the same as in the first example. We may therefore say that the phase systems are equally selective in the two cases. However, a poor separation would be obtained in the latter case if the phase volumes were equal. We have to use an upper phase volume which is 100 times larger than the lower phase in order to get an equally good separation as in the first case. Maximal separation of two substances having the partition coefficients K_1 and K_2 is obtained when the product of their partition ratios is 1, that is,

$$G_1 G_2 = 1 \tag{3}$$

or

$$K_1 K_2 \left(\frac{V_t}{V_b}\right)^2 = 1 \tag{4}$$

Applications of Single-Step Partition

Proteins. Several applications have been published where proteins, mainly enzymes, have been partially purified by phase partition involving one or a few partition steps (Tables 6.1 and 9.9). The dextran–PEG phase systems have been used either alone or combined sequentially.

TABLE 6.1 Proteins Which Have Been Partly Purified by One or a Few Polymer Two-Phase Partition Steps[a]

Enzyme	Reference
Chromatin proteins	Bidney and Reeck (1977); Gineitis et al. (1984); Gaziev and Kuzin (1973)
Phosphatidylserine synthetase	Raetz and Kennedy (1974)
DNA polymerase (*B. subtilis*)	Falaschi and Kornberg (1966); Okazaki and Kornberg (1964)
DNA polymerase (thermophilic bacteria)	Stenesh and Roe (1972)
DNA polymerase (sea urchin)	Loeb (1969)
RNA polymerase (Qβ)	Eoyang and August (1968); Shapiro et al. (1968)
RNA polymerase (*E. coli*)	Babinet (1967)
Leucyl–tRNA synthetase	Hayashi et al. (1970)
Initiator protein C (F2)	Hertzberg et al. (1969)
Polypeptide chain elongation factors	Iwasaki and Kaziro (1979)
Polypeptide chain-termination factors	Klein and Capecchi (1971)
Deoxyribonuclease (*Diplococcus pneumoniae*)	Vovis and Buttin (1970)
tRNA-nucleotidyltransferase	Miller and Philipps (1971)
Excision nucleases	Kaplan et al. (1969)
Exonucleases (λ)	Little et al. (1967); Radding (1966)

[a]See also Table 9.3, Chapter 9.

Partition in the dextran–PEG system is particularly useful for removing nucleic acids and cell debris from enzymes. In a dextran–PEG system with 0.1 *M* NaCl, for example, nucleic acids and membrane vesicles favor strongly the lower dextran phase while some enzymes have a more even partition, the partition coefficient varying between 0.1 and 10.

By using suitable salt composition and volume ratio, an efficient separation of proteins from nucleic acids and membrane particles can therefore be obtained by one or a few partitions.

If NaCl is added up to high concentrations (2–5 *M*), almost all proteins favor the upper phase (Fig. 5.3), while nucleic acids favor the lower phase. This was used to separate various enzymes (Table 6.1) from nucleic acids and other cell components. After the upper PEG phase was collected and dialyzed, ammonium sulfate was added to form a new, ammonium sulfate–PEG phase system in which the enzyme collected, is in the lower, ammonium sulfate, phase. By this procedure PEG is removed. The enzyme can then be further purified by chromatography.

Phase systems containing detergents have been used to purify membrane-bound proteins [Albertsson, 1973; Albertsson and Andersson, 1981; Salach, 1977)] and by affinity-partition a considerable purification may be obtained by a single partition step (Kopperschläger and Johansson, 1982). See also Chapter 9.

The salt–PEG systems have been used to purify several enzymes on a large scale (see Chapter 9).

The advantage of these procedures is that the enzymes can be freed rapidly from several impurities before subsequent column procedures. The columns can then be better utilized. Partition of an enzyme is usually independent of its concentration, the presence of other substances, and the total scale. Both small samples, on a milliliter scale, and large volumes, of several liters, may therefore be handled in this way. In these respects partition is far superior to precipitation methods.

A partition step can therefore be used as a general first step in enzyme purification. The enzymes are thereby separated from nucleic acids, ribosomes, and other cell fragments. Subsequent column operations then remove the phase polymers from the enzyme. By a combination of partition, including affinity partition, and chromatography, both simple and effective large-scale purifications of enzymes should be possible.

Nucleic Acids. The partition behavior of nucleic acids were described in Chapter 5, Figure 5.18. The partition coefficients change drastically within narrow changes in phase composition. Different nucleic acids can be selectively transferred from one phase to the other by small changes in the ionic composition of the dextran–PEG phase system. This is a characteristic behavior of large molecules and has been exploited for separation of nucleic acids (Table 6.2).

Single-stranded and double-stranded DNA have very different K values (Fig. 5.20) and can be efficiently separated from each other in one or a few partition steps. This has been demonstrated by Alberts (1967), who also used the same phase system to separate, in a single step, a small fraction of cross-linked, reversibly denaturable DNA molecules from the single-stranded products predominating in normal denatured DNA samples. The same technique was used by Summers and Szybalski for detection and quantification of single-strand breaks in DNA (1967). The addition of solid CsCl to the upper (polyethylene glycol rich) phase selectively salts out the polyethylene glycol. DNA in the upper phase can be examined and characterized by direct equilibrium sedimentation in a CsCl density gradient (Alberts, 1967).

A combination of selective melting and partition was used by Patterson and

TABLE 6.2 Applications on Separation of Nucleic Acids and Nucleoproteins by Phase Partition

Nucleic Acid	Reference
Single-stranded from double-stranded DNA	Alberts (1967); Albertsson (1965)
Satellite DNA, sea urchin	Patterson and Stafford (1970)
Cross-linked DNA	Cohen and Crothers (1970)
Circular plasmid DNA, *E. coli*	Ohlsson et al. (1978)
DNA–RNA hybrids	Mak et al. (1976)
Deoxyribonucleoproteins	Turner and Hancock (1974)
Chromosomes	Pinaev et al. (1979); Bandyopadhyay and Pinaev (1983)

Stafford (1970) to isolate a satellite DNA from sea urchin sperm. The main DNA having a lower melting point first was heat denatured at a temperature that the satellite DNA remained native. The two DNA's were then separated by a partition in a dextran–polyethylene glycol system. The satellite was purified 700-fold by this step alone and was then further purified by CsCl density gradient centrifugation. The main advantage of the partition step was that it allowed a large scale isolation; 1.3 mg of satellite was isolated from 1 g of DNA initially present.

Partition in the dextran–polyethylene glycol was used to remove from DNA the enzymes used for modification of the DNA (Lindahl and Edelman, 1968).

The difference in partition between single-stranded and double-stranded DNA was utilized to separate cross-linked, low-molecular-weight DNA from noncross-linked DNA (Cohen and Crothers, 1970). Since the cross-linked DNA renatures easily, the two types of DNA could be separated by melting and cooling, followed by partition in a dextran–polyethylene glycol system. Although the separation factor for the lower-molecular-weight DNA ($M_w < 10^6$) was smaller than for high-molecular-weight DNA, a satisfactory separation could be obtained in two partition steps only.

Circular plasmid DNA was isolated from cleared lysates of *Escherichia coli* (Ohlsson et al., 1978). The rapid reannealing of the plasmid DNA was used to separate it from denatured chromosomal DNA and RNA in a dextran–PEG phase system. The method was found to be more rapid than the conventional dye-centrifugation technique, and plasmid DNA of comparable purity and yield was obtained.

Mak et al. (1976) utilized the difference between double- and single-stranded nucleic acids for selecting DNA–RNA hybrids from unhybridized RNA. The top phase of a dextran–PEG system contained the DNA and DNA–RNA hybrid, while unhybridized RNA was confined to the bottom phase. The hybridized RNA was then separated from DNA by melting and subsequent chromatography.

The use of partition chromatography and affinity partition for DNA was described in Chapter 5.

Chromosomes and Nucleoproteins. Fractionation of chromosomal deoxyribonucleoproteins by the dextran–PEG system was described by Turner and Hancock (1974). A partial separation of nucleoproteins with different nonhistone/DNA ratios may be obtained by a single-step partition. By countercurrent distribution a spectrum of fractions of deoxyribonucleoproteins was obtained.

Partition of chromosomes was studied by Pinaev et al. (1979). Using charged

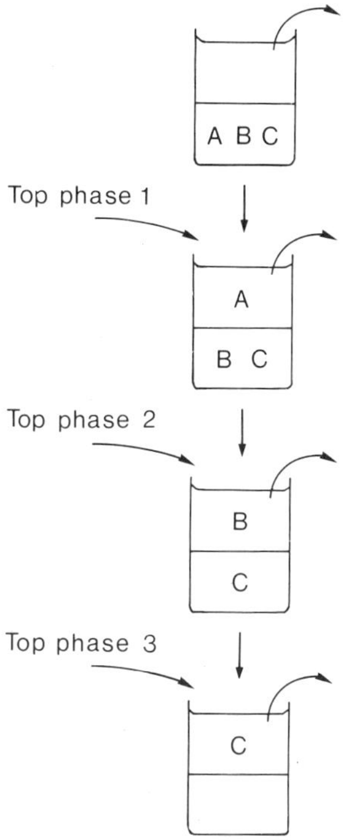

FIGURE 6.1. Gradient extraction. Principle of repeated batch extraction with changing composition of the phase system. The mixture of the components A, B, and C is first partitioned in a phase system where all collect in the lower phase. A new top phase 1 is added with concomitant change in the phase system so that A is transferred to the top phase. A new top phase is added to extract B, and so forth. The change in phase system composition can be made in small or large steps at will.

PEG (negative S-PEG or positive TMA-PEG) chromosomes could be transferred from one phase to another. A certain selectivity was found. Countercurrent distribution in combination with multiple sedimentation (Bandyopadhyay and Pinaev, 1983) separated a preparation of HeLa chromosomes into several subpopulations.

REPEATED BATCH EXTRACTIONS—GRADIENT EXTRACTION

By this procedure one phase is extracted several times with the other phase. By changing the composition of the extracting phase different components of a mixture can be extracted selectively one after the other (Fig. 6.1). Shifts in the partition coefficients can be induced by all the different factors which determine the partition coefficient, as described in Chapter 4, that is, ionic composition, polymer concentration, molecular weight of the polymer of the extracting phase, and temperature. Charged PEG- or PEG-bearing hydrophobic or biospecific groups

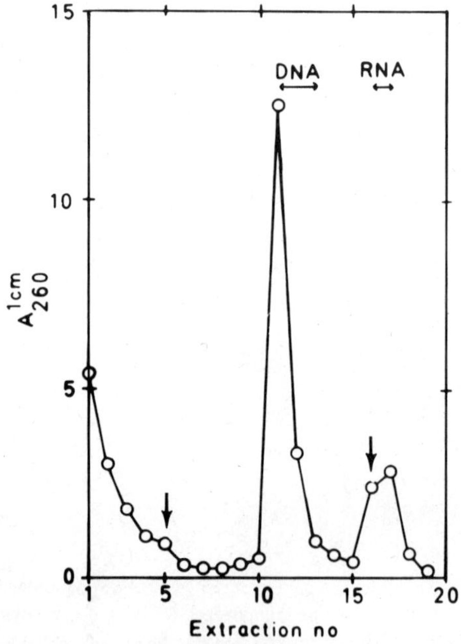

FIGURE 6.2. Isolation of DNA from *E. coli* by gradient extraction. Vertical arrows indicate shifts in ionic composition of the extracting top phase. First arrow: shift to 0.005 *M* Na_2HPO_4. Second arrow: shift to 0.005 *M* Na_2HPO_4, 0.005 *M* N_3PO_4 (Rudin and Albertsson, 1967).

on the polymer of the extracting phase (usually PEG) also can be used. In the case of membrane proteins detergents have been used.

Figure 6.2 shows an extraction profile from a bacterial extract. An *E. coli* extract was first partitioned in a dextran–PEG system with high NaCl concentration. Nucleic acids partition into the lower phase and proteins into the upper phase. After four extractions a top phase without NaCl is used and the NaCl is decreased after each new extraction (NaCl partitions almost equally between the phases). DNA and high-molecular-weight RNA are then extracted by appropriate ionic composition of the upper phase (Albertsson, 1965).

Figure 6.3 shows an extraction profile for chloroplasts which were fractionated using a detergent, Triton X-100, which partitions in the upper phase (Albertsson and Andersson, 1981). The procedure starts with a phase system without detergent and with a salt composition in which the membranes are in the lower phase. The upper phase is then removed and replaced by a new upper phase containing a certain low concentration of detergent. After shaking and phase separation by a

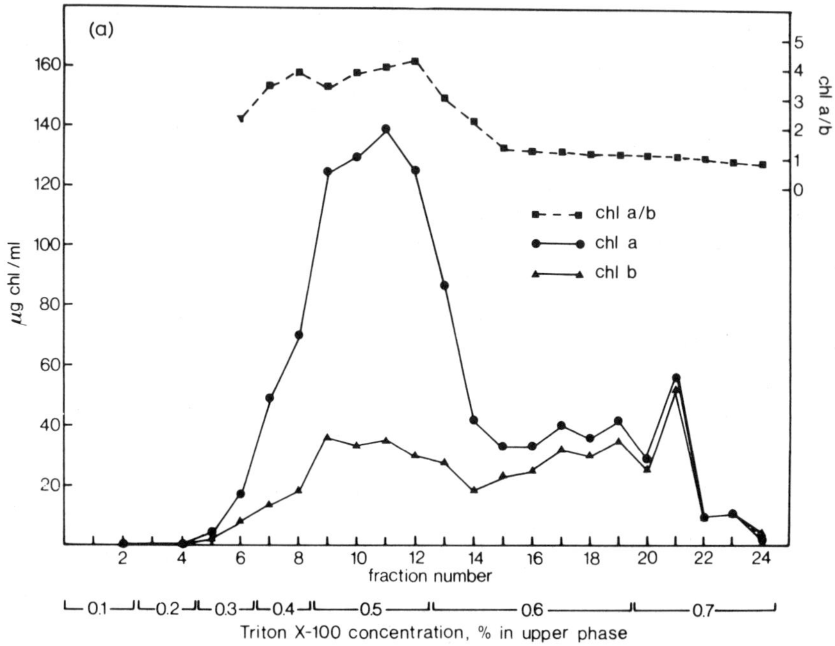

FIGURE 6.3. Gradient extraction of thylakoids by the detergent Triton X-100. In a phase system of 7% Dextran 500, 4.4% PEG 6000 and 0.1 *M* NaCl the thylakoids are concentrated at the interface or in the lower phase. By extraction with top phases having increasing concentration of the detergent, components of the membrane are selectively extracted (Albertsson and Andersson, 1981).

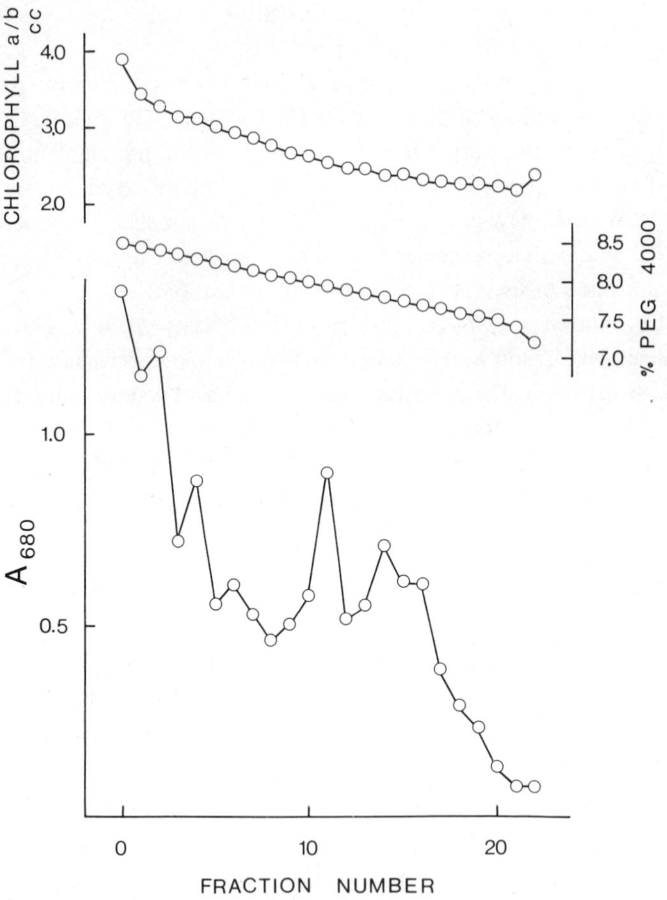

FIGURE 6.4. Gradient extraction of thylakoid vesicles obtained by mechanical press treatment. In this example the top phase was made lower and lower in PEG concentration (middle curve) so that the composition of the phase system approaches that of the critical point. The lower curve shows the amount of vesicles in the upper phase after each extraction. The upper curve showed the chlorophyll *a*/*b* ratio. The first fractions (0 to 8) contain right-side-out vesicles while the last fractions (15 to 20) contain inside-out fractions.

short, low-speed centrifugation, the upper phase is removed and replaced by a new upper phase containing a higher concentration of detergent, and so on. The detergent concentration can be increased successively to obtain a differential extraction. (If a detergent is used which partitions into the lower phase, this has to be the extracting phase and the upper phase has to contain the membrane material at the start.)

Studies with different detergents, Triton X-100, Tween, octylglucoside, and others, have shown that completely different extraction profiles are obtained, indicating a considerable specificity in the effect of the detergent action (Svensson et al. 1985).

Figure 6.4 shows an experiment where membrane vesicles from thylakoids were separated by repeated extractions with a dextran–PEG phase system. In this experiment the polymer concentration was changed successively. First, 6% of each polymer was used. Subsequent partitions were carried out with PEG solutions with decreasing concentration after each step. In this way the composition of the phase system changed more and more toward that of the critical composition. As a result of this gradient extraction, vesicles with different compositions were extracted at different concentrations of PEG. The right-side-out vesicles, originating from the stroma membranes (high chlorophyll *a*/*b* ratio) were first extracated (Fraction No. 0-5) and the inside-out vesicles originating from the grana partitions (low chlorophyll *a*/*b* ratio) were extracted last (Fraction No. 12-20).

Gradient extraction has several advantages. It is very simple experimentally. You need only a centrifuge tube for the partition and a standard laboratory centrifuge. One can try several ways of changing the partition and thereby influencing selectivity (salt, polymer composition, hydrophobic and biospecific effects, and so forth). The procedure is fairly quick. Each extraction only requires about 3 min, and it can easily be scaled up.

COUNTERCURRENT DISTRIBUTION

The method of countercurrent distribution has been treated in a number of books and review articles, of which those of Craig (1956), Hecker (1955), and King and Craig (1962) may be mentioned.

In this method a large number of extraction steps are carried out in order to separate substances having different partition coefficients. A typical extraction scheme is the one developed by Craig. It can best be explained with an example (Fig. 6.5).

A substance is distributed in a two-phase system in which its partition coefficient is 1. Also, the volume ratio is 1. After attainment of equilibrium, the amount of substance in the upper phase equals that in the lower phase. If the total amount is 100, a 50/50 distribution is obtained (tube No. 0 in Figure 6.5). The upper phase is now transferred to a new tube, No. 1, containing fresh lower phase of equal volume. To the bottom phase of tube No. 0 fresh top phase of equal volume is added. The two tubes are shaken, and after equilibrium the

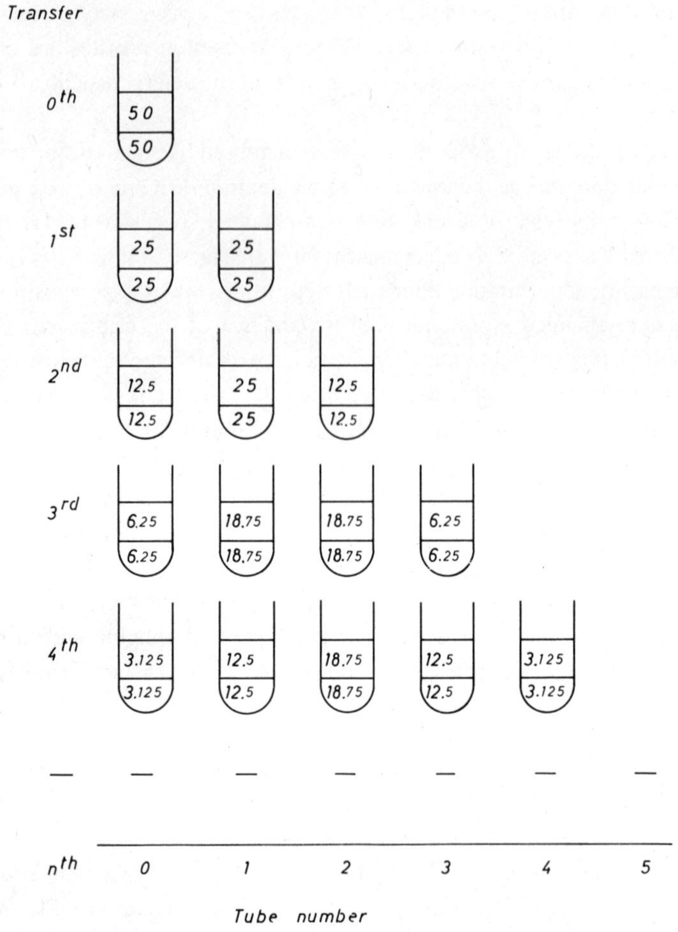

FIGURE 6.5. Scheme for countercurrent distribution, according to Craig.

substance will be distributed as in the second row of Figure 6.5, that is 25% of the original amount of the substance is in each phase. One transfer is now completed. Then the upper phase of tube No. 1 is transferred to tube No. 2 containing fresh lower phase; the upper phase of tube No. 0 is transferred to tube No. 1; and fresh top phase is added to tube No. 0. After equilibration the material will distribute according to the third row of Figure 6.5. The second transfer is thereby completed. The procedure then may be repeated several times. With each transfer, the series of tubes is increased by 1. The upper phases are each transported to the adjacent tube with higher tube number.

The numerical values in Figure 6.5 show how the substance is distributed among the tubes during the process. If concentration of material is plotted versus tube number, a symmetrical peak with its maximum in the center of the diagram is obtained. Had the partition ratio been different from 1, the peak would have been displaced from the center and would be asymmetrical.

The mathematical treatment of countercurrent distribution will not be treated here, the reader is referred to Craig (1956) and Hecker (1955); only a few points will be summarized here.

1. For a large number of transfers, each solute, behaving ideally, gives rise to an approximately symmetrical peak whose position is determined by the number of transfers and the partition ratio (which is the product of the partition coefficient and the volume ratio). Figure 6.6 shows examples of some theoretical distribution curves. The following relation holds between the partition ratio G of the solute in the tubes, the number of transfers n, and the number of the tube (r_{max}) at peak maximum.

$$r_{max} = \frac{nG}{1 + G} \qquad (5)$$

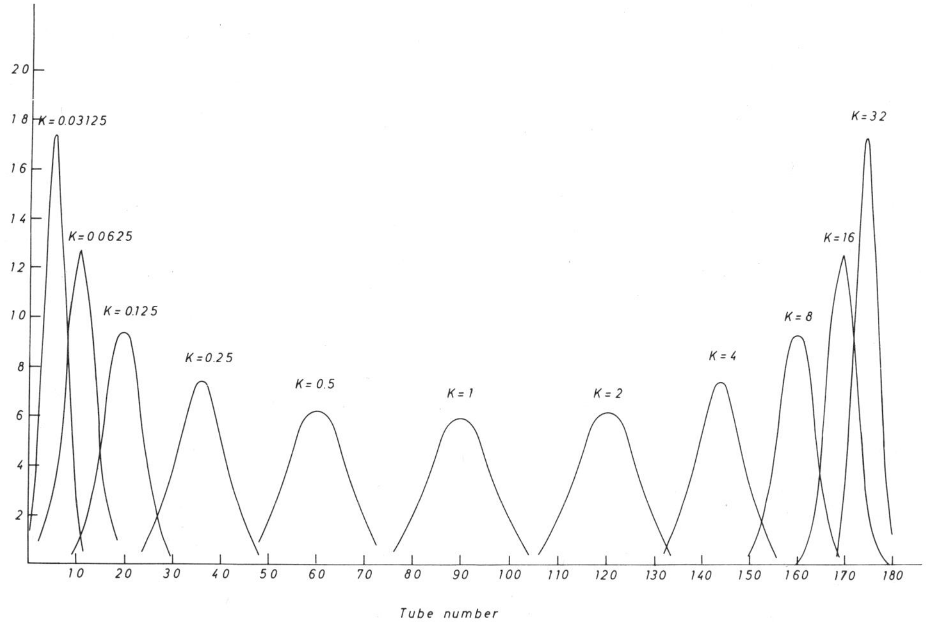

FIGURE 6.6. Theoretical curves obtained in countercurrent distribution of substances having different K values. The volume ratio is 1 and the number of transfers is 180.

2. The width of the distribution curve increases with $\sqrt{n}$ while the r_{max} increases with n. Two substances are therefore more efficiently separated by increasing the number of transfers.

3. For a given number of transfers maximal separation of two substances is obtained if their peaks are symmetrically located on opposite sides of the center of the diagram, that is, $G_1G_2 = 1$.

Countercurrent distribution is, in theory, very similar to chromatography. An outstanding feature of countercurrent distribution is that it is based upon several steps each involving an equilibrium between two liquid phases. This is particularly advantageous when macromolecules are involved, since for these it is thought to be easier to achieve an equilibrium between two liquid phases than between a liquid and a solid phase.

In the diagram obtained by countercurrent distribution each substance gives rise to a peak which may be compared with a theoretical curve. Such a comparison gives information as to whether the distributed substance is pure or behaves nonideally. Countercurrent distribution is therefore an important analytical tool. Since different substances are separated from each other, it may also be used for preparative purposes. It is, therefore, of the greatest importance, to develop suitable phase systems, which may be used to characterize and fractionate macromolecules, such as proteins and nucleic acids, and particles in a countercurrent apparatus.

It is of particular interest that countercurrent distribution can be applied on cell particles and thereby allow analysis and separation of complex particle mixtures such as suspension of cells or cell organelles.

Liquid–Interface Countercurrent Distribution

The distribution scheme of Figure 6.5 is based on partition between the two bulk phases. This is the normal procedure for soluble substances. In the case of particles, however, partition usually takes place between one bulk phase and the interface. We have found that also in this case the countercurrent principle can be applied. One may therefore distinguish between two types of countercurrent distribution (Fig. 6.7).

The first (Fig. 6.7*a*) involves the distribution of the substance between the two phases, no significant adsorption of the solute taking place at the interface. This is the conventional way of distribution in a countercurrent experiment and will be referred to as liquid–liquid countercurrent distribution. Each substance is characterized by its partition coefficient K, and is transported along the dis-

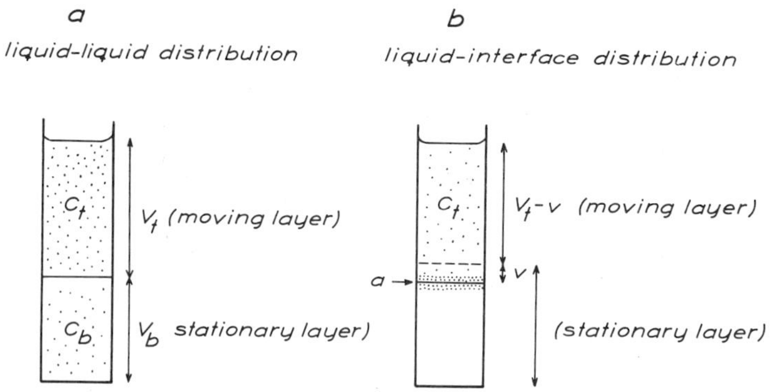

FIGURE 6.7. (*a*) Liquid–liquid countercurrent distribution and (*b*) liquid–interfacial countercurrent distribution. In (*a*) no significant adsorption at the interface takes place. The entire top phase is the moving layer in the countercurrent apparatus. In (*b*) the distribution takes place between the top phase and the interface. The bottom phase, the interface, and a small layer above the interface form the stationary layer in the countercurrent apparatus.

tribution train according to its distribution ratio G [Eq. (1)], which is the ratio between the amount of substance in the top phase and the bottom phase.

The other distribution type (Fig. 6.7*b*), which is called liquid–interface countercurrent distribution, involves the distribution of the solute between one of the phases and the interface. This is used mainly for particles. In Figure 6.7*b* the distribution takes place between the upper phase and the interface. The bottom phase, together with the interface and a small layer of the top phase just above the interface, is kept stationary in the apparatus while the rest of the top phase is the moving phase. Let the volume of the top phase be V_t, the volume of the layer of the top phase which is kept stationary in the distribution train be v, and let C_t be the concentration of the solute in the top phase and a the amount of the solute adsorbed at the interface. The ratio which is of interest for the countercurrent distribution is then

$$G_i = \frac{(V_t - v)C_t}{vC_t + a} \tag{6}$$

When the substance behaves ideally, that is, the adsorption at the interface is reversible and proportional to C_t, it will be transported along the distribution train according to the G_i value, and, for calculation of theoretical curves, this may be used in the same way as the G value in liquid–liquid countercurrent distribution.

In a similar manner, a distribution between the bottom phase and the interface may be used for countercurrent distribution. The top phase, together with the interface and a small layer below it, is then the moving layer in the countercurrent experiment while the remaining bottom phase is stationary.

Apparatus

Several different types of apparatus have been constructed for countercurrent distribution. For reviews see King and Craig (1962) and Hecker (1955). The most widely known design is the all-glass machine of Craig and Post (1949). This can be used for polymer two-phase systems both for liquid–liquid and liquid–interface countercurrent distribution (Albertsson and Nyns, 1959; Albertsson and Baird, 1962).

A drawback with polymer-phase systems when using the Craig glass machine is the long time needed for phase separation and the high viscosity of the phases. A special countercurrent distribution apparatus for use with polymer two-phase systems has therefore been constructed (Albertsson, 1965, 1970). See Figures 6.8 and 6.9. In this, the phases form thin layers, about 2 mm deep, which reduces the settling time considerably. Thus the settling time for the dextran–PEG system can be reduced to 1–5 min from the 10–30 min necessary with the conventional types of apparatus used so far.

The countercurrent procedure takes place in a partition cell block which is

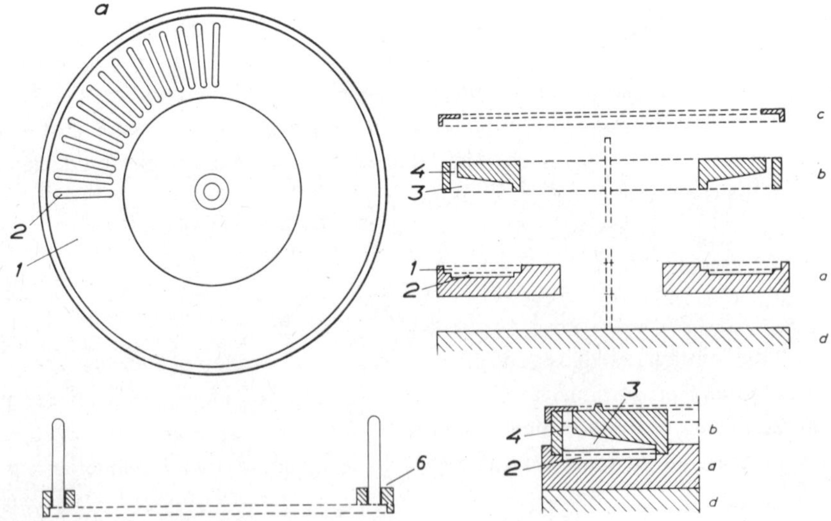

FIGURE 6.8. Drawing of thin-layer countercurrent distribution apparatus. For explanation see text.

made up of two cylindrical plates made of plexiglass (Fig. 6.8). The lower plate (a) has a shallow annular groove (1), which is concentric with the vertical cylindrical axis. A number of shallow cavities (2) in this groove form the bottom parts of the partition cells and contain the lower phases. The depth of these cavities is 2 mm. The upper plate (b) rests in the groove of the lower plate. The upper plate can rotate about its axis and is guided by the inner and outer edges of the groove. The lower surface of the upper plate, which is in contact with the groove, also has cavities (3) of the same horizontal cross section as the bottom plate cavities. The upper plate cavities contain the top phase.

By turning the upper plate relative to the lower plate each cavity of the rotor can be successively brought into coincidence with each cavity of the lower plate. A circular sequence of partition cells is thus formed. The circular arrangement of the cells, also used in Craig and Post's first apparatus, allows recycling procedures so that the number of transfers can be increased to allow maximum separation.

The upper cavities are deeper than the lower cavities to allow air space and efficient mixing in each partition cell during shaking. Each upper cavity is provided with a hole (4) in the top for filling and emptying. The upper part of the inner surface of each top cavity is sloping to facilitate emptying. A circular cover (c) is used to close the holes.

The partition block rests on a horizontal shaking table (d) with a driving motor and device for variation of frequency of shaking (Fig. 6.9).* In the center of the table there is also a unit which drives the rotation of the upper plate. All procedures including shaking, settling and phase transfer are guided automatically by a control unit.

After a countercurrent distribution experiment the content of each cell can be recovered by means of a fraction collector (e) (Fig. 6.8). It consists of a ring with holes, the same number as there are cavities. In each hole a plastic tube (6) (centrifuge tubes) can be placed to receive the contents of the corresponding cell. For emptying, the fraction collector is placed on top of the rotor so that each tube covers a hole in the rotor. Upon inversion of the setup, the contents of each cell flows into the fraction collector tubes (Fig. 6.8).

Åkerlund (1984) has designed an apparatus for countercurrent distribution in centrifugal accelerating field (Fig. 6.10). With this apparatus the time needed for phase separation is reduced to about 1 min by using a centrifugal accelerating field of 100g. Including mixing, transfer, and deacceleration each countercurrent

*The apparatus is manufactured for sale by the workshop of the Chemical Center, University of Lund, Chemical Center, P.O. Box 124, S-221 00 Lund, Sweden.

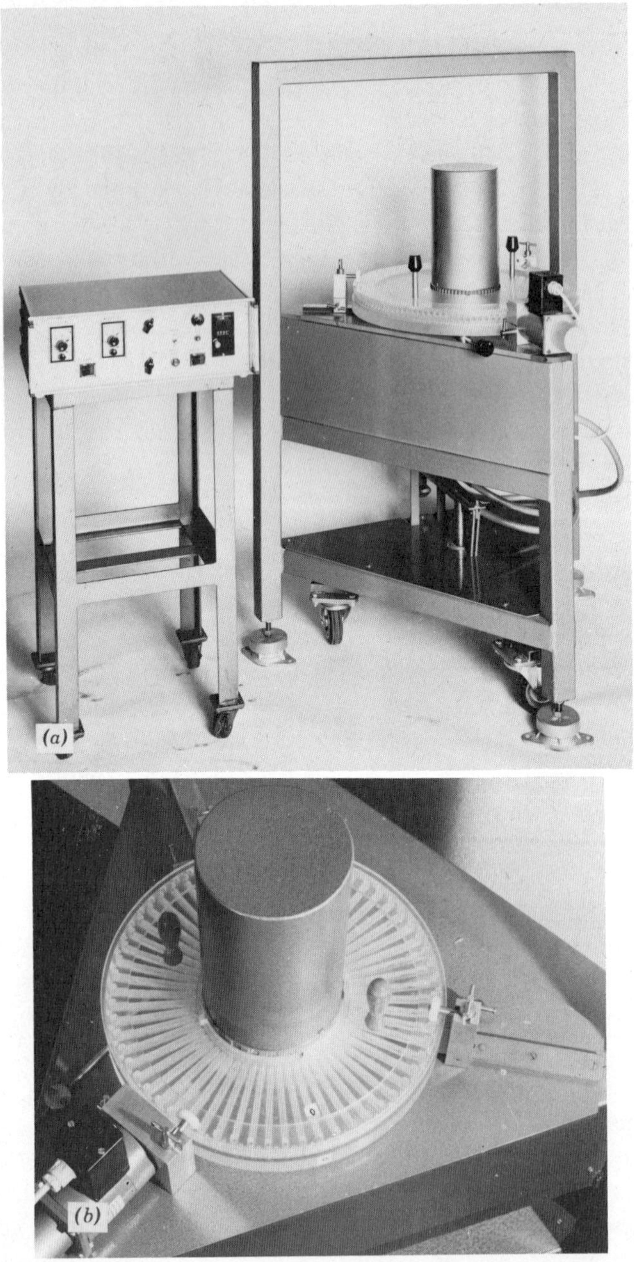

FIGURE 6.9*a*, *b*. (*a*) Automatic version of the thin-layer countercurrent distribution apparatus. (*b*) The partition cell block consists of two cylindrical plates made of plexiglass, resting on the shaking table. The cylinder in the center houses the driving unit for rotating the upper plate during transfer. The main box below the shaking table houses a driving motor and a device for the variation of the shaking frequency.

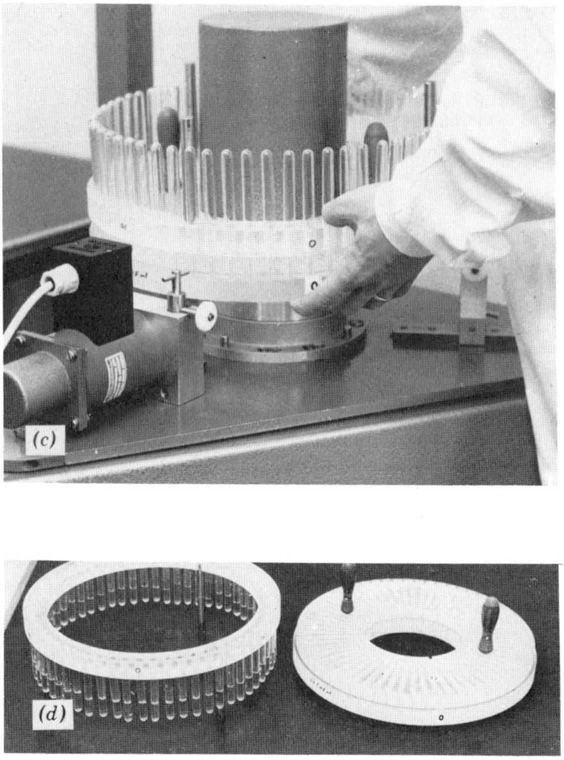

FIGURE 6.9c, d. (*c*) The fraction collector is being placed on top of the upper plate. (*d*) The partition block together with fraction collector is turned upside down to allow the contents of each partition cell to flow into the fraction collector tubes.

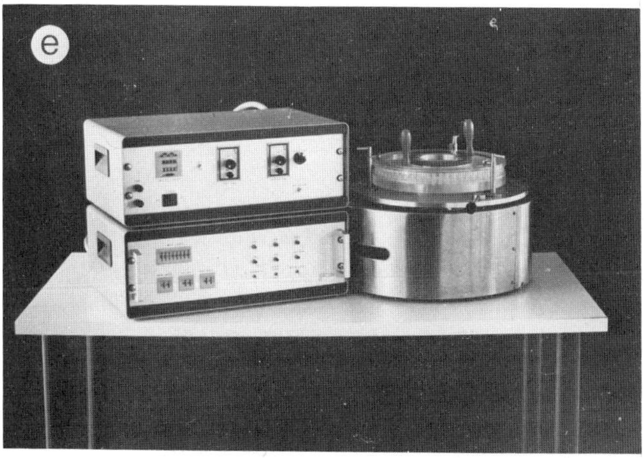

FIGURE 6.9e. Bench model of thin layer countercurrent distribution apparatus.

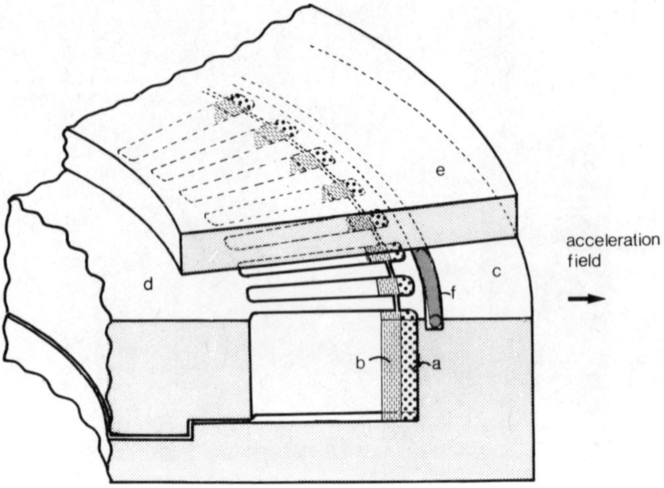

(a)

FIGURE 6.10a. Section of the separation unit for the centrifugal CCD apparatus. It is composed of four units: (c) the outer ring with cavities for the lower phase; (d) the inner ring with cavities for the upper phase; (e) the lid ring; (f) an O-ring for sealing. The position of the two-phase system during centrifugation, with (b) the upper phase and (a) the lower phase is shown.

distribution step takes about 2 min. A 50-transfer experiment therefore takes less than 2 hr. An additional advantage with this apparatus is that even highly viscous and slowly settling phase systems can be used.

Countercurrent Distribution of Proteins

Countercurrent distribution of proteins has mainly employed the dextran–polyethylene glycol system. This system has a comparatively short settling time and the upper, moving phase, has a comparatively low viscosity. The partition coefficient can also be adjusted to a suitable value by the ionic composition or by including charged PEG (TMA-PEG or PEG-S) or ligand-PEG. Figure 6.11 shows a countercurrent distribution of a purified protein, phycocyanin (Albertsson and Nyns, 1959). The experimental curve fits perfectly the theoretical one showing an ideal behaviour of this protein, that is, the partition coefficient is independent of concentration, and equilibrium is established before each transfer. Any deviation from the theoretical curve indicates that the protein is either not pure or does not behave ideally.

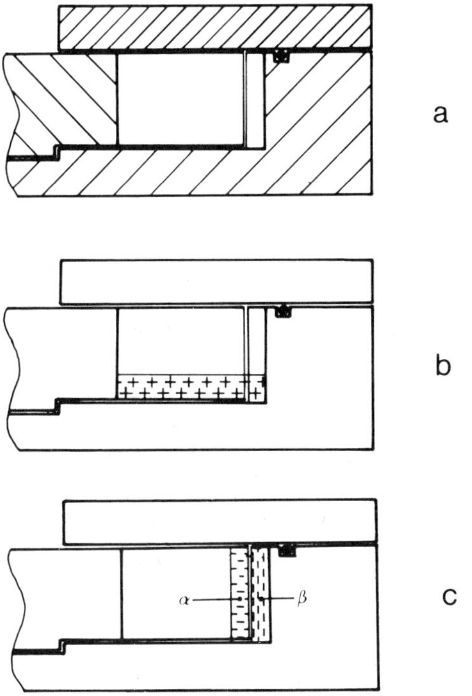

(b)

FIGURE 6.10*b*. Side view of a radial cut through one cavity. *a*, Empty; *b*, with phase system during mixing; *c*, during centrifugation with the phases separated; α, upper phase; β, bottom phase (Åkerlund, 1984).

To construct a theoretical curve for one component is fairly simple (King and Craig, 1962). If a mixture of components is present the experimental curve looks much more complicated. One can then assume a certain number of theoretical curves, and by the help of a computer one can find out how many components and what relative amount gives the best fit with the experimental curve. Such an approach has been described (Blomquist and Wold, 1974) and is illustrated in Figures 6.12 and 6.13. Figure 6.12 illustrates a countercurrent distribution of the enzyme enolase from baker's yeast. Two main peaks are obtained. By assuming two protein components a theoretical curve, as shown in Figure 6.12*a*, is obtained. It fits the experimental curve well except in the valley between the two peaks. A much better fit over the entire diagram is obtained if three components are assumed instead (Fig. 6.12*b*). Indeed, enolase from baker's yeast is

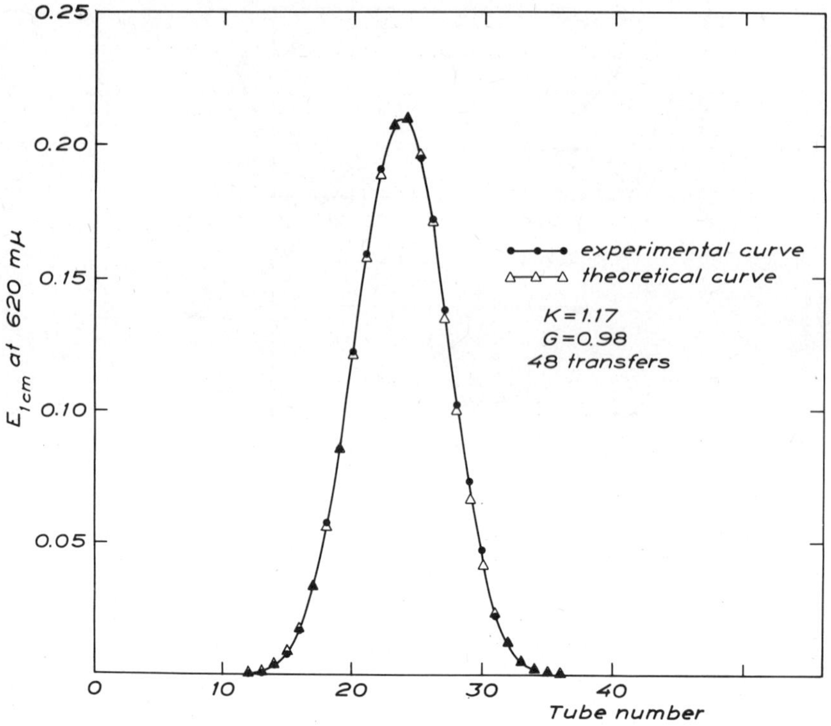

FIGURE 6.11. Countercurrent distribution of phycocyanin (Albertsson and Nyns, 1959).

known to consist of three isoenzymes. Figure 6.13 shows a similar experiment with lactic dehydrogenase. This enzyme is composed of five isoenzymes. In the countercurrent distribution diagram these components overlap, and it might seem to be difficult to compare the experimental curve with a theoretical one. By using a computer program however, it can be tested whether a best fit is shown for four or five components (Fig. 6.13*a* and *b*). As seen, a curve composed of five components gives an almost perfect fit.

Countercurrent Distribution of Cells and Cell Organelles

Examples of liquid–interface countercurrent distribution of cell organelles and cells are found in Chapter 7 and 8. In the case of particle separation we do not know if a single population of identical particles will give a CCD diagram in perfect agreement with a theoretical curve. Therefore the calculation of theoretical curves for a particle mixture is not so meaningful at the moment.

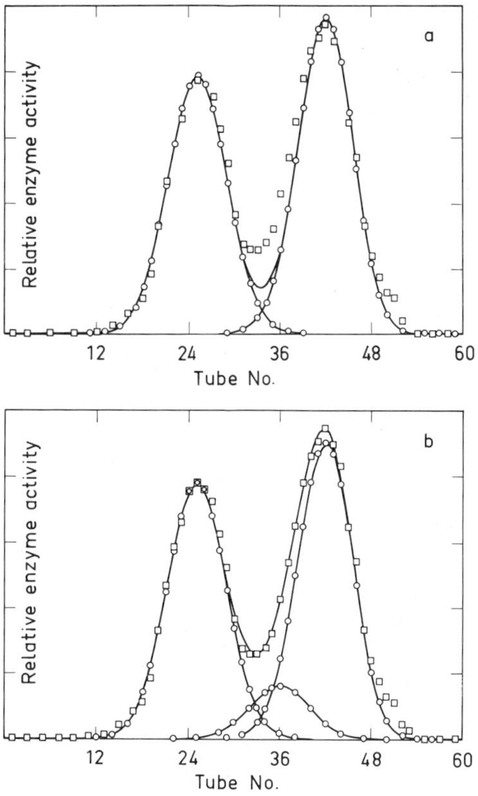

FIGURE 6.12. Countercurrent distribution of enolase from baker's yeast (□) represents experimental curve, (○) represents theoretical curve, (——) represents the sum of theoretical curves. In (*a*) two components, and in (*b*) three components are assumed to be present in the enzyme sample when calculating the theoretical curves (Blomquist and Wold, 1974).

LIQUID–LIQUID PARTITION COLUMNS

Several liquid–liquid partition columns are described in the literature. For a review see the books by Treybal (1951) and Hecker (1955) and the article by Craig (1956). The basic principle behind all these columns is that two immiscible phases move in opposite direction under alternating mixing and phase separation. Mixing is accomplished by different means, for example, by pulsation or by stirring rods. Phase separation is accomplished either by settling under the gravitational field or by centrifugation. Most columns have been applied for separations on a large scale in industry and they have found very little use in research laboratory work.

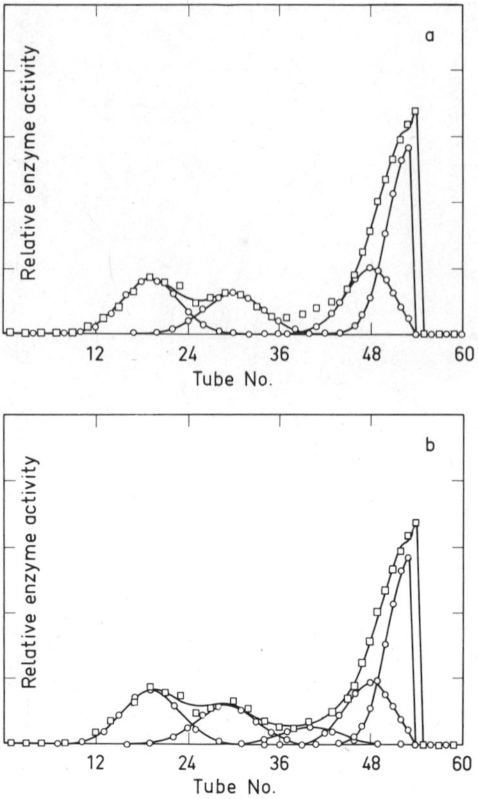

FIGURE 6.13. Countercurrent distribution of lactic dehydrogenase from pig liver. Symbols as in Figure 6.12. In (*a*) four components and in (*b*) five components are assumed when calculating the theoretical curves (Blomquist and Wold, 1974).

Of particular interest are the recent column designs developed by Ito, Sutherland, and collaborators (Ito and Bowman, 1973; Ito and Conway, 1984; Ito et al., 1966; Sutherland et al., 1984). By using narrow helical tubes, with a large number of turns (in some cases up to 17,000 turns), very efficient liquid–liquid partition columns are obtained. The number of theoretical plates is several thousand. These columns should offer extremely powerful tools for analyzing macromolecules and particles.

In the author's laboratory a number of different column designs of another type have been tested. A schematical description of three of them is shown in Figure 6.14.

The column shown in Figure 6.14*a* is a modification of the column described by Rometsch (1950) and Scheibel (1950). The column consists of a series of chambers, made of Plexiglas, separated by sieve plates which allow transport of the phases. Every second chamber is a mixing chamber and every other second chamber is a settling chamber. The heavier phase is pumped in from above and pumped out from below while the lighter phase is pumped in from below and pumped out from above. In each settling chamber there is a phase separation. The lighter phase moves up to the mixing chamber above with concomitant flow of phase mixture from this chamber to the settling chamber, while the heavier phase sinks to the next lower mixing chamber with concomitant flow of phase mixture from this chamber to the settling chamber. The net result is that the

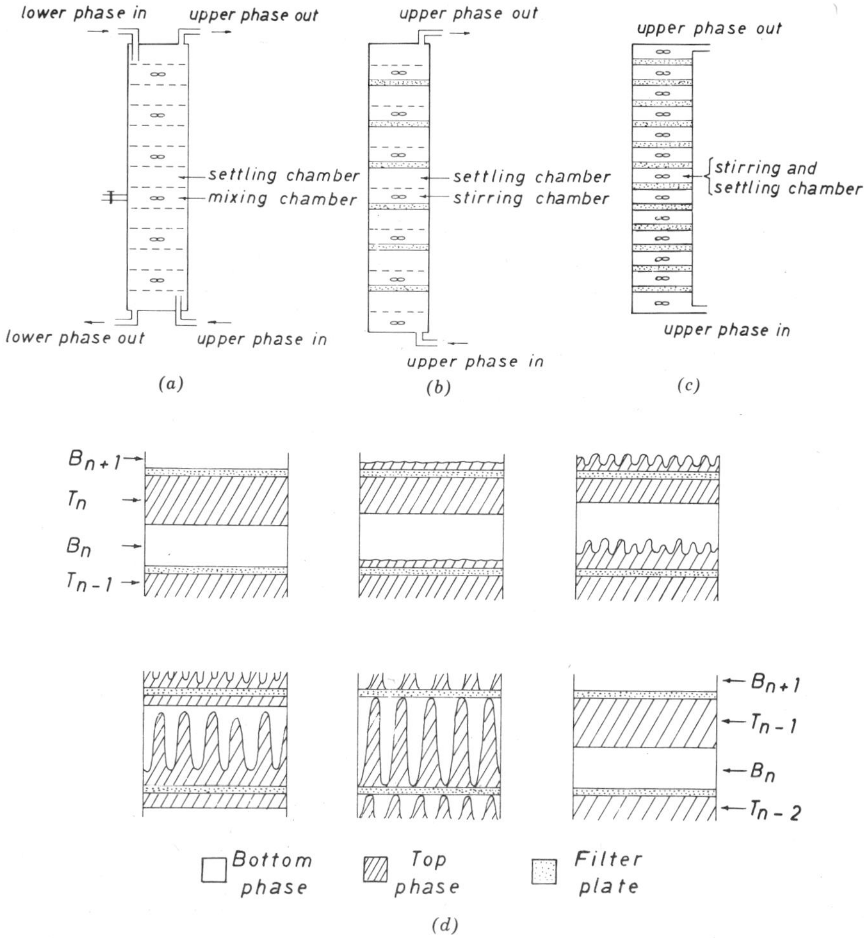

FIGURE 6.14. (*a*)–(*c*) Different designs of liquid–liquid partition columns (half natural size). (*d*) Principle of phase transport in the column shown at (*c*). For explanation see text.

heavier phase moves down and the lighter phase moves up with intervals of contacting and mass transfer in the mixing chambers. The sample to be separated either can be introduced, continuously, together with one of the input phases, or into one of the chambers of the column. In this way continuously operating separation process is obtained. After a while a stationary state is obtained, substances with low *K* values will follow the heavier phase out of the column at the bottom while substances with high *K* values will leave the column at the top. Figure 6.15 shows such an experiment.

The sample also may be added once as a small volume, either in one of the input phases or in one of the chambers of the column. In this way a zonal separation is obtained and the different substances of the sample will move upward or downward with speeds depending on the *K* values and the relative volume flow of the phases. Figure 6.16 shows an experiment where native and denatured DNA have been separated according to this procedure.

The mathematical treatment of this column is described in the article by Rometsch (1950).

A special case of the column in Figure 6.14*a* is obtained if the inlets and

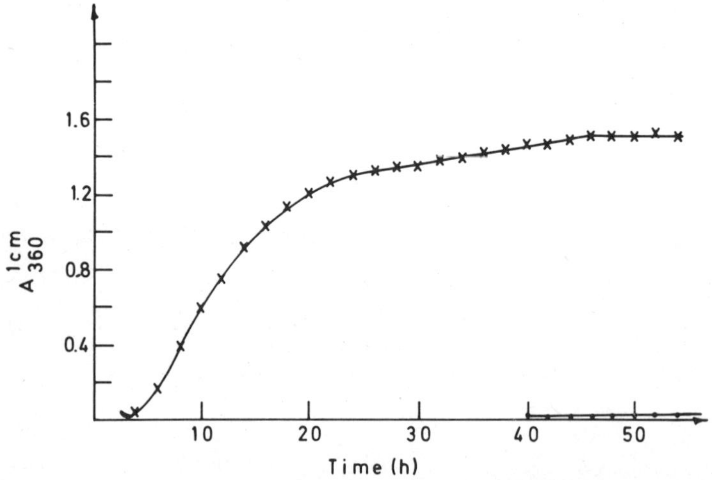

FIGURE 6.15. An experiment with DNP-methionine in the dextran–polyethylene glycol system with the column shown in Figure 6.14*a*. Top phase was pumped in from below with a speed of 0.11 mL/min. Bottom phase was pumped in from above with a speed of 0.096 mL/min. DNP-methionine dissolved in top phase was pumped at a speed of 0.01 mL/min in a mixing chamber in the middle of the column with 10 mixing and 10 settling chambers above and below. After about 45 hr a steady state was obtained. The upper and lower curves show the concentration of DNP-methionine leaving the column at the top and bottom, respectively. The number of theoretical plates calculated according to Rometsch (1950) was 31 in this experiment (Blomquist and Albertsson, 1972).

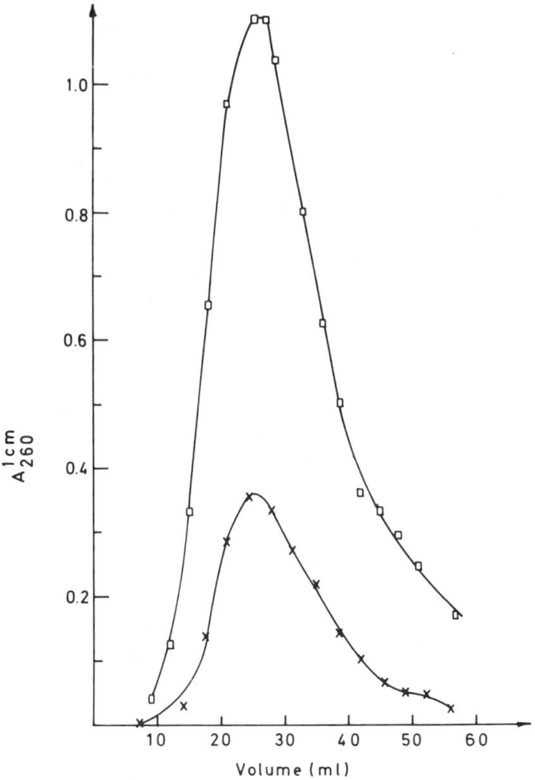

FIGURE 6.16. Separation of native from denatured DNA in the column of Figure 6.14*a*. A pulse of the mixture was injected in the middle of the column. Native DNA, having a high *K* value, leaves the column with the top phase (□), while denatured DNA, having a low *K* value, leaves at the bottom (×). Phase system: 5% w/w Dextran 500; 4% w/w PEG 6000; 5 m*M* Na_2HPO_4, 5 m*M* Na_2HPO_4. Top and bottom phase moved with the same speed (Blomquist and Albertsson, 1972).

outlets of heavier phase are closed. The heavier phase is then stationary in the column and held up by the flow of the lighter phase. (The reversed case where the lighter phase is stationary and the heavier phase is moving is also possible by closing the inlets and outlets of the lighter phase.) Such a column is very similar to a chromatographic column. Another type of column is shown in Figure 6.14*b*. Here every second plate is a sieve of the same type as in the column in (*a*). Every other second plate is a filter which has such properties that it allows pumping of liquid through it, but it does not allow exchange of heavier and lighter phase by gravity, that is, a heavier phase may rest on top of a lighter phase if they are separated by the filter. Every unit between two filters consists of a lower mixing unit and an upper settling unit. This column may be operated in two different ways, (a) continuous and (b) step-wise.

(a) In this procedure the lighter phase is pumped in from below. Each unit between filters works as an extraction unit, that is, in the mixing chambers the two phases are being mixed while in the settling chamber the phases separate and the lower phase goes back to the mixing chamber underneath, while the upper phase continues to the next mixing chamber above. Thus the lower phase does not leave the column; it acts like a stationary phase in chromatography. If the flow of lower phase stops, the phase in each unit separates and stays there (in contrast to the column in Figure 6.14*a*, the lower phase does not settle in the lower part of the entire column because the bottom phase cannot pass through the filters). An experiment with this column is shown in Figure 6.17.

If only one mixing chamber and one settling chamber of the column shown in Figure 6.14*b* are used, an extractor of the type described by Blomquist and Albertsson (1969) is obtained.

(b) The step-by-step variant is operated as follows: The column is filled so

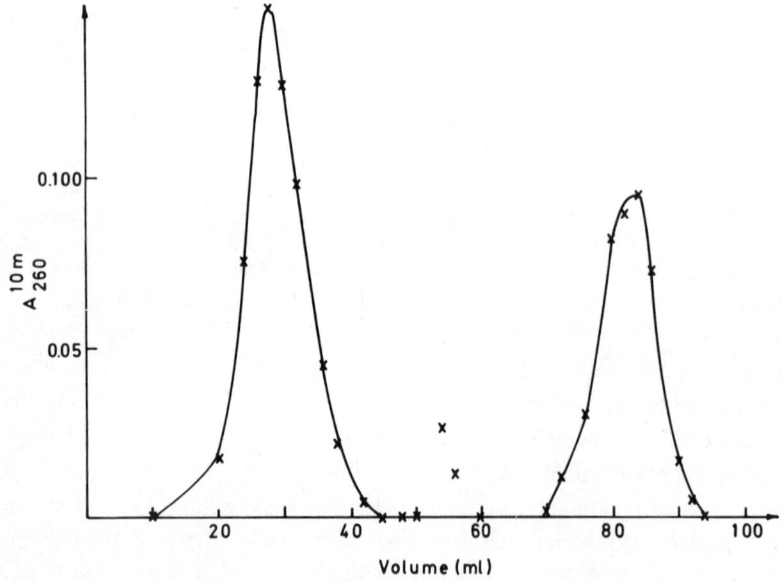

FIGURE 6.17. Separation of native and denatured DNA in a column of the type shown in Figure 6.14*b*. A sample of the mixture in top phase was injected from below. Phase system as in Figure 6.16. Native DNA, which has a high *K* value in this system, leaves the column first. Denatured DNA, which has a low *K* value, stays in the column. By changing the buffer in the top phase to 10 m*M* Na_2HPO_4 denatured DNA can be eluted from the column (Blomquist and Albertsson, 1972).

that the stirring chamber contains heavier phase, and the settling chamber contains lighter phase. A volume of top phase equivalent to the volume of the settling chamber (or less) is pumped into the column with concomitant stirring. A phase mixture will then leave each mixing chamber, and since it is denser than the upper phase, it will form a boundary below the upper phase and lift the upper phase of each unit so that it is pressed into the adjacent mixing chamber above. After the volume equivalent to (or less than) the settling chamber has been pumped into the column, stirring and pumping is stopped and the phases are allowed to settle. Since the phases can separate and settle freely through the sieve plate of each unit, lower phase will collect in the mixing chamber and upper phase in the settling chamber. After settling has finished, a cycle has been completed and may be repeated. After each cycle the upper phase has moved from one unit to the unit above, while the lower phase remains in the unit. Alternatively, the lower phase instead of the upper phase may be pumped in from above, or they may be pumped in alternatively. The sample may thus be fed in from below or from above or into some of the units along the column, either as a single sample, giving rise to a zonal separation, or repeatedly, giving rise to a continuously operating multistage extractor, in much the same fashion as the column shown in Figure 6.14*a*.

In the column shown in Figure 6.14*c* each chamber is both a mixing and settling chamber. They are separated by filter plates which allow transport of liquid by pumping but not by gravity alone. Each chamber contains both lighter and heavier phases. After mixing and phase separation a volume of top phase equal to or less than the volume of the upper phase is pumped into the column from below. As a result, the same volume of lighter phase in each chamber is pressed through the filter plates into the adjacent chamber above. In this chamber the lower phase is first pressed by the entering lighter phase from below, which then rises through the heavier phase and comes in contact with the lighter phase which is leaving this chamber. To prevent such a contact and mixing between lighter phases from two adjacent chambers the pumping speed has to be so high that the time required for the lighter phase to move to the adjacent chamber is less than the time required for the lighter phase to rise through the heavier phase layer (Fig. 6.14*d*). This time is very short, less than 1 sec for a thin layer of conventional phase systems. For polymer phase systems, however, this time is much longer because of the high viscosity and the low density difference. It is of the order of a few seconds when the phase layer is about 5 mm deep. This makes it feasible to pump the lighter-phase volume into the column and thereby displaces the top phase of each chamber into the adjacent chamber without mixing the two phases.

The column also may be operated with the top phase stationary and the bottom phase moving down by pumping it in from above. Also, top and bottom phase may be alternately pumped into the column. The sample may be fed into the column gradually, either from below, above, or somewhere in the middle, giving a steady-state process, or the sample may be fed once into the column giving a zonal separation.

An example of the application of the column shown in Figure 6.14*c* is given in Figure 6.18. It shows a separation of two microorganisms, namely *E. coli* from *Chlorella pyrenoidosa*. *E. coli* leaves the column first as it partitions mainly in favor of the moving top phase. Chlorella remained in the column at first, because it favors the interfaces and the bottom phase more under the prevalent conditions. It was, therefore, later eluted by a shift in the ionic composition.

Comparison between Thin-Layer Countercurrent Distribution and Column Methods

The partition columns have the advantage of a rather simple experimental setup. The relatively high viscosity of polymer phases however, makes the flow through the columns rather slow. Therefore, thin-layer countercurrent distribution is gen-

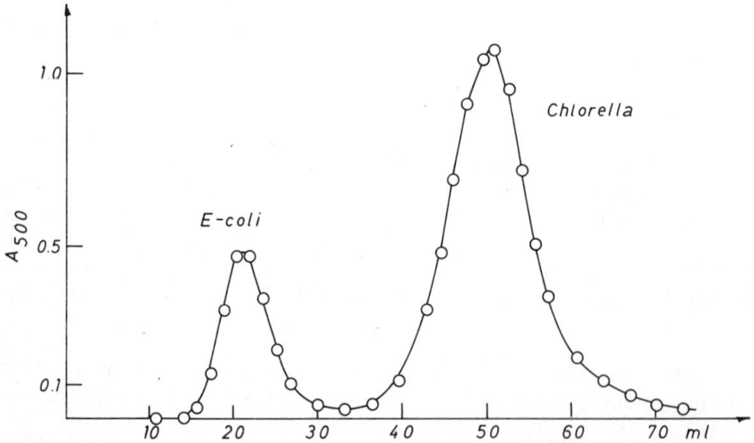

FIGURE 6.18. Separation of *E. coli* from *Chlorella pyrenoidosa* by a column shown in Figure 6.14*c*. A mixture of the two organisms was injected in a phase system of 5% w/w Dextran 500; 4% w/w PEG 6000; 0.01 *M* Na phosphate, pH 6.8; 0.1 *M* NaCl. The chlorella cells stay at the interface in this system and remain in the bottom chambers, while the *E. coli* goes to the moving top phase and leaves the column at the top. Then after 11 transfers, chlorella was eluted by injecting a top phase with a composition as above, except that NaCl was omitted. The whole column consisted of 20 chambers, each 5 mm deep. Each chamber holds 1.6 mL, and at each displacement 0.8 mL top phase was injected during a time of 6 sec, followed by 30 sec mixing and 5 min settling time. The top/bottom volume ratio was 1.

erally a quicker method in the present case. Another advantage is that after a thin-layer countercurrent distribution experiment, the phases in all the partition cells can be recovered rapidly and all material can be accounted for. Column support material can cause irreversible adhesion of cells. Therefore, in the author's opinion, thin-layer countercurrent distribution has proved to be superior to the column methods, at least with column constructions used so far. Most multistage experiments described in this book have been carried out with countercurrent distribution. I have still included the section on columns in this chapter because it might be useful for special cases and may also form the basis for further improvements. Particularly, the procedure of phase transport according to Figure 6.14*d*, has, to my knowledge, not yet been exploited and might prove useful for other phase systems too.

PARTITION CHROMATOGRAPHY

If one of the phases of a liquid–liquid two-phase system can be bound to a solid support, this may be used as column material for partition chromatography. In order to test which phase has the most affinity for a solid material this material is shaken with the two-phase system. If the solid particles then preferably collect in one phase, this is used as the stationary phase while the other phase is used as moving phase.

Morris (1963) has used celite and synthetic calcium silicates as solid support for chromatography of proteins with the dextran–polyethylene glycol system. The lower dextran-rich phase was the stationary phase. The proteins tested were ovalbumin, serum albumin, cytochrome *c* and lysozyme. A separation of lysozyme from bovine serum albumin was demonstrated.

Anker (1971) used agarose gel as the stationary support. If the dextran–polyethylene glycol system is prepared using a low-molecular-weight dextran (Dextran 40), which can penetrate the agarose beads, the dextran-rich phase can be held stationary by the agarose beads.

The most systematic and thorough study on partition chromatography with polymer phase systems has been carried out by Müller and co-workers (Müller and Eigel, 1981; Müller and Kütemeier, 1982; Müller et al., 1979). Microgranular cellulose was found to be a good support for partition chromatography of DNA fragments. The ratio of mobile to stationary phase, the ionic composition, and the temperature were optimized. Figures 6.19 and 6.20 show examples of partition chromatography of DNA. The resolution obtained approaches that of polyacrylamide and agarose gel electrophoresis, but the capacity is 200 to 500 times higher.

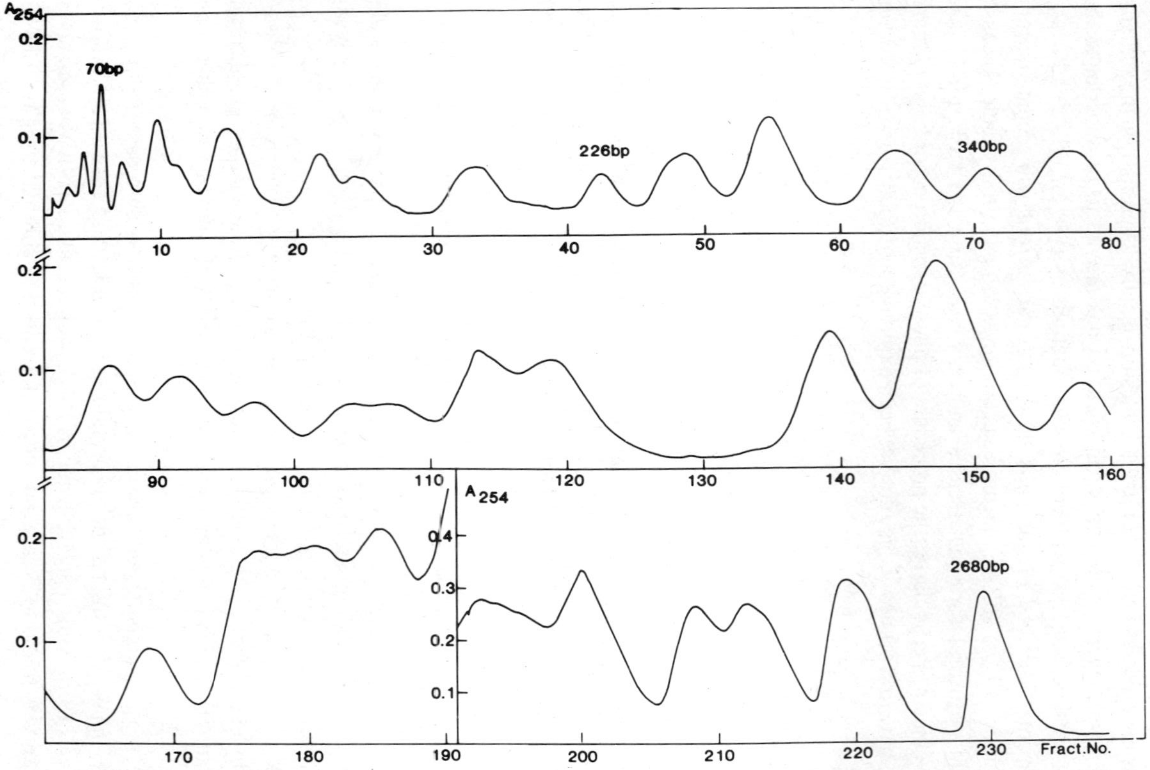

FIGURE 6.19. Partition chromatography of fragments of a Hae III digest of calf thymus DNA with a dextran-PEG phase system on a column with cellulose as support. Number of base pairs, determined by gel electrophoresis, is given on the peaks (Müller and Kütemeier, 1982).

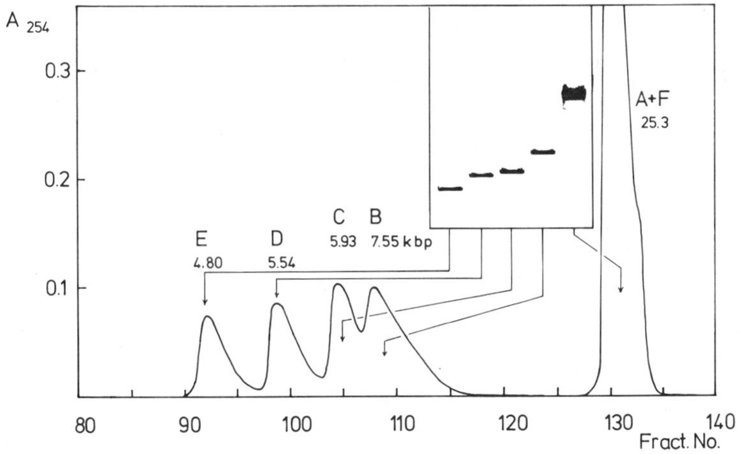

FIGURE 6.20. Fractionation of an EcoR I digest of 0.5 mg of λ DNA on a PEG–dextran column. Column size, system, support, coating, and temperature as for Figure 6.4; gradient, linear (800 mL) from 240 m*M* LiOAc to 80 m*M* Li_2SO_4; flow rate, 9 mL/h; fraction size, 3 mL. The homogeneity of the peaks eluted is documented by the gel pattern shown in the insert obtained from an agarose gel (0.7%) on which aliquots of the fractions indicated by the arrows were run in 40 m*M* tris, 20 m*M* NaOAc, 2 m*M* EDTA, pH 7.8 at 2.5 V/cm (Müller and Kütemeier, 1982).

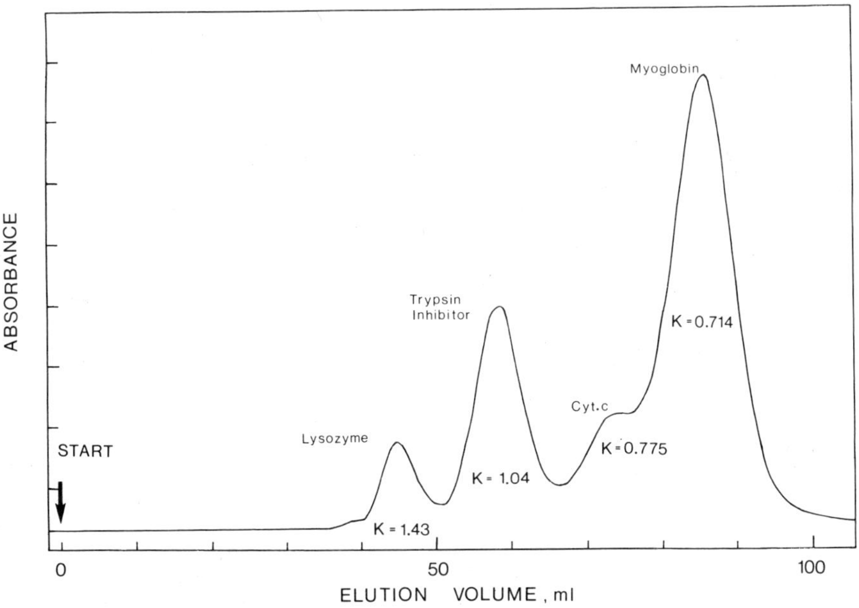

FIGURE 6.21. Partition chromatography of a mixture of myoglobin, cytochrome *c*, trypsin inhibitor, and lysozyme on a column of Altex (43 × 1.5 cm). Flow rate, 15 mL/h; Phase system, Dextran–PEG (37°C), pH 6.2; 0.2 *M* K acetate (Müller, unpublished).

Müller (personal communication) has also separated proteins by partition chromatography using AcA44 as solid support. See Figure 6.21.

Earlier work on partition chromatography suffered from low resolution and a low flow rate due to the high viscosity of the mobile phase. With the introduction of microgranular supports and high-pressure columns the prospects for partition chromatography with aqueous phase systems have been considerably improved.

REFERENCES

Åkerlund, H.-E. (1982). *Meth. Biochem. Anal.*, **28,** 115–150.

Åkerlund, H.-E. (1984). *J. Biophys. Biochem. Meth.*, **9,** 133–141.

Alberts, B. (1967). *Biochemistry,* **6,** 2527.

Albertsson, P.-Å. (1965). *Biochim. Biophys. Acta,* **103,** 1.

Albertsson, P.-Å. (1965). *Anal. Biochem.*, **11,** 121.

Albertsson, P.-Å. (1970). *Science Tools,* **17**(3), 56.

Albertsson, P.-Å. (1971). *Partition of Cell Particles and Macromolecules,* 2nd ed., Almqvist & Wiksell, Stockholm, and Wiley, New York.

Albertsson, P.-Å. (1973). *Biochemistry,* **12,** 2525–2530.

Albertsson, P.-Å., and Andersson, B. (1981). *J. Chromatogr.*, **215,** 131–141.

Albertsson, P.-Å., and Baird, G.D. (1962). *Exptl. Cell Res.*, **28,** 296.

Albertsson, P.-Å., and Nyns, Ed. J. (1959). *Nature,* **184,** 1465–1468.

Albertsson, P.-Å., Andersson, B., Larsson, C., and Anker, H.S. (1971). *Biochim. Biophys. Acta,* **229,** 290.

Babinet, C. (1967). *Biochem. Biophys. Res. Commun.*, **26,** 639.

Backman, L., and Johansson, G. (1976). *FEBS Lett.*, **65,** 39.

Bandyopadhyay, D., and Pinaev, G. (1983). *Exptl. Cell. Res.*, **144,** 313.

Bidney, D.L., and Reeck, G. (1977). *Biochemistry,* **16,** 1844–1849.

Blomquist, G., and Albertsson, P.-Å. (1969). *Anal. Biochem.*, **30,** 222.

Blomquist, G., and Albertsson, P.-Å. (1972). *Biochim. Biophys. Acta,* **73,** 125–133.

Blomquist, G., and Wold, S. (1974). *Acta Chem. Scand.*, **B28,** 56–60.

Blomquist, G., Hartman, A., Shanbhag, V., and Johansson, G. (1974). *Eur. J. Biochem.*, **48,** 63–69.

Capecchi, M.R. (1967). *Proc. Nat. Acad. Sci. USA,* **58,** 1144.

Cohen, R.J., and Crothers, D.M. (1970). *Biochemistry,* **9,** 2533.

Craig, D. (1956). In A. Weissberger, Ed., *Technique of Organic Chemistry,* Vol. III, Part I, 2nd ed., Interscience, New York.

Craig, L.C., and Post, H.O. (1949). *Anal. Chem.*, **21,** 500.

Cuzin, F., Kretchmer, N., Greenberg, R.E., Hurwitz, R., and Chapeville, F. (1967). *Proc. Nat. Acad. Sci., USA,* **58,** 2079.

Eoyang, L., and August, J.T. (1968). *Meth. Enzym.* **12B,** 530.

Falaschi, A., and Kornberg, A. (1966). *J. Biol. Chem.*, **241,** 1478.

Favre, J., and Pettijohn, D.E. (1967). *Eur. J. Biochem.*, **3,** 33.

Frederick, E.W., Maitra, U., and Hurwitz, J. (1969). *J. Biol. Chem.*, **244,** 413.

Gaziev, A.J., and Kuzin, A.M. (1973). *Eur. J. Biochem.*, **37,** 7–11.

Gineitis, A.A., Suciliene, S.P., and Shanbhag, V.P. (1984). *Analyt. Biochem.*, **139,** 400–403.

Gordon, J., Lucas-Lenard, J., and Lipman, F. (1971). *Meth. Enzym.* **20,** 281–291.

Hayashi, H., Knowles, J.R., Katze, J.R., Lapointe, J., and Söll, D. (1970). *J. Biol. Chem.*, **245,** 1401.

Hecker, E. (1955). *Verteilungsverfahren in Laboratorium, Monographien zu Angewandte Chemie und ChemiIngenieur-Technik,* No. 67, Verlag Chemie, GMBH, Weinheim/Bergstr., Germany.

Hertzberg, M., Lelong, J.C., and Revel, M. (1969). *J. Mol. Biol.*, **44,** 297.

Ito, Y., and Bowman, R.L. (1973). *Science,* **182,** 391–393.

Ito, Y., and Conway, W.D. (1984). *Anal. Chem.*, **56,** 534A.

Ito, Y., Weinstein, M.A., Aoki, I., Harada, R., Kimura, E., and Nunogaki, K. (1966). *Nature,* **212,** 985.

Iwasaki, K., and Kaziro, Y. (1979). *Meth. Enzym.*, **60,** 657–676.

Johansson, G., Andersson, M., and Åkerlund, H.-E. (1984). *J. Chromatogr.*, **298,** 483–493.

Kaplan, J.C., Kushner, S.R., and Grossman, L. (1969). *Biochemistry,* **3,** 144.

King, T.P., and Craig, L.C. (1962). *Meth. Biochem. Anal.*, **10,** 201–228.

Klein, H.A., and Capecchi, M.R. (1971). *J. Biol. Chem.*, **246,** 1055.

Kopperschläger, G., and Johansson, G. (1982). *Anal. Biochem.*, **124,** 117.

Lentz, K.E., Skeggs, L.T., Hochstrasser, H., and Kahn, J.R. (1963). *Biochim. Biophys. Acta,* **69,** 263.

Lindahl, T., and Edelman, G.M. (1968). *Proc. Nat. Acad. Sci. USA,* **61,** 680.

Little, J.W., Lehman, I.R., and Kaiser, A.D. (1967). *J. Biol. Chem.*, **242,** 672.

Loeb, L.A. (1969). *J. Biol. Chem.*, **244,** 1672.

Mak, S., Oberg, B., Johansson, K., and Philipson, L. (1976). *Biochemistry,* **15,** 5754–5761.

Miller, J.P., and Philipps, G.R. (1971). *J. Biol. Chem.*, **246,** 1274.

Mok, C.C., Grant, C.T., and Taborsky, G. (1966). *Biochemistry,* **5,** 2517.

Morris, C.J.O.R. (1963). In H. Peeters, Ed., *Protides of Biological Fluids,* Vol. 10, p. 325. Pergamon, Oxford.

Müller, W., and Eigel, A. (1981). *Anal. Chem.*, **118,** 269–277.

Müller, W., and Kütemeier, G. (1982). *Eur. J. Biochem.*, **128,** 231–238.

Müller, W., Scheutz, H.J., Guerrier-Takada, C., Cole, P.E., and Potts, R. (1979). *Nucleic Acid. Res.*, **7,** 2483–2499.

Ohlsson, R., Hentschel, C.C., and Williams, J.G. (1978). *Nucleic Acid. Res.*, **5,** 583–590.

Okazaki, T., and Kornberg, A. (1964). *J. Biol. Chem.*, **239,** 259.

Patterson, J.B., and Stafford, D.W. (1970). *Biochemistry,* **9,** 1278.

Pettijohn, D.E. (1967). *Eur. J. Biochem.*, **3,** 25.

Pettijohn, D.E. (1969). *Biochem. Biophys. Res. Commun.*, **34,** 541.

Pinaev, G., Bandyopadhyay, D., Glebov, O., Shanbhag, V.P., Johansson, G., and Albertsson, P.-Å. (1979). *Expt. Cell Res.*, **124,** 191–203.

Radding, C.M. (1966). *J. Mol. Biol.*, **18,** 235.

Raetz, C.R.H., and Kennedy, E.P. (1974). *J. Biol. Chem.*, **249,** 5038–5045.

Rometsch, R. (1950). *Helv. Chimica Acta,* **33:28,** 184.

Rudin, L. (1967). *Biochim. Biophys. Acta,* **134,** 199–202.

Rudin, L., and Albertsson, P.-Å. (1967). *Biochim. Biophys. Acta,* **134,** 37–44.

Sacks, L.E., and Alderton, G. (1961). *J. Bacteriol.,* **82,** 331.

Salach, J.I. (1977). *Meth. Enzymol.,* **53,** 495–501.

Scheibel, E.G., and Karr, A.E. (1950). *Ind. Eng. Chem.,* **42:6,** 1048.

Shanbhag, V., Blomquist, G., Johansson, G., and Hartman, A. (1972). *FEBS Lett.,* **22,** 105–108.

Shapiro, L., Franze de Fernandez, M.T., and August, J.T. (1968). *Nature,* **220,** 478.

Skeggs, L.T., Lentz, K.E., Hochstrasser, H., and Kahn, J.R. (1963). *Exptl. Med.,* **118,** 73.

Stenesh, J., and Roe, B.A. (1972). *Biochim. Biophys. Acta,* **272,** 156–166.

Summers, W.C., and Szybalski, W. (1967). *J. Mol. Biol.,* **26,** 227.

Sutherland, I.A., Heywood-Waddington, D., and Peters, T.J. (1984). *J. Liq. Chrom.,* **7,** 363–384.

Svensson, P., Schröder, W., Åkerlund, H.-E., and Albertsson, P.Å. (1985). *J. Chromatogr.,* **323,** 363–372.

Tavel, P.V., and Signer, R. (1956). *Advances in Protein Chem.,* **11,** 237.

Turner, G., and Hancock, R. (1974). *Biochem. Biophys. Res.. Comm.,* **58,** 437–445.

Treybal, R.E. (1951). In *Liquid Extraction,* McGraw-Hill, New York.

Vovis, G.F., and Buttin, G. (1970). *Biochim. Biophys. Acta,* **224,** 29.

Westrin, H., and Backman, L. (1983). *Europ. J. Biochem.,* **136,** 407–411.

7 CELL ORGANELLES AND MEMBRANE VESICLES

Phase partition has found many applications for separation of cell organelles and membrane vesicles both for analytical and preparative purposes. These applications have led to many important discoveries.

Unlike soluble molecules such as proteins and nucleic acids, which partition between the two bulk phases, particles such as cell organelles and cells usually partition between one of the phases and the interface. This can be explained by considering the surface-free energies of the particle in the two phases and the liquid–liquid interface. As shown in Chapter 3, the tendency for adsorption of particles at the interface increases the larger the particles and the larger the liquid–liquid interfacial tension. Also the adsorption at the interface is favored the more equal the affinity of the particles for the two phases. The affinity for the two phases in turn depends on many factors, such as type and molecular weight of polymer and ionic composition, and so forth, qualitatively in much the same way as was described for proteins and nucleic acids.

FACTORS DETERMINING PARTITION OF CELL ORGANELLES

Molecular Weight of the Polymers

In general, if one polymer of a phase system is replaced by the same type of polymer having a lower molecular weight, then the particles will favor more the phase containing the polymer with the lower molecular weight. Thus, if PEG

6000 is replaced by PEG 4000 of the dextran–PEG system, particles will favor more the PEG 4000-containing phase. The strongest tendency for particles to partition into the upper phase is therefore found with a phase system of high-molecular-weight dextran, such as Dextran 2000 and low-molecular-weight PEG, such as PEG 1500. Conversely, the strongest tendency for particles to partition into the lower phase is found for a system with a low-molecular-weight dextran, such as Dextran 40 and a high-molecular-weight PEG, such as PEG 40000.

Polymer Concentration

Increasing the polymer concentration of a phase system will remove it more from the critical point composition and also increase the interfacial tension. As a result, adsorption of particles at the interphase is favored when the polymer concentration is increased. Figure 7.1 shows the percentage of cell organelles found in the upper phase as a function of the dextran and PEG concentration. Close to the critical point all particles are in the upper phase. As the polymer concentration is increased the particles are removed from the upper phase and are found at the interface and/or the lower phase. The effect is selective, so by choosing a suitable polymer concentration a phase system with optimal separation effect can be obtained. By comparing different cell organelles from different sources one can now make a general statement concerning their affinity for the upper phase (Albertsson et al., 1982). The affinity for the upper phase increases in the following order:

For plant cells

inside-out thylakoid < intact chloroplasts mitochondria, peroxisomes
< endoplasmic reticulum < thylakoids
< Golgi < multiorganelle complex
= protoplasts = plasma membrane.

For animal cells

endoplasmic reticulum < mitochondria
< lysosomes < Golgi < plasma membranes.

Since the phase diagram may be slightly different for different batches of polymers, a plot of the partition as a function of the polymer concentration may not be exact compared to experiments in different laboratories. Therefore it is desirable to report also the polymer concentration in relation to the concentration

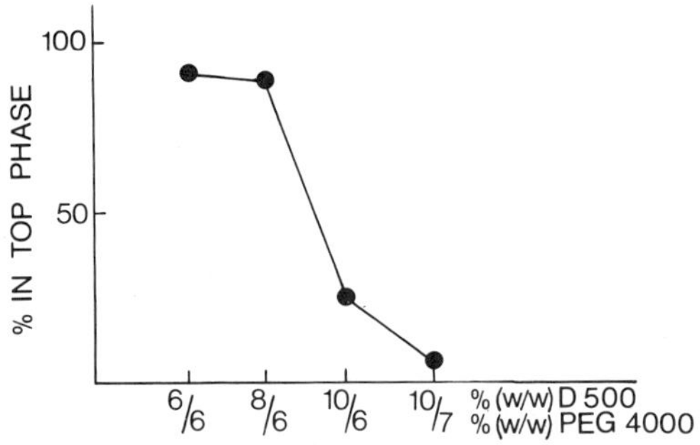

FIGURE 7.1. Partition of chloroplasts in the dextran–polyethylene glycol system as a function of polymer concentration. The ionic composition is kept constant while the concentration of both polymers is increased. The phase system is thereby more-and-more removed from the critical point.

at which the system turns from a one-phase to a two-phase system. If we denote these concentrations of dextran and PEG as C^0_{PEG} and C^0_D, respectively, then we should plot partition as a function of $C_D - C^0_D$ or $C_{PEG} - C^0_{PEG}$ in order to allow a better comparison between different experiments (see Figure 7.2).

Cell organelles and cells usually have specific requirements regarding the ionic composition and tonicity of the medium. Keeping the ionic composition constant and using the polymer concentration as a variable is therefore a convenient way to select a suitable phase system for fractionation.

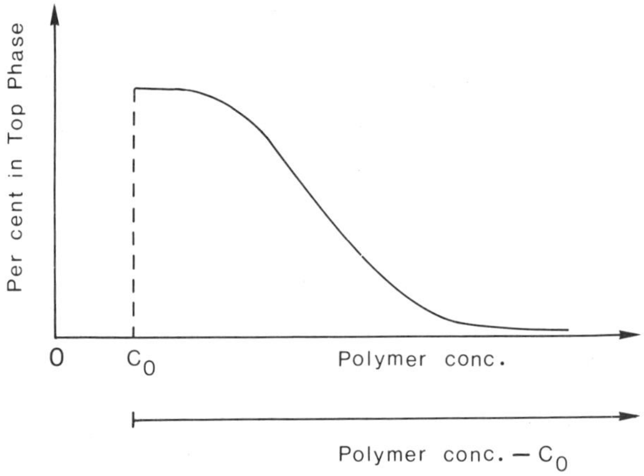

FIGURE 7.2. Different batches of the polymers may give slightly different phase diagrams. This in turn may influence a curve such as that of Figure 7.1. In order to obtain a more reproducible plot it is better to plot the percentage particles in the upper phase versus the concentration of the polymer minus its concentration (C_0) when the mixture shifts from a one-phase to a two-phase system.

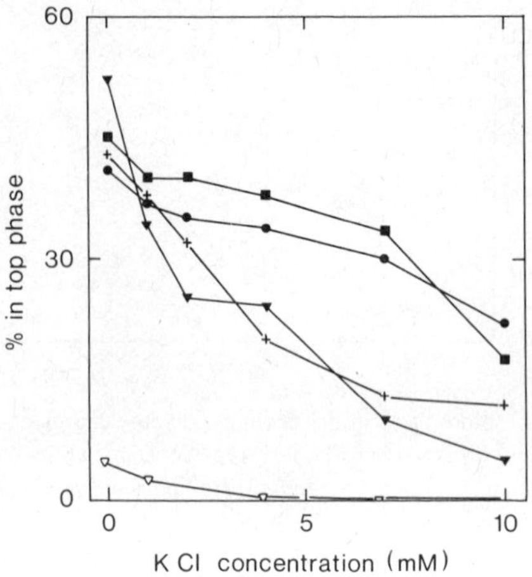

FIGURE 7-3. Effect of salt composition on the partition of rat liver microsomes in a dextran–PEG phase system; 6% w/w Dextran 500, 6% w/w PEG 4000, 15 m*M* Tris-sulfate, pH 7.8. Marker enzymes in top phase as function of KCl. Symbols represent (marker enzyme in parentheses): ■, plasma membrane (5′-nucleotidase); ●, lysosomes (acid phosphate); ▼, Golgi (*N*-acetylglucose amine galactosyltransferase); ▽, endoplasmic reticulum (NADPH-cytochromic reductase); +, protein (Gierow et al., 1985).

Ionic Composition of the Phase System

The same qualitative effects for proteins holds for particles, that is, for negatively charged particles positive ions push the particles into the top phase, in the order $Cs^+ \sim K^+ < Na^+ < Li^+$ and the negative ions in the order $Cl^- \sim H_2PO_4^- < HPO^{2-}{}_4 \sim SO^{2-}{}_4 \sim$ citrate. To get a maximal distribution into the upper phase one should therefore use Li_2HPO_4, Li_2SO_4, or $LiC_6H_7O_7$, and to get a maximal distribution into the lower phase one may use NaCl or KCl. Figure 7.3 shows curves illustrating the salt effect.

Affinity Partition

Both hydrophobic and biospecific affinity partition may be applied on cell organelles and membrane vesicles. By using hydrophobic groups coupled to PEG (PEG esterified with fatty acids or deoxycholate) selective separation of chloroplasts with the dextran–polyethylene glycol system was obtained (Westrin et al., 1976; Johansson and Westrin, 1978). Biospecific affinity partition has been applied on membrane vesicles from nervous tissues (Flanagan et al., 1976; Johansson et al., 1981).

MITOCHONDRIA

Rat Liver Mitochondria

A conventional preparation of mitochondria from rat liver can be separated into two populations by countercurrent distribution (Fig. 7.4). These subpopulations are not the same as the ones obtained by centrifugation in sucrose gradients, since both populations obtained by countercurrent distribution can be further subdivided by gradient centrifugation. No structural differences between the two countercurrent distribution populations has been detected by electron microscopy. The chemical composition was somewhat different. The different types of mitochondria may be considered as isomitochondria (Eriksson, 1974).

Mitochondria from developing young rats have been compared with those from adult rats. As seen in Figure 7.4, the mitochondria from a $1\frac{1}{2}$-day-old rat gives a single peak more to the right in the CCD diagram. With age this peak diminishes and is gradually replaced by the two peaks to the left of the diagram characteristic of adult rats.

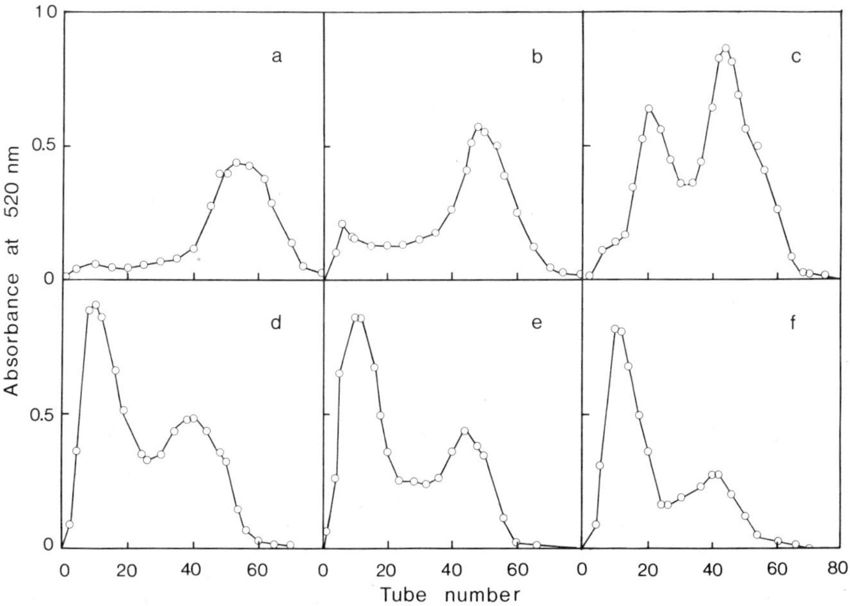

FIGURE 7.4. Countercurrent distribution of liver mitochondria from developing young rats. Age of rats: (*a*) 1.5 days, (*b*) 7 days, (*c*) 15 days, (*d*) 22 days, (*e*) 35 days, (*f*) full grown (mother) (Ericson, 1974a).

Rat Brain Mitochondria

Preparations of rat brain mitochondria by centrifugation are contaminated with synaptosomes, which include mitochondria. Free mitochondria can be effectively separated from synaptosomes by partition since their surface properties are entirely different.

Figure 7.5 shows a countercurrent distribution experiment demonstrating a complete separation of synaptosomes from free mitochondria. Since the mitochondria partition essentially to the lower phase plus the interface, and the synaptosomes to the upper phase, both fractions can be obtained by a simple and rapid batch procedure (Lopez-Perez et al., 1981).

Plant Leaf Mitochondria

To isolate leaf mitochondria from chlorophyll-containing membranes is a real challenge. Differential centrifugation alone gives a preparation with a protein to chlorophyll ratio of about 15. Such a preparation can be purified further by phase partition to a protein to chlorophyll ratio of about 100. About the same degree of purification is reached by density-gradient centrifugation. The mitochondria preparation is still green, however, and it is only by a combination of all three methods, differential centrifugation, density gradient and phase partition, that a

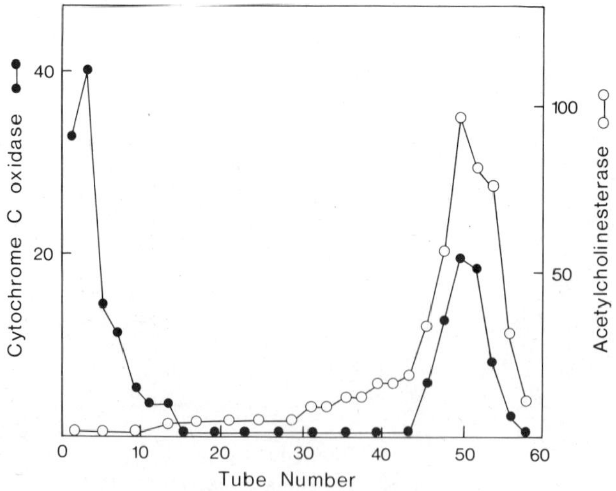

FIGURE 7.5. Separation of free rat brain mitochondria (left peak) from synaptosomes (right peak) by countercurrent distribution (Lopez-Perez et al., 1981).

chlorophyll-free preparation can be obtained (Gardeström et al., 1978; Bergman et al., 1980).

Submitochondrial Particles

Sonicated mitochondria from *Arum maculatum* have been subjected to countercurrent distribution with a phase system of dextran–PEG and hexamethylene diamine coupled to PEG (Möller et al., 1981). The mixture displayed a broad heterogeneity in the CCD diagram. An enrichment of inside-out particles was obtained in the lower phase. The hexamethylene diamine–PEG seemed to bind preferentially the right-side-out vesicles. This experiment is therefore an example of affinity parti-partition.

CHLOROPLASTS

Chloroplasts isolated by differential centrifugation can be resolved into three populations by countercurrent distribution (Fig. 7.6). The left peak (I) represents intact chloroplasts surrounded by their envelope and containing stroma, the middle peak (II) represents naked thylakoid membranes (class II chloroplasts) while the peak to the right (III) represents multiorganelle complexes, which are composed of one or more intact chloroplasts surrounded by a membrane-enclosed cytoplasmic layer including mitochondria and peroxisomes. The amount of multiorganelle complexes varies between 0.5 and 5% of total chlorophyll in chloroplast preparations from spinach. Cotelydons and young leaves give the highest yield of multiorganelle complexes. Prolonged homogenization increase their yield. The membrane enclosing the multiorganelle complexes is probably identical to the plasma membrane. They show a very similar partition behavior to protoplasts or isolated plasma membranes.

The results of Figure 7.6 demonstrate the usefulness of combining two different separation methods, centrifugation and phase partition. The particles of peak I and III are of about the same size, but since they differ in their surface properties they are effectively separated by partition. It also demonstrates the usefulness of CCD in discovering new types of cell particles. The multiorganelle complexes of peak III were not noticed earlier on electron microscopy pictures of conventional chloroplast preparations due to their low concentration. However, in peak III they are selectively enriched and constitute almost 100% of the population.

All three chloroplast populations also can be obtained by simple batch procedures which are useful for rapid isolation before metabolic studies. Fixation

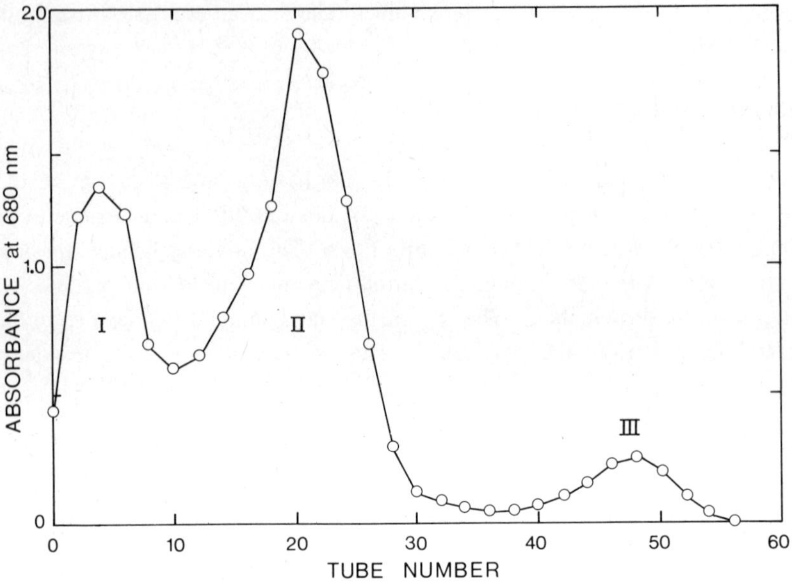

FIGURE 7.6. Countercurrent distribution and chloroplasts from spinach. (Left peak) intact chloroplasts surrounded by their envelope (class I chloroplasts); (middle peak) naked thylakoid membranes (class II chloroplasts); (right) multiorganelle particles containing chloroplasts, mitochondria, peroxisomes, cytoplasm, and surrounded by plasma membrane.

studies on $^{14}CO_2$ have shown that the intact chloroplasts are very pure and only minute amounts of ^{14}C are incorporated into sucrose or amino acids. In contrast, the multiorganelle complexes can make sucrose and amino acids as a result of cooperation between the chloroplasts and the surrounding cytoplasmic layer. Contamination of conventional chloroplast preparations by the multiorganelle complexes constitutes a problem in studies on chloroplast metabolism and on the localization of enzymes. This can be overcome, however, by removing the multiorganelle complexes by phase partition. Chloroplasts purified by phase partition are therefore particularly well suited for studies on compartmentation of metabolic pathways, as demonstrated by Larsson and Albertsson (1974) and Buchholz et al. (1979) for amino acids, and by Bickel et al. (1979) for plastoquinone and α-tocopherol.

Chloroplasts from both spinach and from mesophyll protoplasts of the C_4 plant *Digitaria sanguinalis* (Hallberg and Larsson, 1981) have been purified by phase partition.

Hydrophobic affinity partition has been used to purify intact chloroplasts from spinach (Westrin et al., 1976; Johansson and Westrin, 1978). PEG–caprate was used to selectively transfer intact chloroplasts to the upper phase while naked thylakoids stayed in the lower phase of the dextran–PEG phase system.

Thylakoid Vesicles—Inside-Out Vesicles

The internal lamellae system of the chloroplast is composed of paired membranes, the thylakoids. These either can occur as single pairs, stroma lamellae, or packed into stacks (grana). Each granum is composed of apposed regions where stacking occurs and nonapposed regions of the end membranes and the margins. Mechanical disintegration of thylakoids yields a mixture of membrane vesicles with different size, chemical composition, and photosynthetic activity. By differential centrifugation this can be separated into size classes of vesicles. These centrifugal fractions can be further separated by partition into classes of vesicles that have different surface properties (Fig. 7.7). Within each centrifugal fraction there is a considerable heterogeneity in surface properties. The material to the left in the

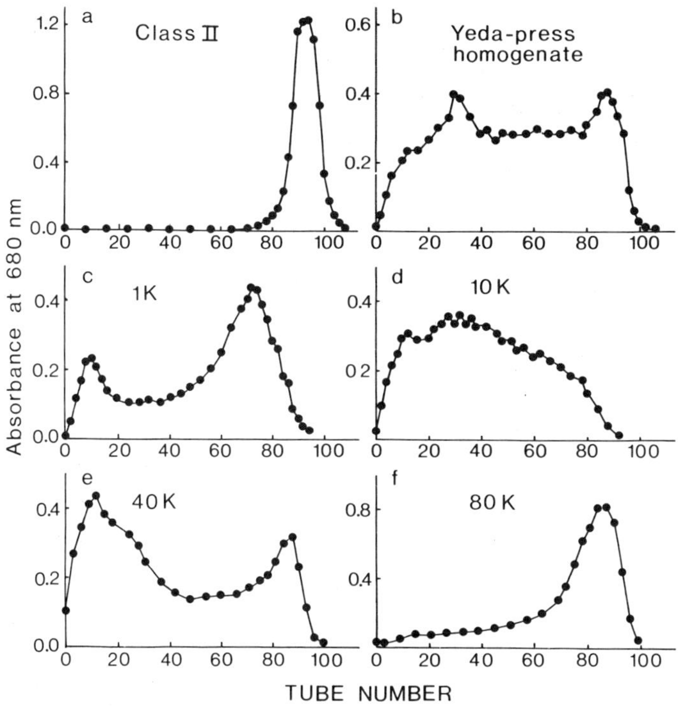

FIGURE 7.7. Countercurrent distribution of different centrifugal fractions from thylakoid vesicles obtained by press treatment. *K* symbolizes 100*g*. (Andersson et al., 1976).

countercurrent distribution diagram is enriched in photosystem II while the material to the right is enriched in photosystem I (Andersson et al., 1976; Åkerlund et al., 1976).

Later studies revealed that the material to the left in the countercurrent distribution diagram consists of inside out vesicles (Andersson et al., 1977; Andersson and Åkerlund, 1978). The heterogeneity in surface properties of the thylakoid vesicles obtained by mechanical disintegration under stacking conditions is therefore in part due to variations in sidedness of the vesicles. Figure 7.8 shows a separation of inside-out from right-side out vesicles. The mechanism for the formation of inside-out vesicles is shown in Figure 7.9.

Inside-out thylakoid vesicles can be disintegrated further by sonic treatment. This yields smaller vesicles which can be separated by countercurrent distribution. Some of the vesicles have turned back to normal sidedness and are found to the right in the countercurrent distribution diagram (Fig. 7.10).

PLASMA MEMBRANES AND CELL WALLS

Plasma membranes from several species have a high partition ratio compared with intracellular membranes. Conditions can therefore be found where the plasma membranes partition mainly into the upper phase while mitochondria, microsomes, chloroplasts, and other cell organelles are found at the interface or in the

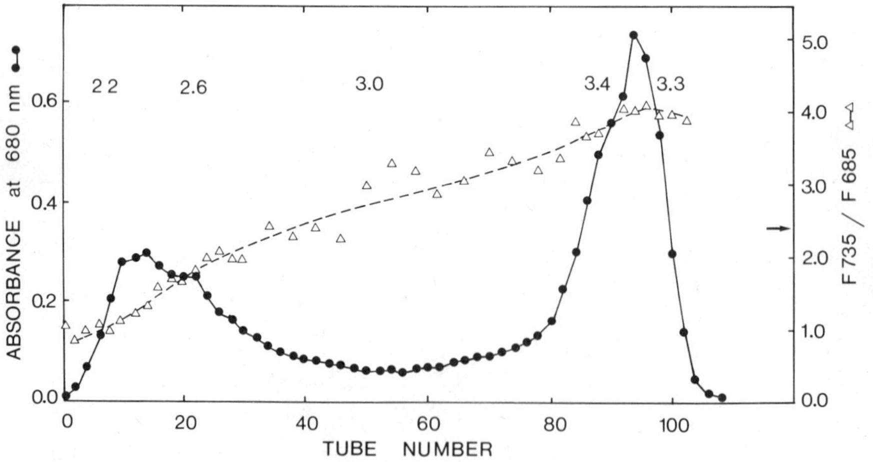

FIGURE 7.8. Separation of inside-out from right-side-out thylakoid vesicles by countercurrent distribution (Åkerlund et al., 1976; Andersson et al., 1977).

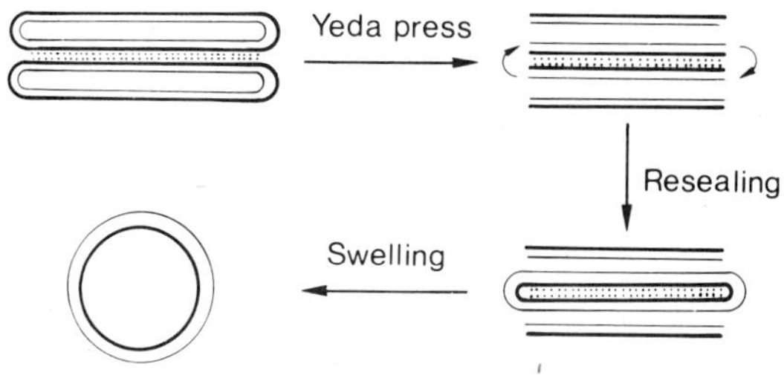

FIGURE 7.9. Mechanism of formation of inside-out vesicles from thylakoids by press treatment (Andersson et al. 1980).

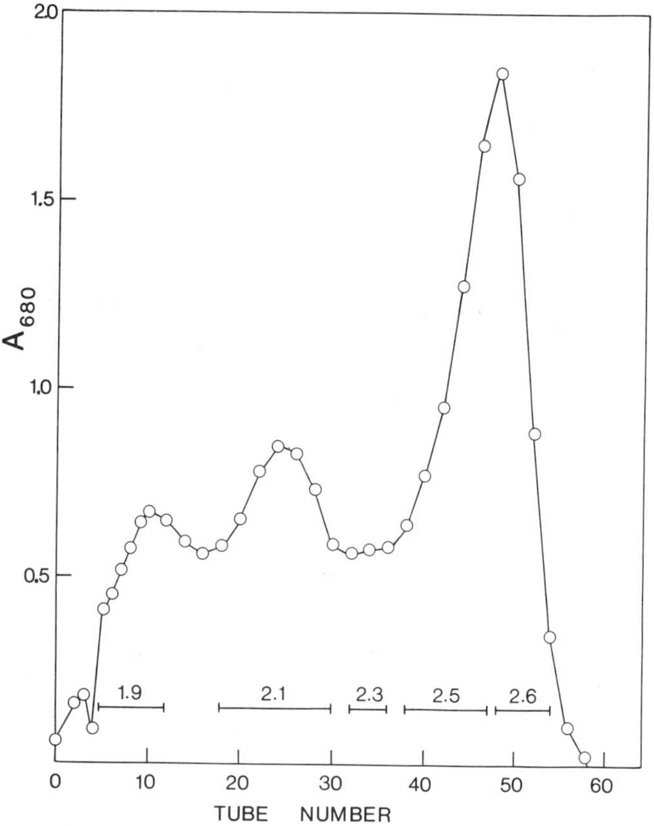

FIGURE 7.10. Further fractionation by countercurrent distribution of inside-out vesicles (peak to the left in Fig. 7.8) after sonication. (Left peak) vesicles which remain inside-out after sonication; (middle) vesicles of unknown sidedness; (right) vesicles which upon sonication have turned back to the right-side-out conformation.

bottom phase. This difference between plasma membranes and other membranes has been found for all species tested so far and can be utilized as a very simple, fast, and effective purification of plasma membranes.

From Plant Cells

Plasmalemma has been isolated from several plant tissues. Figure 7.11 shows a countercurrent distribution diagram on a microsomal fraction from oat roots (Widell et al., 1982). The small peak to the right (fraction 40–50) represents plasmalemma with the marker enzyme glucan synthetase while most of the remaining membranes are found in fractions 0–30. Since the difference in partition between the plasmalemma and other membranes is relatively large a batch pro-

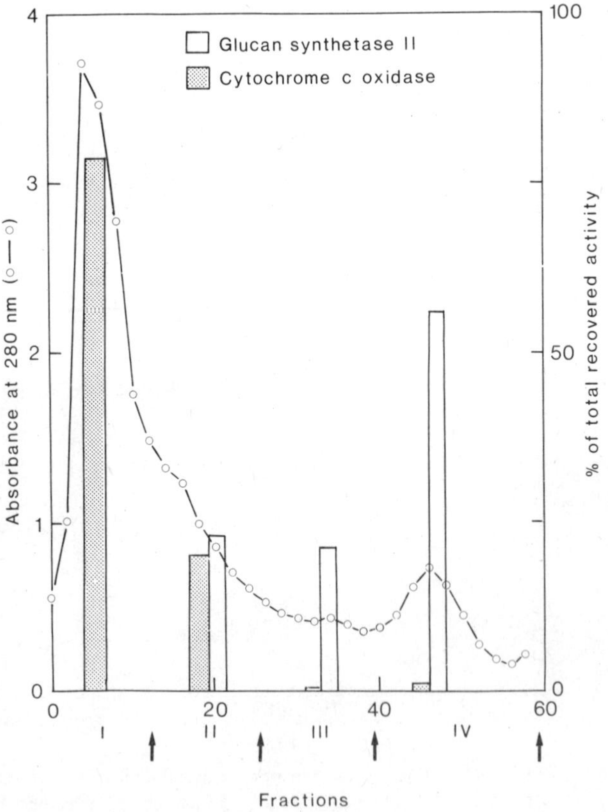

FIGURE 7.11. Countercurrent distribution of a microsomal fraction from oat root. The peak to the right represent plasma membranes. (Glucan synthetase II is a marker for plant plasma membrane) (Widell et al., 1982).

cedure involving only a few partition steps can be used for preparative isolation of plasmalemma.

Since the plasma membranes are isolated by a surface-dependent method, they are relatively homogeneous with respect to surface properties and thereby also with respect to sidedness. Latency studies indicate that the vesicles are sealed and right-side out (Larsson et al., 1984).

Of particular interest is that phase partition can be used also for isolation of plasmalemma from the microsomal fraction of green leaves. Here the contamination by chlorophyll-containing membrane vesicles is substantial and offers a tough problem in plasmalemma isolation. A batch procedure involving five partition steps gives a good yield of plasmalemma free from chlorophyll, mitochondria, and most of the microsome vesicles (Kjellbom and Larsson, 1984).

From Animal Cells

Plasma membranes from animal cells can be isolated by essentially similar procedures used for plant cells (Lesko et al., 1973; Brunette and Till, 1971; Gruber et al., 1984; Miller et al., 1974). The partition of rat liver plasma membranes was studied in detail by Gierow et al. (1985). Crude membranes obtained by centrifugation were resolved by countercurrent distribution into two peaks. One peak distributed with most of the contaminating membranes while the other was highly enriched in plasma membranes (Fig. 7.12). A batch procedure was developed for isolation of plasma membranes. Plasma membranes, originating either from the blood sinusoidal or the bile *canalicular domani,* were obtained in high purity and yield by the batch procedure.

Inside-out vesicles from the erythrocyte membrane can be isolated by one partition step (Steck, 1974). The partitions of erythrocyte ghosts, inside-out vesicles, and right-side-out vesicles were compared with the partition of erythrocytes. Both ghosts and right-side-out vesicles differed from the erythrocytes in their partition (and hence in surface properties). Inside-out vesicles had a lower partition coefficient than right-side-out vesicles, which are very heterogeneous and could even give a binodial CCD distribution. They show, therefore, a different partition behavior compared with whole cells (Walter and Krob, 1976).

Cell Walls

Cell walls from the green algae chlorella were among the first subcellular particles purified by phase partition (Albertsson, 1956, 1958). Both a system of potassium phosphate–PEG–water and a dextran–PEG system were used. The latter was found to give the purest preparation. A similar procedure was used by

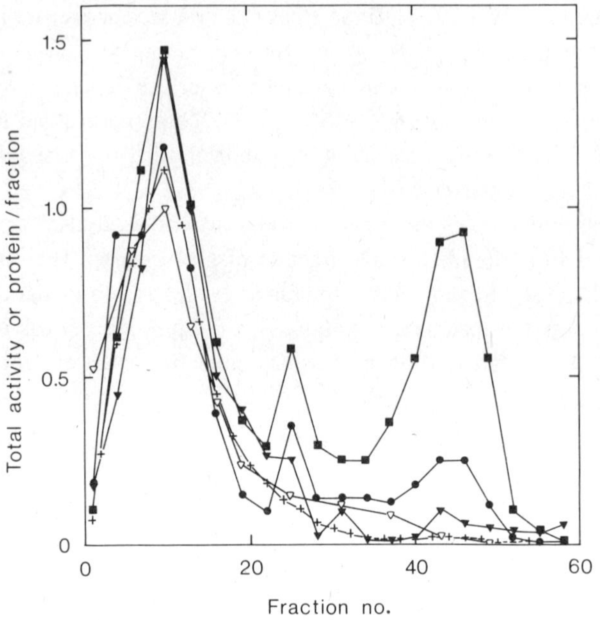

FIGURE 7.12. Countercurrent distribution of a rat liver microsomal fraction. Symbols as in Figure 7.3 (Gierow et al., 1985).

Burczyk (1973) to isolate cell walls from *Scenedesmus obliquus*. Cell walls from Aerobacter were isolated by a potassium phosphate–PEG system (Albertsson, 1958).

PROTOPLASTS

Intact protoplasts can be obtained from several plant species by digestion of leaf segments with cellulase and pectinase. Such preparations, however, contain broken protoplasts, free chloroplasts, and other cellular contaminants. They can be purified further by phase partition. Kanai and Edwards (1973) purified protoplasts from C_3, C_4, and CAM plants by a phase system composed of dextran and polyethylene glycol. Dextrans of different molecular weights were compared in order to optimize the yield and purification, although the size of the protoplasts varied considerably, from 17 μm with *Panicum capillare,* to 119 μm with *Sedum rubrotinctum* the phase partition step was effective. The protoplasts from Sedum are probably the largest particles purified so far by partition. Since protoplasts expose the plasmalemma, their partition is very similar to isolated plasmalemma, and they prefer the upper phase more than do subcellular membranes and cell organelles such as mitochondria and chloroplasts. Figure 7.13 shows a countercurrent distribution of a chloroplast preparation from protoplasts. The residual intact protoplasts have a much higher partition ratio and are therefore found to the right in the distribution train (tubes 80–95), while chloroplasts and thylakoids are found to the left (Hallberg, 1983). A batch procedure for purification of protoplasts from *Digitaria sanguinalis* was developed based on this information and using only two partition steps at room temperature (Hallberg and Larsson, 1981).

NEURAL MEMBRANES

Affinity partition was employed in the purification of membranes rich in cholinergic receptor from *Torpedo californica* (Flanagan et al., 1976). Bis-trimethylamino polyethylene glycol was added to a dextran–polyethylene glycol system

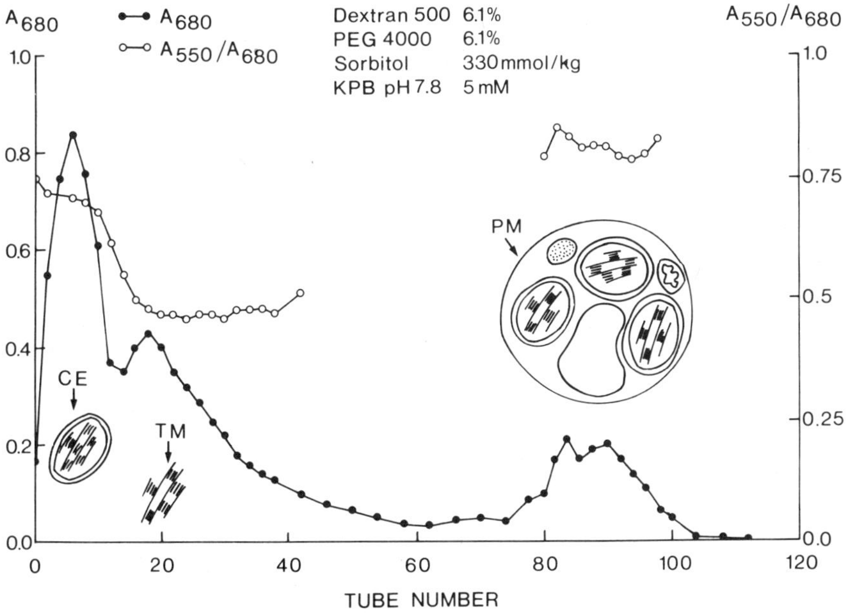

FIGURE 7.13. Separation of chloroplasts from protoplasts by countercurrent distribution. The peak between tubes 0 and 10 contains intact chloroplasts; the peak around tube 20, thylakoids; and the peak to the right, protoplasts (Hallberg, 1983).

to achieve specific partition of the receptor-containing membranes into the upper phase. The specific activity of the upper phase membranes was four to five times higher than the lower-phase membranes, indicating a considerable purification. Countercurrent distribution resolved the membrane fragments into three populations. Two of these were highly enriched in nicotinic receptor (Johansson et al., 1981). The polypeptide composition of affinity purified membranes has been described (Gysing et al., 1981).

Neural membrane vesicles from calf brain were fractionated by countercurrent distribution (Johansson et al., 1984). The partition behavior of the vesicles seemed to change with time, probably due to aggregation or lateral rearrangement of membrane components. This problem could be circumvented, however, by using the centrifugal countercurrent distribution apparatus described in Chapter 6 (Johansson et al., 1984).

Hartman and Heilbronn (1978) used a phase system of dextran and polyethylene glycol for partition of membranes from *T. marmorata*. Negatively charged polyethylene glycol (PEG-sulfonate) was employed to adjust the partition. With increasing concentration of PEG-sulfonate the membranes are transferred from the upper phase via the interface to the lower phase (Fig. 7.14). They could separate nicotinic-receptor-enriched membranes from those containing acetylcholin esterase in a single step. The two types of membranes could be further

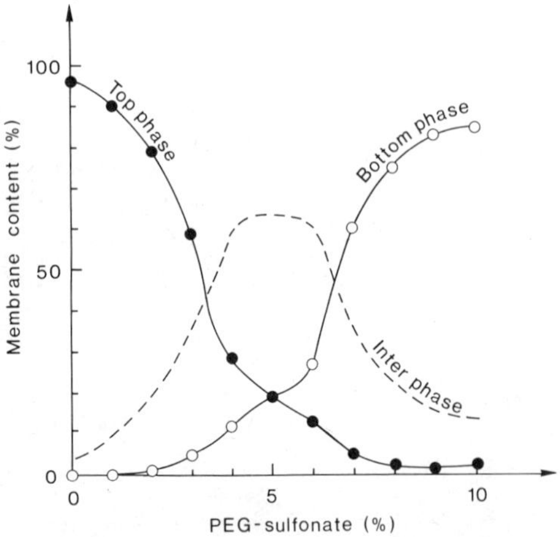

FIGURE 7.14. Partition of membrane vesicles from *Torpedo marmorata* electric organ in a phase system of dextran and PEG + PEG sulfonate. The negatively charged PEG sulfonate partitions into the upper phase and repels the membranes selectively from this phase (Hartman and Heilbronn, 1978).

subfractionated by countercurrent distribution. The acetylcholine-receptor-enriched membranes from *T. marmorata* electroplax purified by phase partition were characterized with regard to polypeptide composition, sidedness, and protein phosphorylation (Heilbronn et al., 1979).

LIVER HOMOGENATE

Rat liver homogenate was analyzed by countercurrent distribution using a small volume partition apparatus capable of 17 transfers in which separation of the phases was enhanced by low-speed centrifugation (Morris and Peters, 1982). Effect of polymer concentration and salt composition was studied. The different organelles and membranes (mitochondria, lysosomes, plasma-membranes, endoplasmic reticulum, nuclei) behaved differently and some marker enzymes showed a binodial distribution. Similar experiments were carried out with a toroidal-coil centrifuge (Heywood-Waddington et al., 1984).

LIPOSOMES

Partition of liposomes has been used to study the contribution of lipids to the partition of membranes. Liposomes made of different phospholipids, having different polar head groups and different fatty acids, were partitioned in the dextran–polyethylene glycol system (Eriksson and Albertsson, 1978). It was found that the polar head groups of the phospholipids play a dominant role in determining partition of liposomes, while the degree of unsaturation is of less importance. This again stresses that partition is a surface-dependent method. If cholesterol was incorporated in liposomes, it reduced the difference in partition between liposomes of various composition.

Also, when hydrophobic affinity partition was employed the polar head group was the main determining factor (Eriksson, 1980). Phosphatidylcholine liposomes with different chain length and saturation of the fatty acids and phosphatidylserine liposomes were partitioned in a dextran–polyethylene glycol system with the addition of increasing amounts of PEG fatty acid ester (Fig. 7.15). The liposomes are thereby transferred from the lower phase to the upper phase. The fatty acid composition of the liposomes have no influence on the extraction curve; it is only the polar head groups of the liposomes and the chain length of the PEG-bound hydrocarbon chain which is of importance.

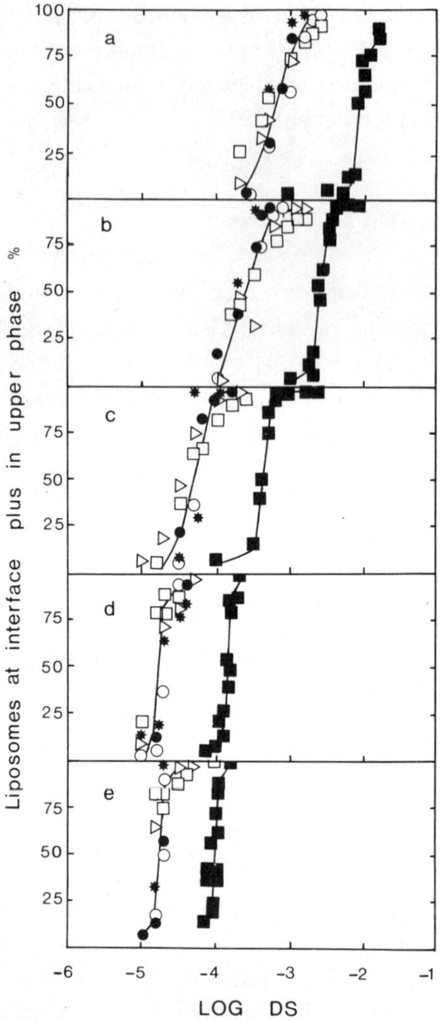

FIGURE 7.15. The percentage of liposomes at interface and in upper phase as function of the fraction of polyethylene glycol end groups esterified with fatty acids. DS-degree of substitution. *, dilauroyl phosphatidylcholine; ○, dimyristoyl phosphatidylcholine; △, dipalmitoyl phosphatidylcholine; □, distearoyl phosphatidylcholine; ●, egg phosphatidylcholine and ■, phosphatidylserine. Polyethylene glycol is esterifed with (*a*) capric acid (C_{10}), (*b*) lauric acid (C_{12}), (*c*) myristic acid (C_{14}), (*d*) palmitic acid (C_{16}), and (*e*) stearic acid (C_{18}) (Eriksson, 1980).

Incorporation of glycolipids of defined carbohydrate composition have shown that the surface carbohydrate affects partition (Sharpe and Warren, 1984).

The picture that emerges from these studies is that the outer surface layer of the liposomes determines the partition behavior. The effect of PEG–fatty acid esters is the result of binding of the hydrophobic tail of the PEG ester to the lipid layer of the liposomes. This binding is stronger the longer the hydrocarbon chain length of the PEG ester but independent of the fatty acid composition of the liposome. The binding is, however, also affected by the polar head group of the liposome.

The effect of liposome charge was also studied. Partition in charged phase systems is able to detect charges on liposomes. However, the charge detected is not always what would be expected from the overall charge of the head group. For example, the phosphatidyl head group is neutral yet partition showed that the surface of the phosphatidyl liposomes was slightly positively charged (Eriksson and Albertsson, 1978).

CHROMOSOMES

Chromosomes from HeLa cells have been fractionated by partition in the dextran–polyethylene glycol system. By adding positively or negatively charged PEG the chromosomes could be transferred from one phase to the other via the interface (Fig. 7.16). Different karyotypes behaved differently. A selective partition of chromosomes was also demonstrated by countercurrent distribution (Pinaev et al., 1979; Bandyopadhyay and Pinaev, 1983). (See Table 7.1).

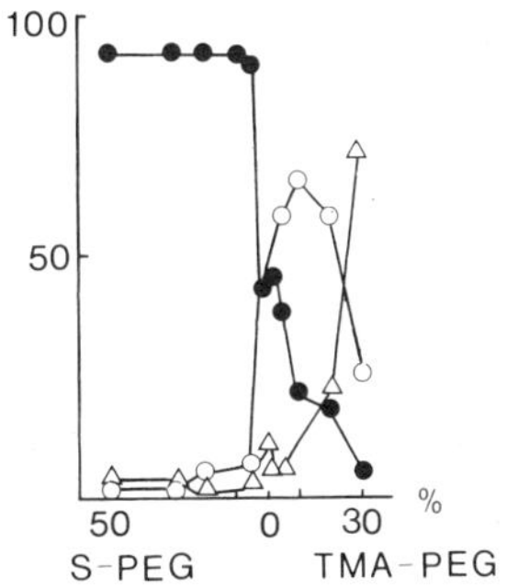

FIGURE 7.16. Partition of chromosomes from HeLa cells in a dextran–PEG phase system as a function of either 0–30% of the PEG as positively charged trimethylamino-PEG (TMA-PEG) or 0–50% of the PEG as the negatively charged PEG-sulfonate (S-PEG). Ordinate: percentage of the chromosomes in top phase (●), bottom phase (△), and interface (○) (Pinaev et al., 1979).

TABLE 7.1 Applications on Cell Organelles and Membrane Vesicles

	Reference
Chloroplasts	Albertsson and Baltscheffsky (1963; Albertsson and Larsson (1976); Karlstam and Albertsson (1969, 1972); Larsson et al., (1971); Johansson and Westrin (1978); Westrin et al. (1976)
Thylakoids	Andersson et al. (1976); Åkerlund et al. (1976)
Inside-out thylakoids	Andersson et al. (1977, 1980); Andersson and Åkerlund (1978); Sundby et al. (1982); Åkerlund and Andersson (1983)
Etioplasts	Treffry (1974)
Mitochondria from leaves	Gardeström et al. (1978); Bergman et al. (1980)
Submitochondrial vesicles	Möller et al. (1981)
Golgi	Hino et al. (1978 a, b)
Peroxisomes	Horie et al. (1979)
Microsomes	Ohlsson et al. (1978); Gierow et al. (1985)
Synaptosomes	Lopez-Perez et al. (1981)
Membrane vesicles from neurons	Flanagan et al. (1976, 1984); Hartman and Heilbronn (1978); Heilbronn et al. (1979); Johansson et al. (1981, 1984)
Cell membranes	
rat liver	Gierow et al. (1985); Lesko et al. (1973)
murine fibrosarcoma	Miller et al. (1974)
corneal endothelial cells	Whikehart and Soppet (1981)
lung fibroblasts	Gruber et al. (1984)
L-cells	Brunette and Till (1971)
kidney brush border	Glossman and Gips (1974)
embryo fibroblasts	Russel and Pastan (1973)
inside-out erythrocyte membrane	Steck (1974); Walter and Krob (1976)
Insulin-secreting granules	Coore et al. (1969)
Rat liver homogenate	Morris and Peters (1982)
Chromosomes	Pinaev et al. (1979); Bandyopadhyay and Pinaev (1983)
Multiorganelle complexes	Larsson et al. (1971); Albertsson and Larsson (1976)
Plasma lemma, plasma membranes	
corn	Widell et al. (1983)
cauliflower	Caubergs et al. (1983)
oat roats	Widell et al. (1982)
wheat	Lundborg et al. (1981)
barley, spinach	Kjellbom and Larsson (1984)
rye	Uemura and Yoshida (1983)
orchard grass	Yoshida et al. (1983)
Poly-β-hydroxybuturate granules	Griebel et al. (1968)

TABLE 7.1 *(continued)*

	Reference
Membranes of *Acholeplasma laidlawi*	Wieslander (1979)
Cell walls from green algae	Albertsson (1956, 1958); Burczyk (1973)
Mitochondria from rat liver	Lopez-Perez et al. (1981)
Mitochondria from rat liver	Ericson (1974a, b)
Mitochondrial inner membrane	Lundberg and Ericson (1975)
	Heywood-Waddington (1984)
Liposomes	Eriksson and Albertsson (1978); Eriksson (1981); Sharpe and Warren (1984)

CROSS PARTITION

The Isoelectric Point of Cell Particles Can Be Determined by Cross Partition

Because the electrical potential between the phases, partition can be used to characterize the charge properties of cell organelles and membrane vesicles. The isoelectric point of the particles can be determined by cross partition. By this method the partition is studied as a function of pH with two different salt media. If the partition coefficient or percentage of particles in the upper phase is plotted against pH, two curves are obtained which cross at the isoelectric point.

Cross partition has been applied to a large number of proteins (see Chapter 5). It can also be used for particles. Figure 7.17 shows the cross partition of chloroplasts, thylakoids, and subthylakoid vesicles, inside-out and right-side out. From these studies the isoelectric point of the different surfaces of the membranes of a chloroplast could be determined (Fig. 7.18). Also, the isoelectric points of mitochondria and the inner and outer membranes of mitochondria have been determined by cross partition (Ericsson, 1974b).

This method was also used to determine the isoelectric points of mitochondria and mitochondrial particles from beef heart (Lundberg and Ericsson, 1975). It was found that ATP, but not ITP, lowered the isoelectric point of the mitochondrial particles, indicating a conformational and/or charge change of the membrane.

Cross partition also gives information on the noncharge contribution to the partition. At the cross point the charge contributions are nullified and the partition coefficient at the cross point therefore reflects factors other than charge which determine the partition. This is illustrated by Figure 7.19 which shows the cross

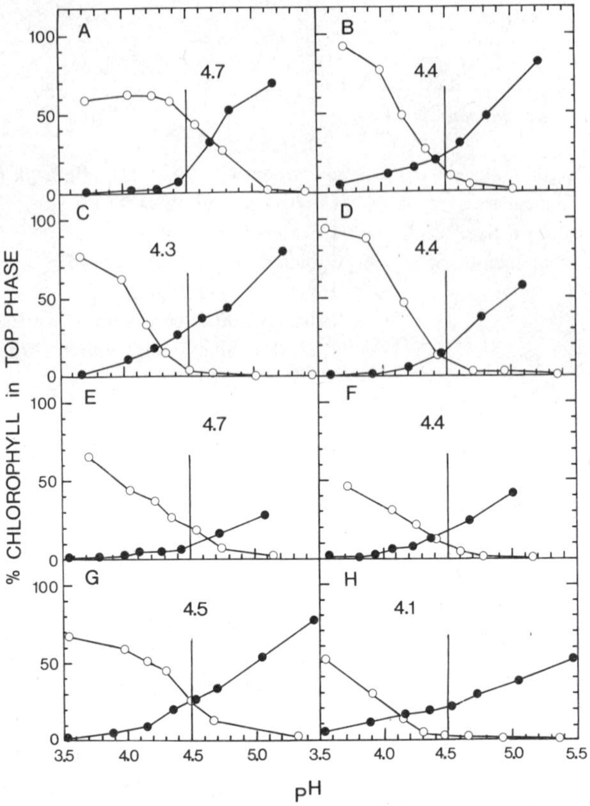

FIGURE 7.17. Cross partition of different thylakoid membrane: (A) chloroplast lamellae (class II chloroplasts); (B) press homogenate of A; (C) 10*K* centrifugal fraction; (D) 40*K* centrifugal fraction; (E) 100*K* centrifugal fraction; (F) low salt fragments; (G) right-side-out vesicles; (H) inside-out vesicles (Åkerlund et al., 1979).

partition of three different particles: intact chloroplasts, thylakoid lamellae, and multiorganelle particles from spinach leaves. These particles expose different types of membranes: chloroplast envelope, thylakoid membrane, and plasma membrane, respectively. They not only have different isoelectric points but their partition characteristics at the isoelectric point are also very different. The multiorganelle particles prefer the upper phase most (80–90% in the upper phase), while the intact chloroplasts are mainly at the interface or in the bottom phase (only 10–20% in the upper phase). The thylakoids have an intermediate partition. These three types of particles can be separated by countercurrent distribution (Fig. 7.6). In the diagram of Figure 7.6 the multiorganelle particles (peak III) are to the right, the intact chloroplasts (peak I) to the left, and the thylakoids in the middle (peak II). Thus the order in the countercurrent distribution diagram

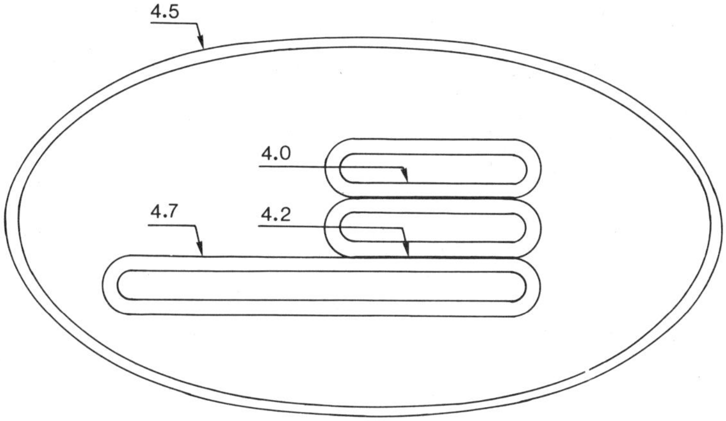

FIGURE 7.18. Isoelectric points of different surfaces of the chloroplast. The isoelectric points were determined by cross partition (Åkerlund et al., 1979).

reflects the partition at the cross point. One can conclude, therefore, that although the separation carried out by countercurrent distribution employed a phase system at pH 7, when all particles are negatively charged, it was the differences in noncharge properties of the particle surfaces, and not differences in charge, which were the main factors causing the separation.

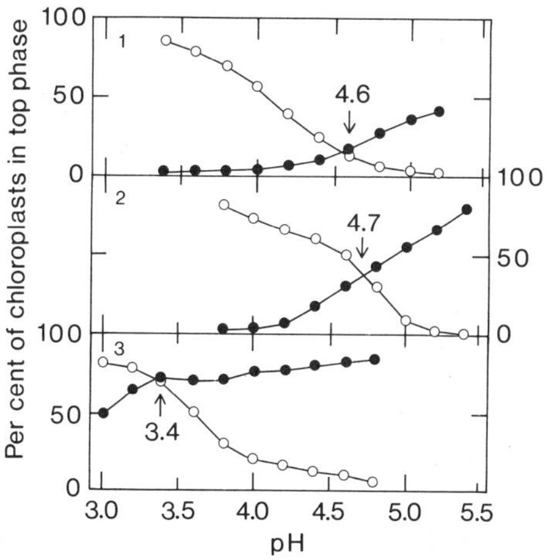

FIGURE 7.19. Cross partition of (1) intact chloroplasts (class I); (2) thylakoid membranes (class II); and (3) multiorganelle particles (peak III in Figure 7.6). Note the difference in partition (percentage in top phase) at the isoelectric point (the cross point) (Westrin et al., 1983).

REFERENCES

Åkerlund, H.-E. (1982). *Meth. Biochem. Analysis,* **28,** 115–150.

Åkerlund, H.-E. (1984). *J. Biophys. Biochem. Meth.,* **9,** 133–141.

Åkerlund, H.-E., and Andersson, B. (1983). *Biochim. Biophys. Acta,* **725,** 34–40.

Åkerlund, H.-E., Andersson, B., and Albertsson, P.-Å. (1976). *Biochim. Biophys. Acta,* **449,** 525–535.

Åkerlund, H.-E., Andersson, B., Persson, A., and Albertsson, P.-Å. (1979). *Biochim. Biophys. Acta,* **522,** 238–246.

Albertsson, P.-Å. (1956). *Nature,* **177,** 771–774.

Albertsson, P.-Å. (1958). *Biochim. Biophys. Acta,* **27,** 378–395.

Albertsson, P.-Å., and Baltscheffsky, H. (1963). *Biochem. Biophys. Res. Commun.,* **12,** 14–20.

Albertsson, P.-Å., and Larsson, C. (1976). *Mol. Cell. Biochem.,* **11,** 183–189.

Andersson, B., Åkerlund, H.-E., and Albertsson, P.-Å. (1976). *Biochim. Biophys. Acta,* **423,** 122–132.

Andersson, B., Åkerlund, H.-E., and Albertsson, P.-Å. (1977). *FEBS Lett.,* **77,** 141–145.

Andersson, B., Sundby, C., and Albertsson, P.-Å. (1980). *Biochim. Biophys. Acta,* **599,** 391–402.

Bandyopadhyay, D., and Pinaev, G. (1983). *Exptl. Cell Res.,* **144,** 313.

Bergman, A., Gardeström, P., and Ericson, I. (1980). *Plant Physiol.,* **66,** 442–445.

Bickel, H., Buchholz, B., and Schultz, G., (1979). In L.Å. Appelqvist and C. Liljenberg, Eds., *Advances in the Biochemistry and Physiology of Plant Lipids,* Elsevier, Amsterdam, pp. 369–375.

Brunette, D.M., and Till, J.E. (1971). *Membrane Biol.,* **5,** 215–224.

Buchholz, B., Rempke, B., Bickel, H., and Schultz, G. (1979). *Phytochemistry,* **18,** 1109–1111.

Burczyk, J. (1973). *Folia Histochem. Cytochem.,* **11,** 119–134.

Caubergs, R., Widell, S., Larsson, C., and De Greef, J.A. (1983). *Physiol. Plant.,* **57,** 291–295.

Coore, H.G., Hellman, B., Pihl, E., and Täljedal, I.-B. (1969). *Biochem. J.,* **111,** 107–113.

Ericson, I. (1974a). Ph.D. Thesis, Umeå University, Umeå, Sweden.

Ericson, I. (1974b). *Biochim. Biophys. Acta,* **356,** 100–107.

Eriksson, E. (1981). *J. Chromatogr.,* **205,** 189–193.

Eriksson, E., and Albertsson, P.-Å. (1978). *Biochim. Biophys. Acta,* **507,** 425–432.

Flanagan, S.D. (1984). "Affinity Phase Partitioning," in J.C. Venter and L.C. Harrison, Eds., *Receptor Biochemistry and Methodology,* Alan R. Liss, New York, 15–44.

Flanagan, S.D., Barondes, S.H., and Taylor, P. (1976). *J. Biol. Chem.,* **251,** 858–865.

Flanagan, S.D., Johansson, G., Yost, B., Ito, Y., and Sutherland, I.A. (1984). *J. Liq. Chromatogr.,* **7**(2), 385–402.

Gardeström, P., Ericson, I., and Larsson, C. (1978). *Plant Sci. Lett.,* **13,** 231–240.

Gierow, P., Jergil, B., and Larsson, C. (1985). in press.

Glossmann, H., and Gips, H. (1974). *Arch. Pharmacol.,* **282,** 439–444.

Griebel, R., Smith, Z., and Merrick, J.M. (1968). *Biochemistry,* **7,** 3676–3681.

Gruber, M.Y., Cheng, K.-H., Lepock, J.R., and Thompson, J.E. (1984). *Anal. Biochem.,* **138,** 112–118.

Hallberg, M., and Larsson, C. (1981). *Arch. Biochem. Biophys.,* **208,** 121–130.

Hartman, A., and Heilbronn, E. (1978). *Biochim. Biophys. Acta,* **513,** 382–394.

Heilbronn, E., Björck, C., Elfman, L., and Hartman, A. (1979). *Adv. Cytopharmacology,* 3.

Hellergren, J., Widell, S., Lundberg, T., and Kylin, A. (1983). *Physiol. Plant.,* **58,** 7–12.

Heywood-Waddington, D., Sutherland, I., Morris, W.B., and Peters, T.J. (1984). *Biochem. J.,* **217,** 751–759.

Hino, Y., Asano, A., and Sato, R. (1978a). *J. Biochem.,* **83,** 925–934.

Hino, Y., Asano, A., and Sato, R. (1978b). *J. Biochem.,* **83,** 935–942.

Horie, S., Ishii, H., Nakazawa, H., Suga, T., and Orii, H. (1979). *Biochim. Biophys. Acta,* **585,** 435–443.

Johansson, G., and Westrin, H. (1978). *Plant Sci. Lett.,* **13,** 201–212.

Johansson, G., Gysing, R., and Flanagan, S.D. (1981). *J. Biol. Chem.,* **256,** 9126–9135.

Johansson, G., Åkerlund, H.-E., and Olde, B. (1984). *J. Chromatogr.,* **311,** 277–289.

Kanai, R., and Edwards, G.E. (1973). *Plant Physiol.,* **51,** 1133.

Karlstam, B., and Albertsson, P.-Å. (1969). *FEBS Lett.,* **5,** 360–363.

Karlstam, B., and Albertsson, P.-Å. (1972). *Biochim. Biophys. Acta,* **255,** 539–552.

Kjellbom, P., and Larsson, C. (1984). *Physiol. Plant.,* **62,** 501–509.

Larsson, C. (1985). "Plasma membranes", in H.F. Linskens and J.F. Jackson, Eds., *Modern Methods of Plant Analysis* Vol. 1, Springer, Berlin.

Larsson, C., and Albertsson, P.-Å. (1974). *Biochim. Biophys. Acta,* **357,** 412–419.

Larsson, C., Collin, C., and Albertsson, P.-Å. (1971). *Biochim. Biophys. Acta,* **245,** 425–438.

Larsson, C., Kjellbom, P., Widell, S., and Lundborg, T. (1984). *FEBS Lett.,* **171,** 271–276.

Lesko, L., Donlon, M., Marinetti, G.V., and Hare, J.D. (1973). *Biochim. Biophys. Acta,* **311,** 173–179.

Lopez-Perez, M.J., Paris, G., and Larsson, C. (1981). *Biochim. Biophys. Acta,* **635,** 359–368.

Lundberg, P., and Ericson, I. (1975). *Biochem. Biophys. Res. Comm.,* **65,** 530–536.

Lundborg, T., Widell, S., and Larsson, C. (1981). *Physiol. Plant.,* **52,** 89–95.

Miller, L.S., Evans, S.B., Rossio, J.L., and Dodd, M.C. (1974). *Prep. Biochem.,* **4,** 489–498.

Möller, I.M., Bergman, A., Gardeström, P., Ericsson, J., and Palmer, J.M. (1981). *FEBS Lett.,* **126,** 13–17.

Morris, W.B., and Peters, T.J. (1982). *Europ. J. Biochem.,* **121,** 421–426.

Ohlsson, R., Jergil, B., and Walter, H. (1978). *Biochem. J.,* **172,** 189–192.

Pestka, S., Weiss, D., and Vince, R. (1976). *Analyt. Biochem.,* **71,** 137–142.

Pinaev, G., Bandyopadhyay, D., Glebov, O., Shanbhag, V., Johansson, G., and Albertsson, P.-Å. (1979). *Exp. Cell. Res.,* **124,** 191–203.

Russel, T., and Pastan, J. (1973). *J. Biol. Chem.,* **248,** 5835–5840.

Sharpe, P.T., and Warren, G.S. (1984). *Biochim. Biophys. Acta,* **772,** 176–182.

Steck, T.L. (1974). In E. Korn, Ed., *Methods in Membrane Biology,* Vol. 2, pp. 245–281.

Sundby, C., Andersson, B., and Albertsson, P.-Å. (1982). *Biochim. Biophys. Acta,* **688,** 709–717.

Treffry, T. (1974). *Eur. J. Biochem.,* **43,** 349–351.

Uemura, M., and Yoshida, S. (1983). *Plant Physiol.,* **73,** 586–597.

Walter, H., and Krob, H. (1976). *Biochim. Biophys. Acta,* **455,** 8–23.

Westrin, H., Johansson, G., and Albertsson, P.-Å. (1976). *Biochim. Biophys. Acta,* **436,** 696–706.

Westrin, H., Shanbhag, V.P., and Albertsson, P.-Å. (1983). *Biochim. Biophys. Acta,* **732,** 83–91.

Whikehart, D.R., and Soppet, D.R. (1981). *Invest. Ophtamol. Vis. Sc.*, **21,** 819–825.

Widell, S., and Larsson, C. (1981). *Physiol. Plant.*, **51,** 368–374.

Widell, S., and Larsson, C. (1983). *Physiol. Plant.*, **57,** 196–202.

Widell, S., Lundberg, T., and Larsson, C. (1982). *Plant Physiol.*, **70,** 1429–1435.

Wieslander, A., Christiansson, A., Walter, H., and Weibull, C. (1979). *Biochim. Biophys. Acta*, **550,** 1–15.

Yoshida, S., Uemura, M., Nicki, T., Sakai, A., and Gusta, L.V. (1983). *Plant Physiol.*, **72,** 105–114.

8 PARTITION OF CELLS AND CHARACTERIZATION OF CELL SURFACES

Several different types of cells have been studied by partition in the dextran–polyethylene glycol system (Tables 8.1 and 8.2). The purpose of these studies has been to

1. Study the factors which determine partition.
2. Gain information regarding the surface properties of cells.
3. Gain information on the heterogeneity of cell population and separate subgroups of cells.
4. Study the change in partition as a function of the cell's stage in its growth cycle.

The surface properties of cells and the factors determining partition of cells can be studied by difference partition as described in Chapter 4, Figure 4.1. That is done by comparing the partition in two different phase systems which differ in one parameter only, for example, in ionic composition or number of hydrophobic groups on the polymers. It should be stressed that one has to carry out the difference partition in order to identify which factor, among all the possible variables, is responsible for the observed partition characteristics. Statements such as "The cells must have a high net charge because they have a high partition

TABLE 8.1 Applications on Microorganisms

Microorganisms	Reference
Bacillus thuringiensis	Delafield et al. (1968)
Bakers' yeast	Johansson (1974)
Chlamydia	Söderlund and Kihlström (1982)
Chlorella	Albertsson and Baird (1962); Burczyk (1975); Walter et al. (1971, 1973)
Escherichia coli	Edebo et al. (1975); Pestka (1977); Albertsson and Baird (1962); Hofsten (1965); Wayne and Walter (1974)
Fungi (mucor, penicillium, rhizopus, trichoderma)	Blomquist et al. (1984)
Neisseria	Magnusson et al. (1979b, c)
Salmonella	Magnusson and Johansson (1977); Magnusson et al. (1977, 1979a); Stendahl et al. (1973a, b, 1974); Leive et al. (1984)
Staphylococci	Miörner et al. (1980, 1982); Collen et al. (1980); Gerson (1980)
Streptococci	Miörner et al. (1980, 1982)
Dictyostelium discoideum	Sharpe and Watts (1984); Sharpe et al. (1982)

in a charged interfacial potential system,'' which one can find in the literature, are not correct unless the partition is compared to a phase system with less interfacial potential.

FACTORS DETERMINING CELL PARTITION

Essentially the same factors which determine the partition of protein and cell organelles are important for cell partition. These factors are ionic composition, type of polymer, polymer molecular weight, charge, and hydrophobicity of the cell surface. By using charged polymers, hydrophobic polymers, and biospecific ligand–polymers we can apply electrochemical, hydrophobic, and biospecific affinity partition, respectively, to cells, too.

While soluble macromolecules such as proteins or nucleic acids partition between the two bulk phases, particles such as cells or cell particles usually distribute between one of the bulk phases and the liquid–liquid interface. This distribution behavior has been described in detail. The following general effects have been established.

TABLE 8.2 Applications on Animal Cells

Cells	Reference
Erythrocytes	Walter (1982)
Fibroblasts	Andreeva et al. (1982)
HeLa and mouse mast cells	Pinaev et al. (1976)
Hepatocytes	Walter (1973)
Leucocytes and lymphocytes	Walter et al. (1969a, b, 1979, 1980b); Malmström et al. (1978); Nelson et al. (1978); Walter and Nagaya (1975); Gerson (1980); Levy et al. (1981); Michalski et al. (1983)
Lymphoblasts	Kessel (1981)
Melanoma cells	Evans et al. (1977)
Melastatic murine lymphosaroma cells	Miner et al. (1981)
Monocytes	Walter et al. (1980a)
Leukemic cells	Gersten and Bosman (1974); Kessel (1976, 1977)
Mouse mammary cancer cells	Nakazawa et al. (1979)
Rat intestinal epilhelial cells	Walter and Krob (1975)
Human bone cells	Sharpe et al. (1984)
L cells	Petrov et al. (1981, 1982); Petrov and Andreyeva (1982)
Hybrids of mouse and rat cells	Freedlanskaya et al. (1983)

Influence of Molecular Weight of the Polymers

The same general trend as was described for proteins holds for particles too, that is, if the molecular weight of one polymer is reduced, the particles have more affinity for the phase rich in that polymer. Thus, if PEG 6000 is replaced by PEG 4000 in the Dextran–PEG system, particles are pushed upwards, that is, from the bottom phase to the interface, from the bottom phase to the top phase, or from the interface to the top phase.

Distribution in Compositions More or Less Removed from the Critical Point

A general tendency for particles is that if the polymer concentration is increased, that is, the composition of the phase system is farther away from the critical point, the greater the number of particles is adsorbed to the interface.

Influence of Ionic Composition of the Phase System

The same qualitative effects for proteins seem to hold for particles, that is, for negatively charged particles positive ions push the particles into the top phase in the order $Cs^+ \sim K^+ < Na^+ < Li^+$ and the negative ions in the order $Cl^- \sim H_2PO^-{}_4 < HPO^{2-}{}_4 \sim SO^{2-}{}_4 \sim$ citrate. To get a maximal distribution into the upper phase one should, therefore, use Li_2HPO_4 or Li_2SO_4, for example, and to get a maximal distribution into the lower phase one may use NaCl or KCl.

Net Charge of Particles

When DEAE–cellulose particles or carboxymethylcellulose particles were shaken with the dextran–polyethylene glycol system with different salt content it was found that Li salts pushed the carboxymethylcellulose particles up as compared to Na salts while the reverse was true for DEAE–cellulose particles. Thus, for charged particles, too, the difference between positive and negative particles is similar to the difference between positive and negative proteins with regard to their partition dependence on salt composition.

BACTERIA, FUNGI, AND ALGAE

Species Differences

The quantity of cells of various species of chlorella found in the top phase of a dextran–polyethylene glycol system is recorded in Figure 8.1. The remainder of the cells of each of these species is adsorbed to the interface. The distribution bahaviors of the different strains in this system are distinctly different.

Similar results with different bacterial strains are shown in Figure 8.2, which show that different strains of *E. coli* show different distribution behavior.

A comparison of four different microorganisms in the same phase system is shown in Figure 8.3. It is clear from this figure that the behavior of the different cells is very different and that the difference between two cell types varies with the salt composition. Thus, for separation purposes an experiment like that of Figure 8.3 can serve as a guide for selecting a phase system of suitable composition. The results of Figure 8.3 were also used for selecting a suitable system for countercurrent distribution (see below, page 182).

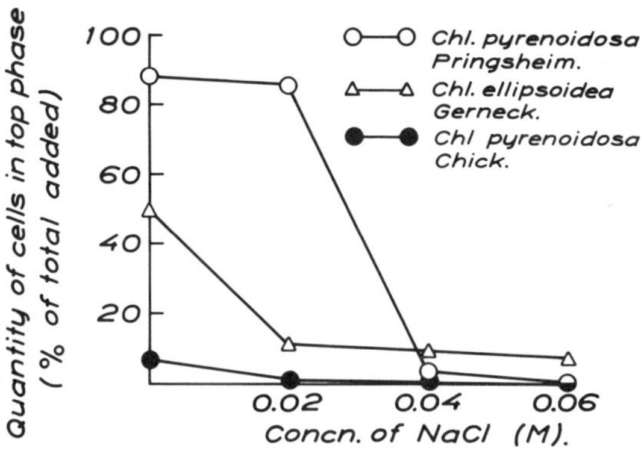

FIGURE 8.1. Partition of different species of chlorella in a dextran–PEG system as a function of NaCl concentration. Phase system composition: 7% w/w Dextran 500, 4.4% w/w PEG 6000, with 0.01 *M* K phosphate, pH 6.9 (Albertsson and Baird, 1962).

Influence of the Quantity of Cells Added

Figure 8.4 shows that for *E. coli* the proportion of cells that partition in favor of the top phase of the dextran–polyethylene glycol system remains constant when the total quantity of cells present in the system is increased. As can be seen from Figure 8.4, approximately 50% of the cells were found in the top

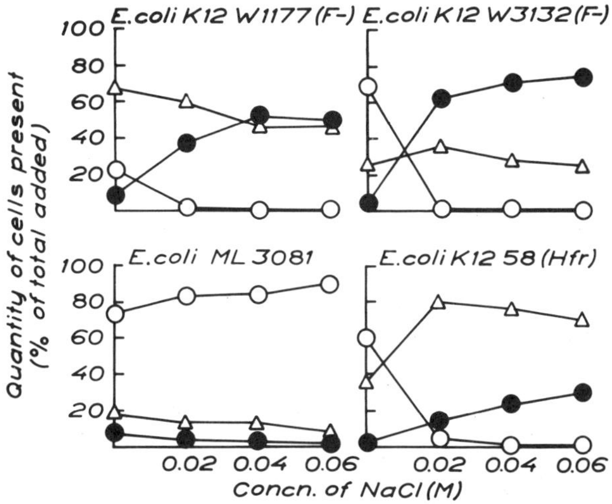

FIGURE 8.2. Partition of different *E. coli* strains as a function of NaCl concentration in the same phase system as Figure 8.3 (○) top phase; (●) bottom phase; (△) interface (Albertsson and Baird, 1962).

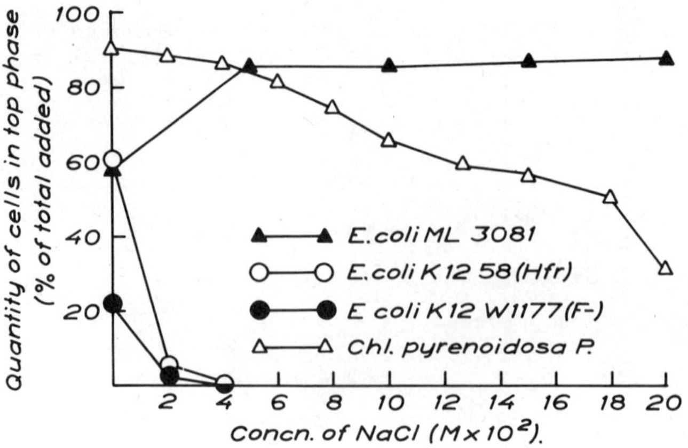

FIGURE 8.3. Partition of different microorganisms in a dextran–PEG system as a function of NaCl concentration. Phase system composition: 5% w/w Dextran 500, 4% w/w PEG 6000, with 0.01 *M* K phosphate, pH 6.9 (Albertsson and Baird, 1962).

phase, irrespective of the total number of cells added. Similar observations have been made in the case of starch particles and of cells of *Chlorella pyrenoidosa* (Albertsson, 1958). In fact, it may be taken as a general rule that variation in the quantity of cells in a phase system does not alter their distribution behavior in the system. In this respect, the partition behavior of suspended particles is similar to that of solutes showing ideal distribution.

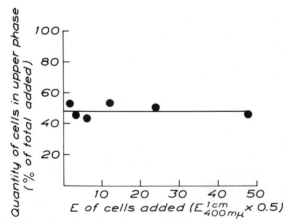

FIGURE 8.4. Partition of the cells of *E. coli* between the upper phase and the interface as a function of cell concentration. Phase system: 5% w/w Dextran 500, 4% w/w PEG 6000, 0.01 *M* K phosphate, pH 6.9, 0.006 *M* NaCl (Albertsson and Baird, 1962).

The fact that the distribution of particles is independent of their concentration, even at fairly high concentrations, indicates a high capacity for particle adsorption of the interface. This is probably due to the very large area of the interface formed during shaking when most of the distribution takes place.

Influence of the Phase Volume Ratio

The effect of the variation of the ratio between the volumes of the top and bottom phases of the dextran–polyethylene glycol system on the distribution of a constant quantity of cells of *E. coli* was also examined (Fig. 8.5). In this experiment, the total volume of the phase system was kept constant at 9.75 mL and the ratio of the volume of the top phase to that of the bottom varied from 0.3 to 3. From Figure 8.5 it is clear that approximately 50% of the cells were present in the top phase of the system under these conditions, irrespective of the volume ratio between the phases. The partition ratio is independent of the volume ratio of the phases also in the case of *C. pyrenoidosa*. This behavior of suspended particles differs from that of soluble substances. Since the ratio of the concentration of soluble substances in the two phases is constant, change in the relative volumes of the two phases in the presence of a fixed quantity of the solute will lead to a change in the proportion of solute in each of the phases.

This unexpected behavior of particles is difficult to explain. It should be pointed out again, however, that the distribution takes place between a phase and the interface mainly during the shaking process when the phases are dispersed

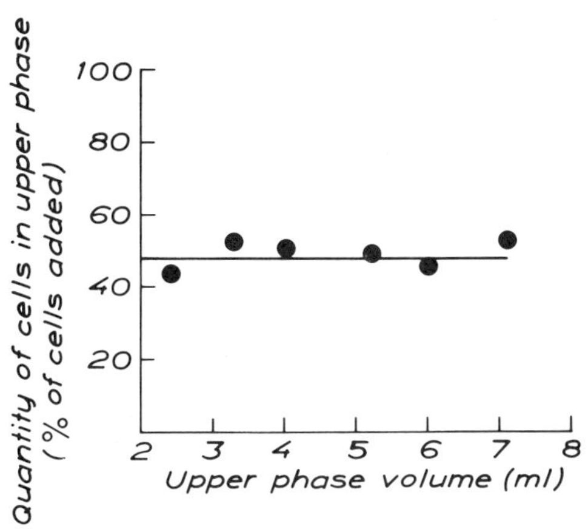

FIGURE 8.5. Similar partition to Figure 8.4, but as a function of the upper-phase volume. Total volume of phase system 9.75 mL (Albertsson and Baird, 1962).

as small drops and the interfacial area becomes very large. The result obtained by this kind of distribution need not, therefore, be similar to what could be expected between a phase and an intact interface which was not broken up and dispersed in this manner.

Effect of Settling Time and Volume Ratio in the Presence of an Emulsion

Since the distribution of particles often occurs between one phase and the interface, rather than between the two bulk phases, the presence of droplets of one phase in the other may influence the partition of particles.

The amount of droplets of one phase dispersed in the other phase depends on the volume ratio. If the upper phase of the dextran–PEG system is small, it will be clear while the lower phase will contain drops of the upper phase. And vice versa, if the lower phase is small, the upper phase will contain droplets. When the two phases are more equal they will both contain droplets. If a particle is attached to the interface, the apparent partition will therefore be highly dependent on the volume ratio (Fig. 8.6*d*).

We can distinguish between the true and the apparent localization of a particle in a phase system that is, a true and an apparent partition. We can locate a particle with the help of a microscope to see if it is freely moving in one of the phases or attached to the interface—this will be the true partition. The apparent partition is obtained by doing a partition experiment, that is, after a certain time of phase separation we take samples from the upper phase and the lower phase. For large particles, such as cells, we cannot allow the phases to settle for too long, since then cells freely suspended in the upper phase may form a sediment at the interface. Therefore, samples have to be taken before all droplets of, for example, the lower phase in the upper phase have settled to the lower phase. Particles attached to these droplets will be counted as belonging to the upper phase and therefore contribute to the apparent partition. The apparent partition of particles therefore depends on the settling time. Moreover, this dependence may be different for different particles. If, for example, different cells are attached to droplets of different sizes, they may reach the horizontal interface at different speeds (Fig. 8.6*a*). Even if two kinds of cells are both attached to the interface (they have the same true partition), their apparent partition may therefore be different (Raymond and Fischer, 1980).

Some particles attached to the interface may stay for a long time in the lower bulk phase. Consider, for example, a top phase droplet in the lower phase (Fig. 8.6*c*). It wants to move upward because of its lower density. However, if, at its surface, there are attached cells which have a higher density than the lower phase

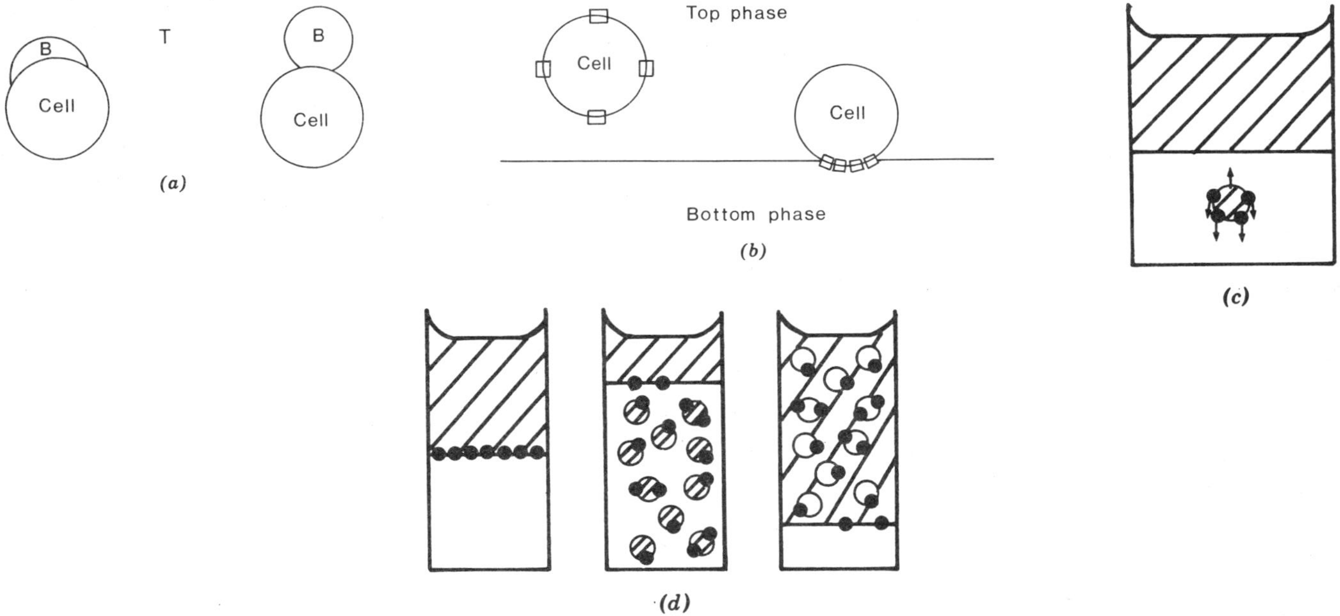

FIGURE 8.6. (*a*) Cells with attached droplets of bottom phase in top phase. Depending on the contact angle the volume of bottom phase attached will vary and thereby also its influence on the sedimentation of the cell. The attached droplets in the two cases will also be differently resistant toward shearing forces during shaking. The bottom-phase droplet to the left will be more firmly bound to the cell than the droplet to the right. (*b*) The surface properties of a cell may change as a result of partition. The domains □ are supposed to have affinity for the bottom phase while the remaining surface has affinity for the top phase. If the cell membrane is fluid, the domains can aggregate and favor adsorption to the interface as shown to the right. (*c*) A drop of upper phase in the lower phase. The particles which are attached to the interface have a higher density than the lower phase and will therefore counteract the upward force on the drop. Thus, the drop will either be pulled up, or down, or be standing still depending upon relative strength of the upward and downward forces. (*d*) Particles which are attached to the interface in a given phase system (left) are partitioned at different volume ratios of that system. In the middle the lower phase is large; therefore, it contains most of the emulsion and the apparent partition will be in favor of the lower phase. (right) The upper phase is large, hence it contains the emulsion and the apparent partition will be in favor of that phase.

these want to pull the droplet down. At a certain ratio between droplet size and amount of particles the result will be that the droplet will stand still. Inspection in a microscope can tell you about these phenomena.

That different types of cells are attached to different droplets may be expected if the contact angles are different (Fig. 8.6). As a result of the shearing forces during shaking it may well be possible that droplets of different size are attached to the cells. One should also consider the heterogeneity of the surface of a cell, that is, only part of the cell surface may attach to the interface. The surface of the cell may also change during shaking and thereby allow more or fewer droplets to be attached to the cell (Fig. 8.6*b*).

COUNTERCURRENT DISTRIBUTION

Countercurrent distributions of bacteria and the algae Chlorella have been carried out with the dextran–polyethylene glycol system (Albertsson and Baird, 1962; Walter et al., 1973). The distribution is between the interface and one or both of the bulk phases. Therefore the liquid–interface countercurrent distribution technique was applied. The microorganisms were originally chosen as suitable

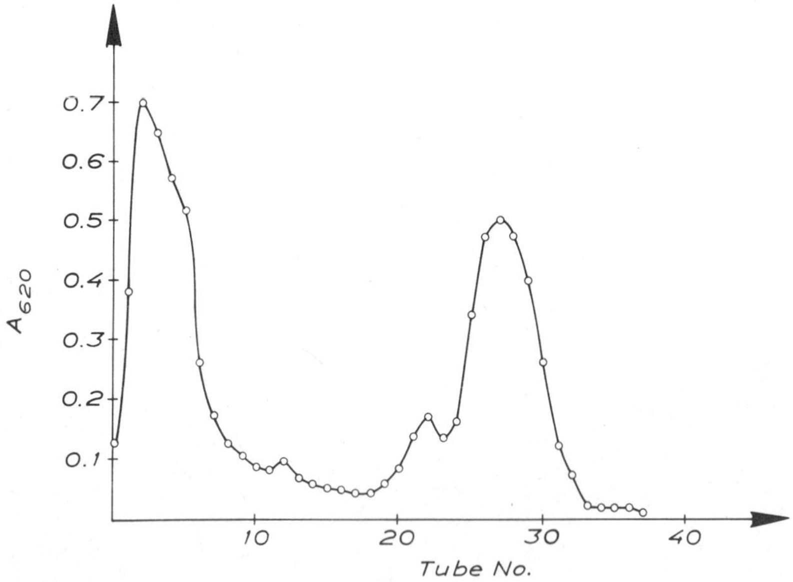

FIGURE 8.7. Separation of a mixture of *Chlorella pyrenoidosa* (left) and *Rhodospirillum rubrum* (right) by liquid–interface countercurrent distribution (Albertsson, 1970).

model particles for studying the liquid–interface countercurrent distribution technique.

Figure 8.7 shows an example where a mixture of two microorganisms has been fractionated by liquid–interface countercurrent distribution (Albertsson, 1970).

A liquid–interface countercurrent distribution experiment with a strain of *E. coli* is shown in Figure 8.8. (The behavior of the strain in an isolated phase pair is shown by Figure 8.3). Two peaks are obtained. In order to determine whether

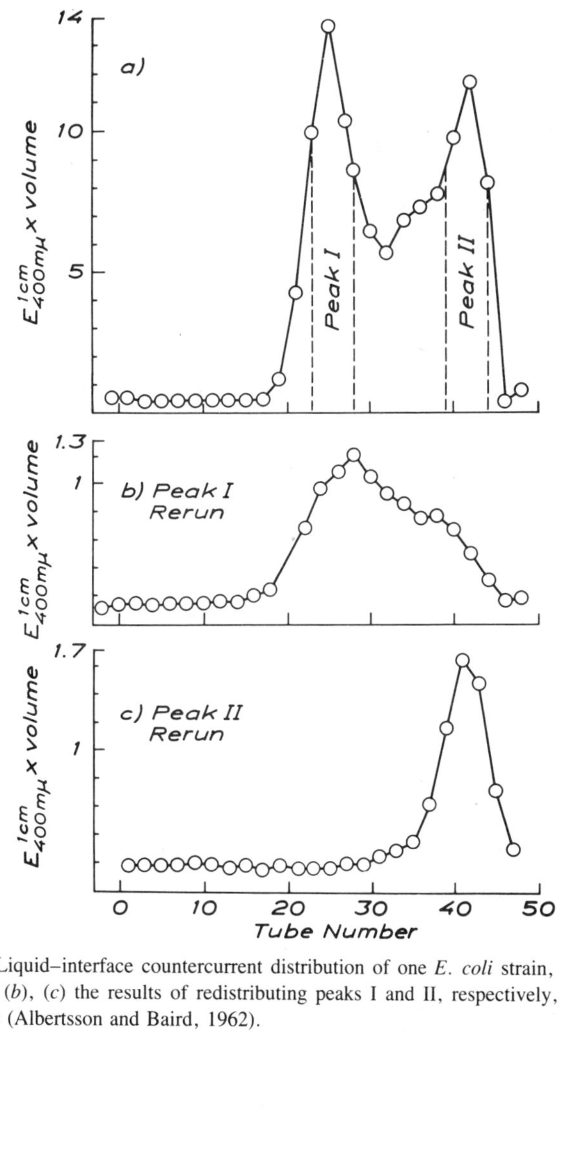

FIGURE 8.8. Liquid–interface countercurrent distribution of one *E. coli* strain, K12 58 (*a*) The original culture; (*b*), (*c*) the results of redistributing peaks I and II, respectively, under the same conditions as (*a*) (Albertsson and Baird, 1962).

the resolution of the cells into two peaks was indicative of genuine heterogeneity or some artifact connected with the treatment of these cells in the phase system the cells under each peak were rerun separately. To this end, the contents of tubes No. 23–28 were combined, as were those of tubes No. 39–44. The cells were then separated from the phases by centrifugation at 3000g and washed twice with 0.01 M phosphate buffer. Subsequently, the two washed cell suspensions were separately redistributed under the same conditions as those pertaining in the original distribution. As may be seen in (*b*) and (*c*), when redistributed in this manner, the cells from each of the two peaks arrive at approximately the same position as they did in the first distribution, (*a*). The peak in (*b*) is, however, somewhat broadened. The fact that the cells from the two peaks, by

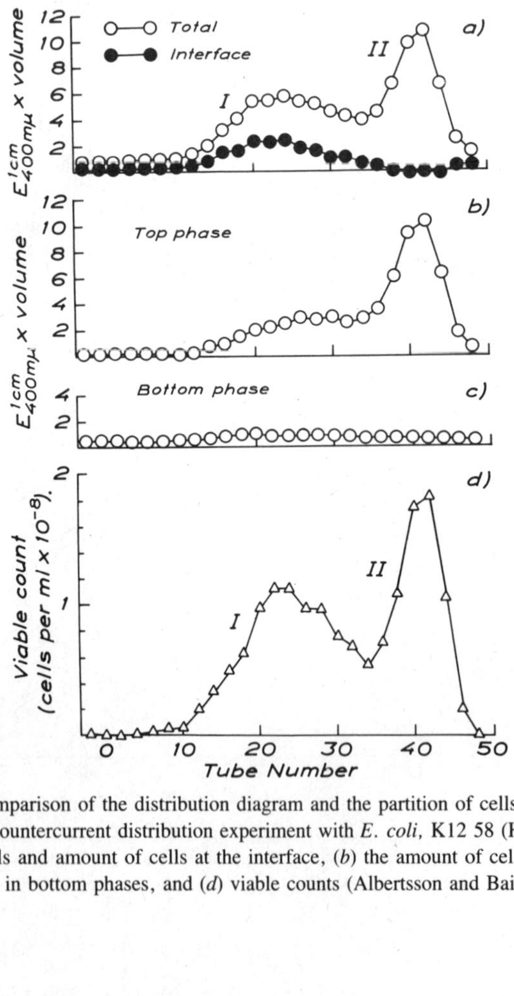

FIGURE 8.9. Comparison of the distribution diagram and the partition of cells in each tube from a liquid–interface countercurrent distribution experiment with *E. coli*, K12 58 (Hfr). (*a*) Shows the total amount of cells and amount of cells at the interface, (*b*) the amount of cells in top phase, (*c*) the amount of cells in bottom phases, and (*d*) viable counts (Albertsson and Baird, 1962).

and large, maintain their integrity of behavior when rerun in this fashion makes it highly probable that they represent two types of cells differing in their surface properties.

Another experiment with the same strain is shown in Figure 8.9. Two peaks were obtained as in the previous case. Here, the quantity of cells in each phase, and therefore, by subtraction, at the interface, was also determined for alternate tubes. The data are plotted in Figure 8.9*a*, *b*, and *c*. From these it can be seen that the cells in peak I are partitioned largely between the top phase and the interface, while those in peak II are found in the top phase alone. That the cells should partition in this fashion in the two peaks is to be expected if they were behaving ideally and if equilibrium were reached before each transfer.

Figure 8.10 shows an experiment in which a mixture of two strains of *E. coli* was distributed in the same phase system as the previous experiments. Three peaks were obtained, and here, too, the distribution of the cells in alternate tubes was determined at the conclusion of the experiment. The quantities of cells present in the bottom and top phases and at the interface are recorded in Figure 8.10*b* and *c*. It may be seen that, as in the previous countercurrent experiment, the

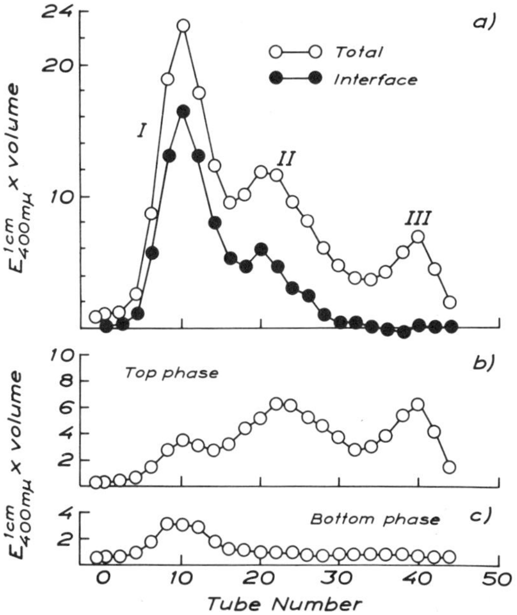

FIGURE 8.10. Similar to Figure 8.9 but with a mixture of two different strains, *E. coli* K12, W1177 and K12, 58/161 (Albertsson and Baird, 1962).

quantity of cell found in the top phase increased with increasing tube number, while the quantity of cells in the bottom phase and at the interface decreased with increasing tube number. Thus, the partition ratio of the distributed cells increases quantitatively with tube numbers as would be expected if the particles distributed ideally. Moreover, a fairly good agreement exists between the partition ratios calculated from the position of a peak maximum and the distribution of cells between the phases in the tubes under the maximum.

These experiments also demonstrate that the distribution behavior of a given type during countercurrent extraction appears not to be influenced by the presence of cells of another type. The separation of mixtures of cells with differing behavior in a given phase system is therefore possible by this technique.

Figure 8.11 shows a separation of a mixture of many different microorganisms. To resolve such a complex mixture a gradient of salt was utilized. This was accomplished by loading some tubes with a different salt composition so that a gradient of salt is automatically developed during the course of the countercurrent distribution procedure. This is useful when the G_i values of the cell types are very different, as in the case here but should not be used for a mixture of cells having G_i close to each other.

Countercurrent distribution of *C. pyrenoidosa* demonstrated that a culture of this organism could be resolved into fractions with cells at different stages in the

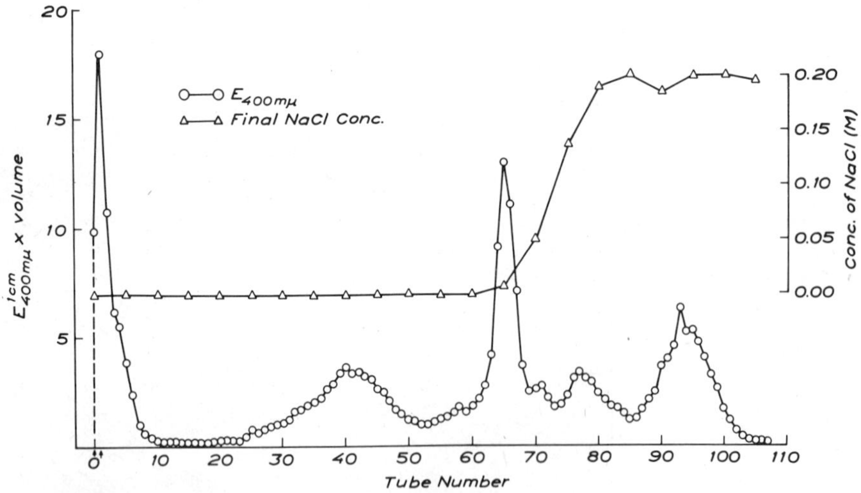

FIGURE 8.11. Separation of a mixture of microorganisms by liquid–interface countercurrent distribution. The peaks represent, from left: yeast cells, *E. coli* K12, W1177, *E. coli* K12, 58, *Chlorella pyrenoidosa* (two small peaks), and *E. coli* ML 3081. A salt gradient was introduced in order to resolve the *E. coli* K12 and the Chlorella peaks (Albertsson and Baird, 1962).

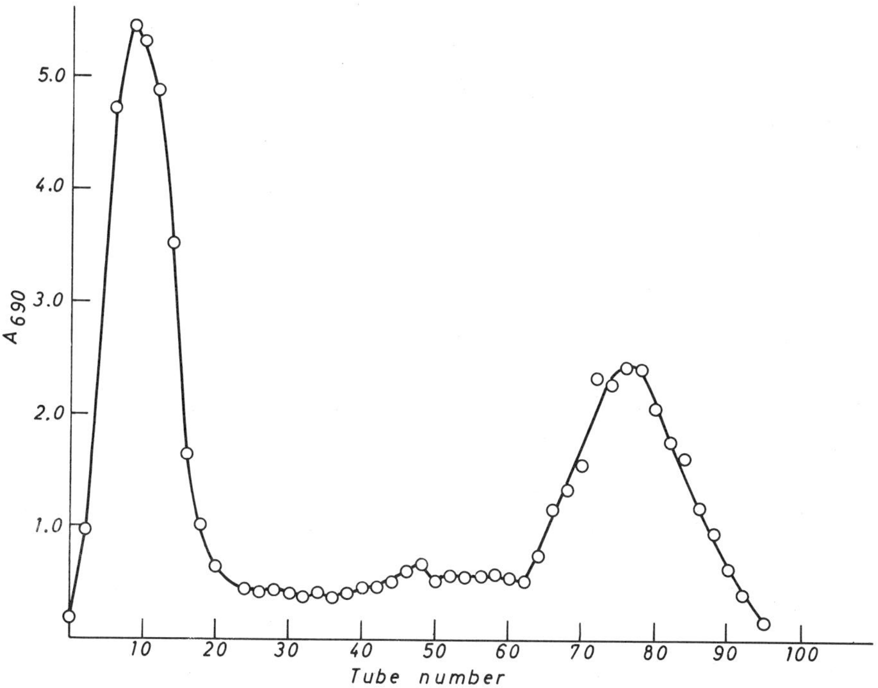

FIGURE 8.12*a*. Separation of different stages in the growth cycle of *C. pyrenoidosa* by countercurrent distribution. The left peak consists of young cells and the right of old cells (Walter et al., 1970).

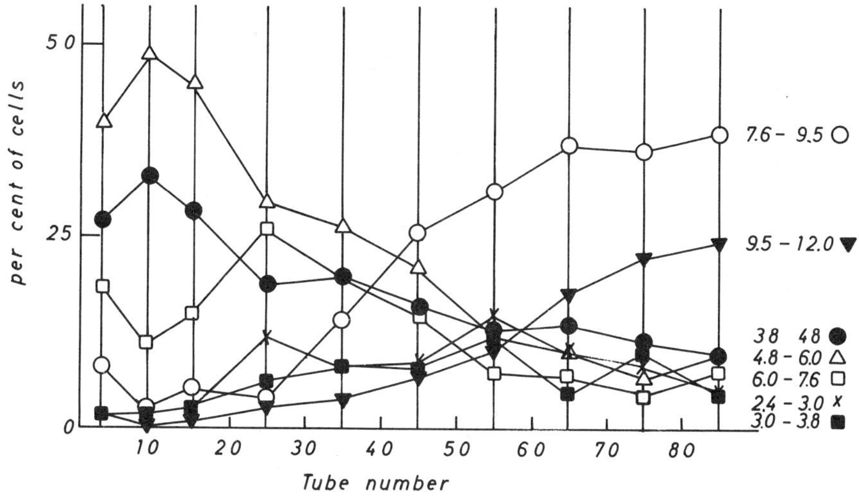

FIGURE 8.12*b*. Size distributions of the cells from the different tubes in the same experiments as (*a*). The numbers to the right indicate size in microns (Walter et al., 1971).

growth cycle (Walter et al., 1971, 1973). When synchronized cells were studied, it was found that the position of the peak depends on the stage in the growth cycle. Figure 8.12 shows the results with an ordinary nonsynchronized culture in the exponential phase. The culture is resolved into two main populations the left peak containing young cells and the right peak old cells.

It is evident from these experiments that, by and large, microbial cells behave in a liquid–interface countercurrent distribution in the way one would expect if the distribution between the top phase and the interface is ideal, that is, reproducible, reversible, and independent of concentration of the cells and the presence of other cells. Thus, liquid–interface countercurrent distribution provides a multiple-stage procedure for separation and analysis of particle mixtures in a fashion similar to countercurrent distribution and chromatography for a mixture of solutes.

Of particular interest is the separation of a culture from a single organism into two or more peaks (Fig. 8.12). This indicates distinct classes of cells in a culture, and would be expected if during the growth cycle the surface of the cells grows in steps rather than continuously. If some of these steps take a longer time than others, then this would lead to an accumulation of certain classes of cells representing different stages in the growth cycle.

In the case of chlorella the different stages in the growth cycle have been analyzed in detail with respect to size, morphology, and metabolism. The countercurrent distribution experiments of Figure 8.12 shows that the surface population of the cells also changes drastically during the cell cycle.

Characterization of Mutants of Bacteria

Mutants of *Salmonella typhimurium* differing in the polysaccharide chain length of the cell wall lipopolysaccharide have been studied by countercurrent distribution. R mutants had a low partition ratio, preferring the lower phase, while S mutants preferred the upper phase. These mutants also differ with respect to virulence, immunogenicity, and phagocytic resistance, and a correlation between partition and phagocytic resistance was found (Stendahl et al., 1973a, b).

Figure 8.13 shows a countercurrent distribution diagram of a UDP–gal–Y–epimerase-less mutant grown in a galactose medium for different times. Without galactose this mutant strain cannot synthesize lipopolysaccharide and its partition ratio is low; it gives a peak to the left in a countercurrent distribution diagram. During growth with galactose, lipopolysaccharide is synthesized and the partition shifts in favor of the upper phase, and the peak to the right in the CCD diagram increases. A semirough mutant gives a CCD pattern which is intermediate (bottom

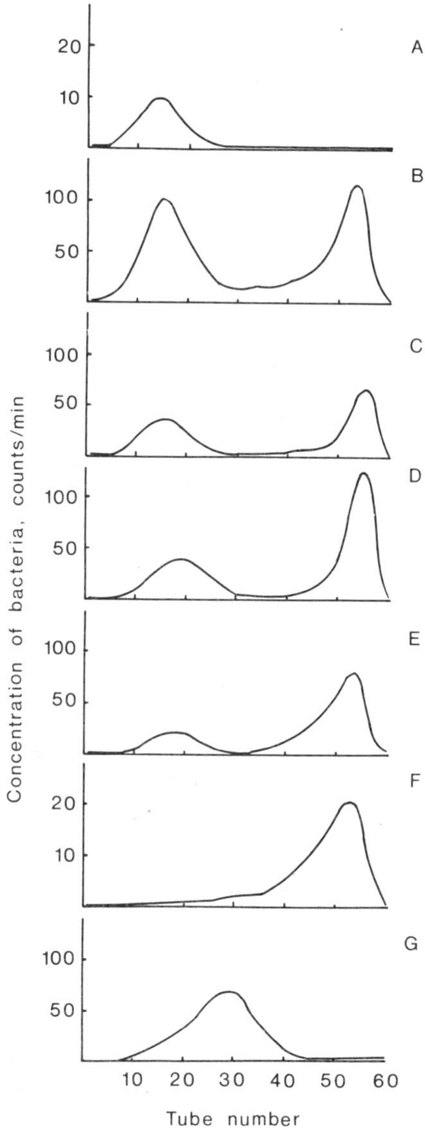

FIGURE 8.13. Countercurrent distribution of an UDP-gal-Y-epimeraseless mutant of *Salmonella typhimurium* grown in a medium with D-galactose for different times. (A) 0 min; (B) 10 min; (C) 20 min; (D) 40 min; (E) 60 min; (F) 180 min; (G) a semirough mutant (Stendahl et al., 1973a).

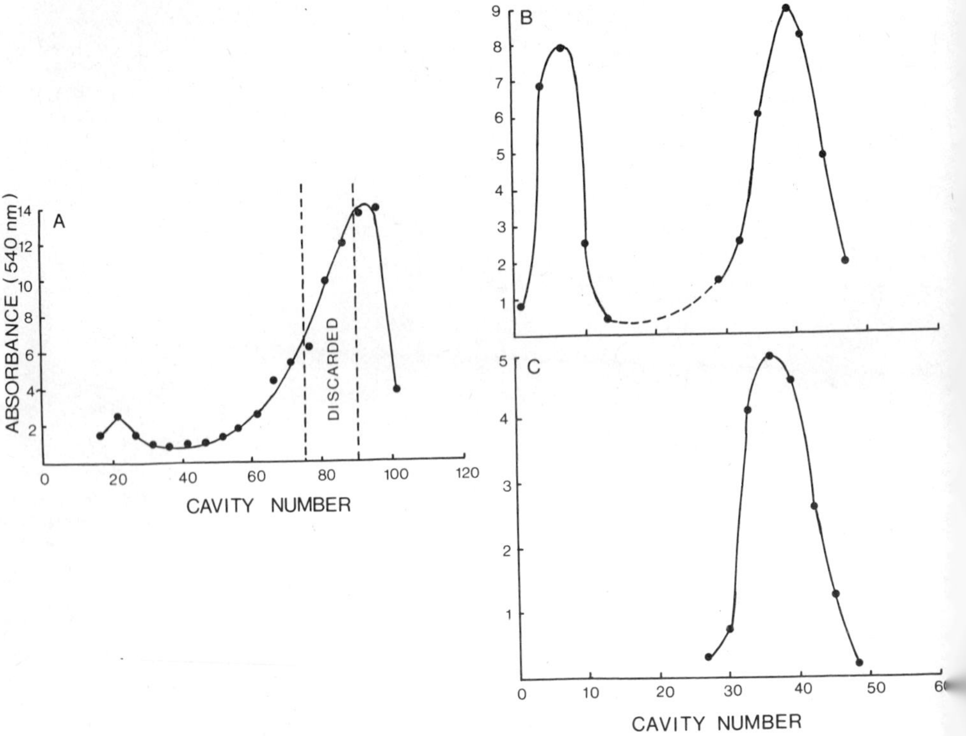

FIGURE 8.14. Selection of mutants by countercurrent distribution: (A) run of an uncloned culture of *E. coli* (120 transfers); (B) rerun of pooled fractions 10–75; (C) rerun of pooled fractions 90–100.

of diagram). Countercurrent chromatography has also been used to separate salmonella strains differing in the surface lipopolysaccharide (Leive et al. 1984).

If bacterial mutants have different partition ratios, one should be able to use countercurrent distribution for separation and selection of mutants. This has been demonstrated with *E. coli* (Wayne and Walter, 1974). A CCD of this strain gave a bimodal curve in the CCD diagram (Fig. 8.14). Clones from the left peak gave a single peak to the left, while cultures from colonies from the right peak gave a single peak to the right. It turned out that the left peak consists of erythromycin-resistant mutants of the parent strain. Also streptomycin-resistant strains of *E. coli* exhibited differences in countercurrent distribution from the parental strains (Pestka et al., 1977).

Different strains of chlorella with different resistance towards lytic agents have also been separated by phase partition (Burczyk, 1975).

Characterization of the Bacterial Surface

Both charge and hydrophobic properties of the bacterial surface has been studied by partition. These properties are of interest in order to understand the virulence and the mechanism of infection of pathogens.

By comparing the partition of bacteria in phase systems differing in the interfacial potential one can gain information on the charge of the bacterial surface. Phase systems with positively or negatively charged PEG (TMA-PEG or S-PEG) have been used in several studies. By cross partition the isoelectric point of the bacterial surface has been determined.

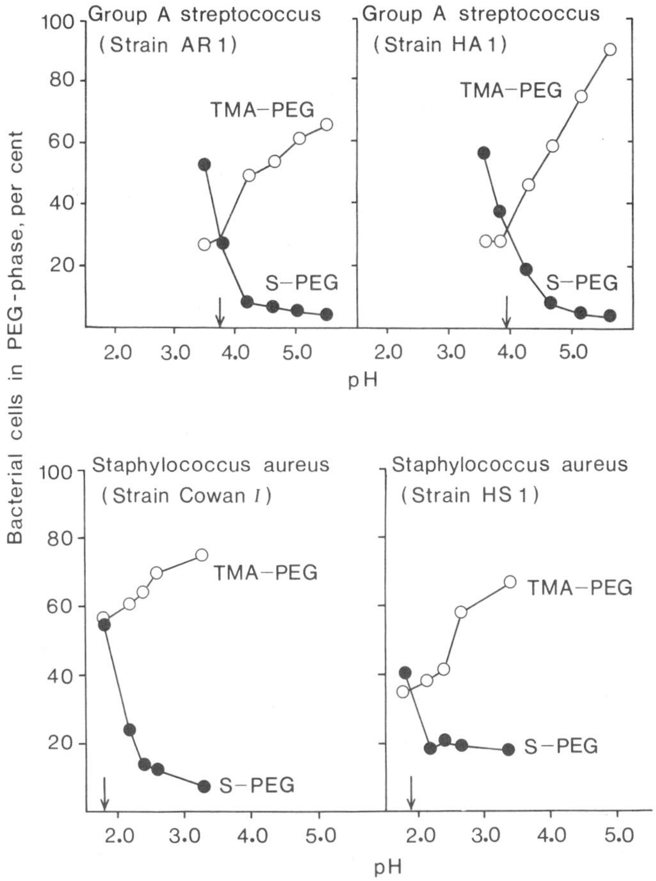

FIGURE 8.15a. Isoelectric points of *streptococcus* and *staphylococcus* strains determined by cross partition (Miörner et al., 1982).

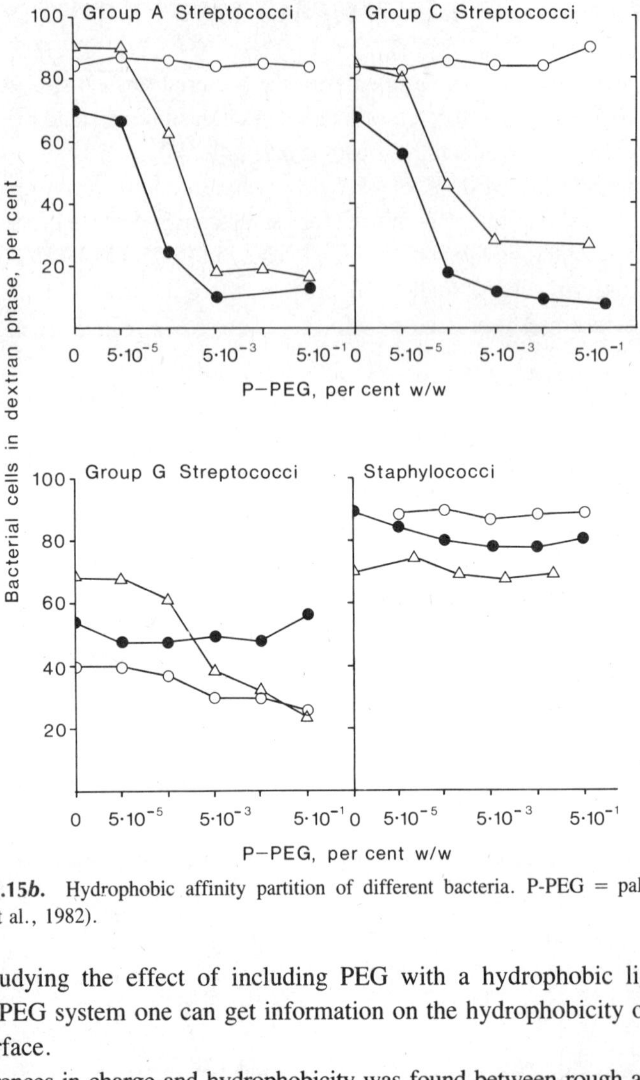

FIGURE 8.15*b*. Hydrophobic affinity partition of different bacteria. P-PEG = palmitoyl-PEG (Miörner et al., 1982).

By studying the effect of including PEG with a hydrophobic ligand in a dextran–PEG system one can get information on the hydrophobicity of the bacterial surface.

Differences in charge and hydrophobicity was found between rough and smooth *S. typhimurium* strain and this difference was correlated with liability to phagocytosis. Thus the phagocytosis-sensitive R mutants have a more charged and a more hydrophobic surface than the smooth phagocytosis-resistant strains (Magnusson et al., 1977; Magnusson and Johansson, 1977).

Thirty-nine streptococcal and 12 staphylococcal strains have been investigated with respect to the isoelectric point, as determined by cross partition, and hydrophobicity as determined by PEG–palmitate (Miörner et al., 1982).

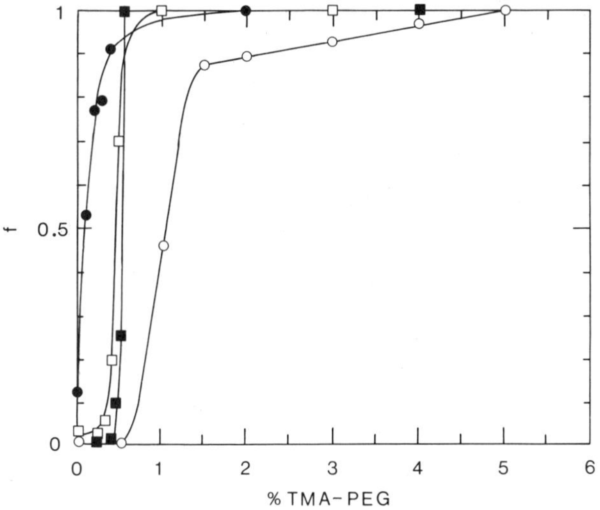

(a)

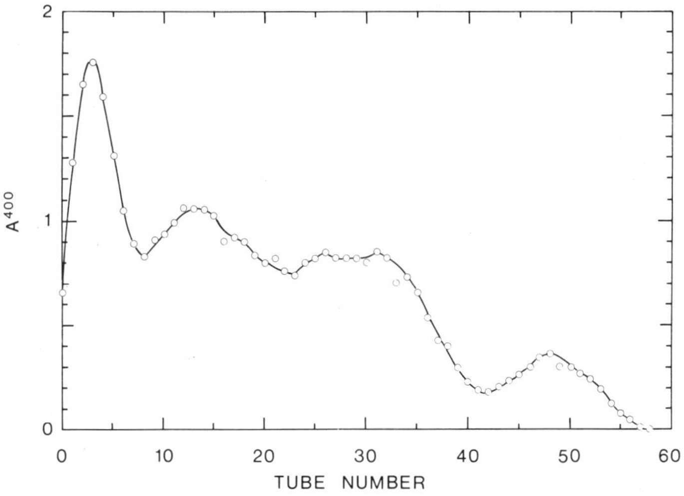

(b)

FIGURE 8.16*a, b*. (*a*) Partition of baker's yeast in dextran–polyethylene glycol system as a function of added TMA-PEG. (8% w/w Dextran 500, 8% w/w PEG 4000, 2 m*M* potassium phosphate, pH 6.8. (*b*) Countercurrent distribution of baker's yeast (59 transfers). The same phase system as (*a*) with 1.1% TMA-PEG.

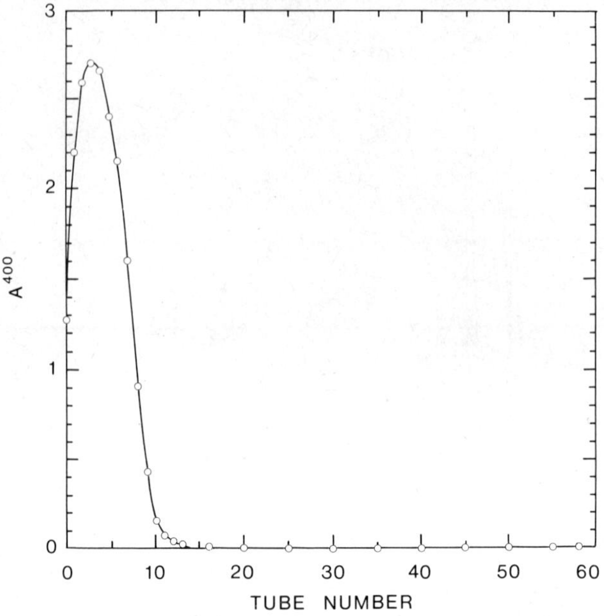

(c)

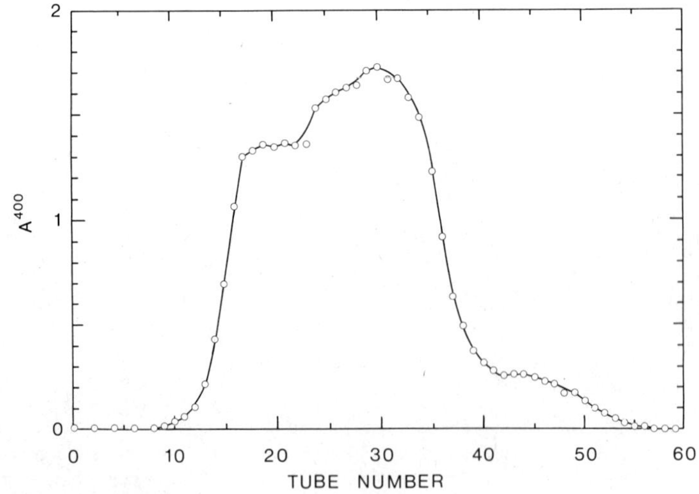

(d)

FIGURE 8.16c, *d*. Same as (*b*), but with two isolated yeast strains (Johansson, 1974).

The isoelectric points of the streptococcal strains were between pH 3.5 and 4.1 while the staphylococcus strain had isoelectric points as low as 1.95 (Fig. 8.15*a*).

All staphylococcus strains displayed little hydrophobicity, that is, they did not change their partition much upon the addition of PEG–palmitate to the phase system. Among the streptococci a large variation in hydrophobicity was observed (Fig. 8.15*b*). The major cell wall component responsible for the surface hydrophobicity of group *A Streptococci* was identified as lipoteichoic acid since there was a correlation between the content of surface lipoteichoic acid and the hydrophobicity of these strains (Miörner et al., 1983).

Partition coefficients of different bacterial cells in the dextran–polyethylene glycol system were compared to the cell surface energies obtained from contact angles of each phase on cell layers (Gerson and Akit, 1980). A linear relation between these two measurements was obtained supporting the importance of contact angles in the partition process described in Chapter 3.

The effect of adsorption of proteins to the bacterial surface has also been studied by partition (Miörner et al., 1980).

Fungi

Figure 8.16 shows a countercurrent distribution diagram of bakers' yeast (Johansson, 1974). These cells have an unusually strong affinity for the dextran phase and the positively charged TMA-PEG (see Chapter 5) has to be added in order to obtain a suitable partition between the upper phase and the interface. Figure 8.16*a* shows an extraction profile when increasing TMA-PEG is added to a dextran–polyethylene glycol phase system. A concentration of 1.1% TMA-PEG was chosen for the CCD experiment in Figure 8.16*b*. As seen, bakers' yeast is divided into several peaks. This is probably due to several strains making up the commercial yeast culture. When one-cell cultures were isolated these gave a more narrow distribution, although these were also heterogeneous, probably because of different stages of the growth cycle (Figs. 8.16*c* and *d*).

Phase partition has been used for purification of fungi belonging to different taxonomic groups (Blomquist et al., 1984). The partition in dextran–polyethylene glycol system differed drastically between the conidia of two penicillium species and the sporangiospores of three Phycomycetes (Fig. 8.17). *Penicillium frequentans* was completely purified from *Mucor racemosus* by means of two partitions. This method was used on air samples from a locality where wood fuel chips were handled.

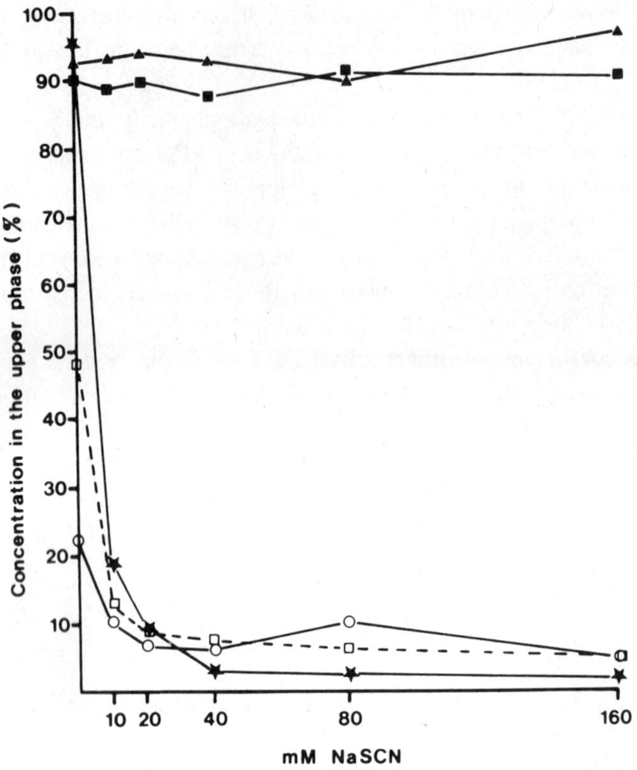

FIGURE 8.17. Partition of *Penicillium chrysogenum* (■), *P. frequentous* (▲), *Phizomucor pucillus* (□), *Mucor rasemosus* (X), and *Rhizopus rhizopodefarmis* (○) in a phase system of 5% Dextran 500 and 4% PEG 6000, 10 m*M* potassium phosphate, pH 6.9, and different concentrations of NaSCN (Blomqvist et al., 1984).

ANIMAL CELLS

Erythrocytes

The erythrocyte is the most extensively studied cell type. Most of the experiments have been made with the dextran–polyethylene glycol system. Both the polymer concentration and the ionic composition has been used to change the partition. The distribution is always between one of the phases and the interface.

Figure 8.18*a* shows the distribution of red blood cells from different species in a dextran–polyethylene glycol system at different ionic compositions. The percentage of the quantity of cells in top phase is plotted against the increase in

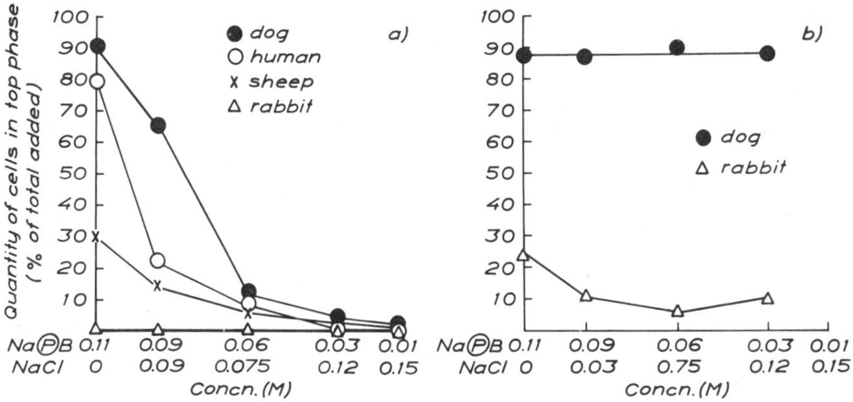

FIGURE 8.18. Partition of erythrocytes from different species in the dextran–polyethylene glycol system at different ionic composition. Phase system (*a*) 5% w/w Dextran 500–4% w/w PEG 6000; (*b*) 5% w/w Dextran 500–3.5% w/w PEG 6000 (Albertsson and Baird, 1962).

NaCl concentration. (In order to vary the ratio between NaCl and Na phosphate without drastically changing the tonicity, the phosphate concentration was decreased by two parts when the NaCl concentration was increased by three parts.) The following conclusions can be drawn from the experiment shown in Figure 8.18*a*.

1. An increase in NaCl (with the concomitant decrease in phosphate concentrations) results in a decrease in the number of cells in the upper phase for all the species tested.
2. The partition of red blood cells is species-dependent.
3. The difference between two species shows up more in some systems than in others. Compare, for example, human and sheep red blood cells. At 0.11 *M* Na phosphate the two types of cells differ much more than at 0.09 *M* Na phosphate + 0.03 *M* NaCl.

Figure 8.18*b* shows a similar experiment but with a lower concentration of polyethylene glycol, that is, with a system closer to the critical point. In this system the percentage of cells found in the upper phase is higher, and the Na phosphate–NaCl ratio does not influence the cell distribution so much. By carrying out experiments of the type shown in Figure 8.18 one can easily find a system which gives a partition suitable for countercurrent distribution. Other salts and other pH may also be tested, for example Li phosphate or Li_2SO_4 in order to increase the partition, or KCl in order to decrease the partition as outlined in Chapter 4. Also, other molecular-weight fractions of the polymers can be used.

Countercurrent Distribution. Figure 8.19 shows a countercurrent distribution experiment with a mixture of rabbit and dog erythrocytes. The two species of cells are very effectively separated. The rabbit cells are mainly at the interface while the dog cells distribute between the top phase and interface. Moreover within the broad peak characterizing the dog cells, the amount of cells in upper phase increases with increasing tube number. This indicates separation of different dog cells within this peak. With erythrocytes as with bacteria (see Figs. 8.9 and 8.10) there is therefore a good agreement between the distribution in each tube and the tube number as would be expected if the countercurrent principle works.

Separation of erythrocytes by countercurrent distribution has been studied in detail by Walter and collaborators (Walter, 1982; Walter et al., 1964, 1965; Walter and Selby, 1966). Of particular interest is the studies on red blood cells of different ages. By injecting the animal with iron radioactive ferrous citrate and then taking blood samples at different times after injection, the position of cells with different ages in the distribution train may be determined.

Figure 8.20 shows a countercurrent distribution diagram of rat erythrocytes with the position of the cells of different ages. The reticulocytes are to the left, tubes 0–4, while the youngest erythrocytes are to the right, around tube 25. Therefore, a drastic change occurs on the cell surface during a short time period

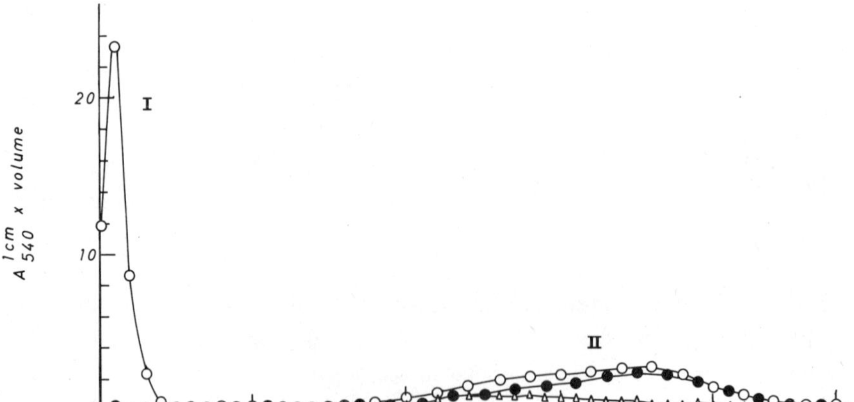

FIGURE 8.19. Separation of a mixture of rabbit (I) and dog (II) erythrocytes by liquid–interface countercurrent distribution (Albertsson and Baird, 1962): ○, total; ●, top phase; △, interface.

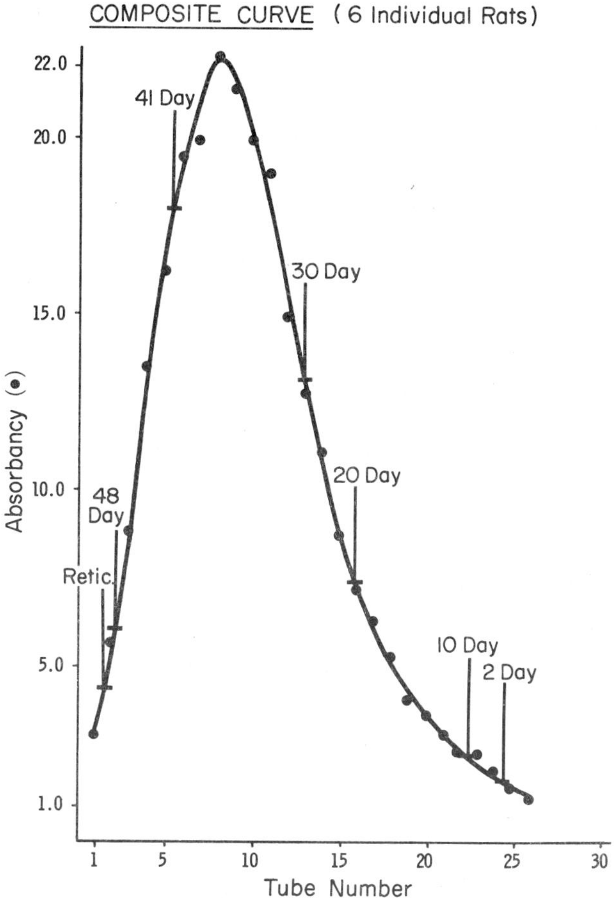

FIGURE 8.20. Countercurrent distribution of rat red blood cells. The approximate position of cells of different ages (Walter and Selby, 1966).

when the reticulocytes are transformed to erythrocytes. As the erythrocytes age there is a gradual change in the distribution towards the left part of the peak.

The separation of red blood cells of different ages by countercurrent has also been demonstrated for cells from chicken, dog, and mouse. However, for human red cells there was no such separation with age (Walter, 1982).

A correlation between partition in the dextran–polyethyleneglycol system and the electrophoretic mobility has been demonstrated in Figure 8.21.

Also, after modification of the cell's surface by neuraminidase, which removes the charge-bearing sialic acid of the cell surface, trypsin, which removes frag-

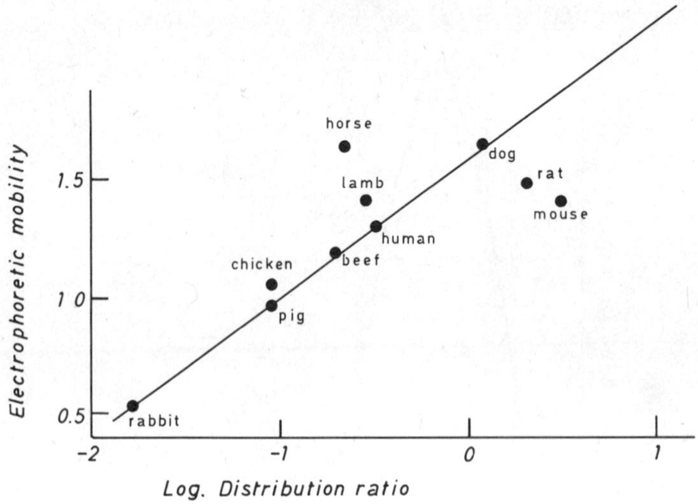

FIGURE 8.21. Relation between electrophoretic mobility of erythrocytes from different species and partition (ratio of number of cells in top phase divided by cells at interface plus bottom phase) in the dextran–polyethylene glycol system.

ments from the protein, or maleic anhydride, which blocks free amino groups, it could be demonstrated that there was a clear correlation between electrophoretic mobility and partition—increasing mobility correlated with a higher partition coefficient (Walter et al., 1983)

These studies employed a phase system which had an intefacial potential, the upper phase being more positive than the lower phase. Thus, one would expect that, other factors being constant, cells carrying more charges would prefer the upper phase.

By comparing red cells from different species in a phase system with a relatively small potential difference, charge differences are minimized and a correlation between the partition and the lipid composition is then demonstrated—cells with higher ratio of polyunsaturated to monounsaturated fatty acids having a higher affinity for the upper phase (Walter, 1982).

Other studies involve comparison between fresh and stored cells and irradiated cells, influence of tonicity, clam red cells, reticulocytes, subgroups of beef red cells, and rat bone marrow cells (Walter, 1982), erythroid cells at different stages for studies of multiple hemoglobin synthesis (Weiser et al. 1976). The effect of settling time on the distribution of cells between the upper phase and the interface has been studied for various species. These studies emphasize the importance of controlling the settling time. It has also been suggested that an enhanced sepa-

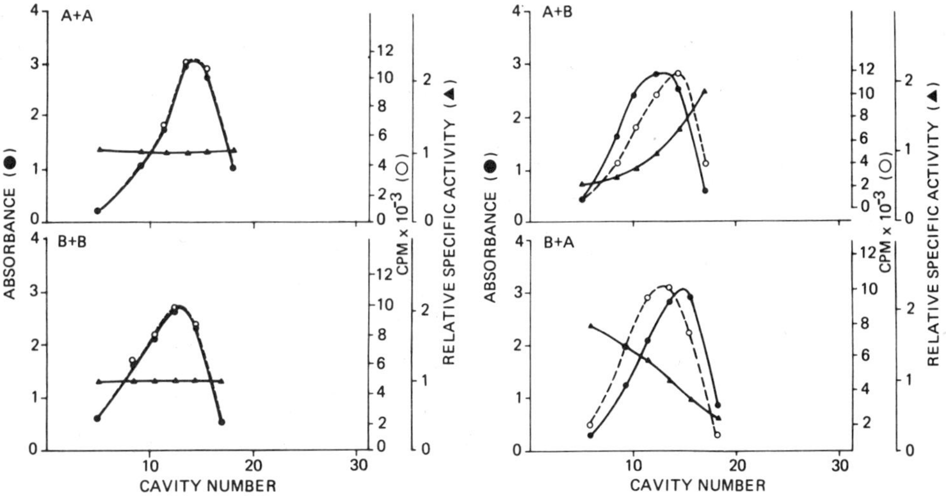

FIGURE 8.22. Countercurrent distribution of human erythrocytes from two individuals A and B. A sample from each of A and B was made radioactive with ^{51}Cr-chromate. Labelled cells were mixed with an excess of unlabelled cells from the other individual (A + B, B + A). As controls, labelled cells were mixed with an excess of cells from the same person (A + A, B + B). Unlabelled cells were measured by hemoglobine absorbance. The two figures to the right show that the cells from B have a lower partition coefficient (curve displaced to the left) than cells from A (Walter and Krob, 1984).

ration can be achieved by collecting the phases prior to reaching the final equilibrium (Fisher and Walter, 1984).

Of great interest are recent experiments where the distribution pattern of erythrocytes from individuals are compared. (Walter and Krob, 1983). By labelling isotopically (with ^{51}Cr-CrO_4) aliquots of erythrocytes from one individual and mixing these with an excess of unlabelled erythrocytes from another individual, the two cell populations can be analyzed independently of each other (by radioactivity and hemoglobin absorption, respectively). Countercurrent distribution of such mixtures revealed that red blood cells from different individuals had small but significant differences in partition, reflecting difference in surface properties of erythrocytes from different individuals (Fig. 8.22). The differences are probably genetic. Erythrocytes from identical twins or triplets displayed no differences.

Affinity Partition. The application of affinity partition to red cells was first demonstrated by Erikson et al. (1976). The affinity of the erythrocyte membrane for hydrohobic substances such as fatty acids or steroids was utilized. Fatty acids

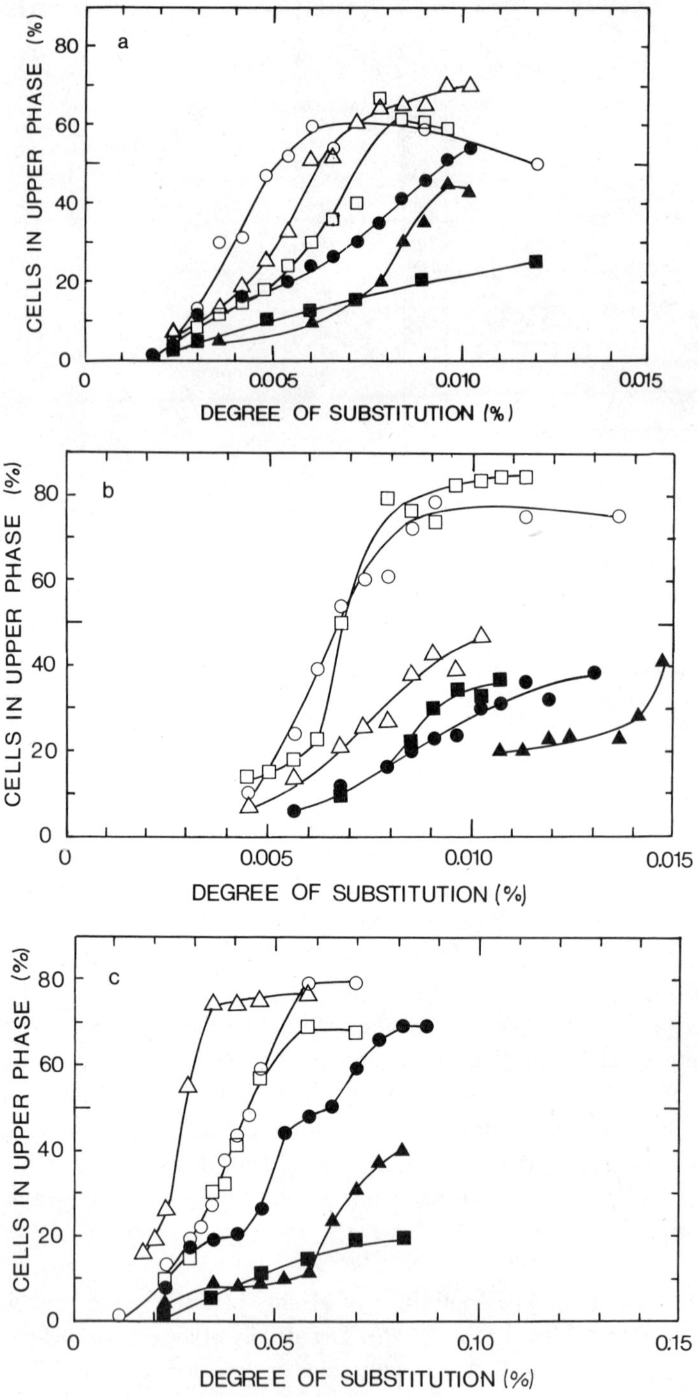

a
CELLS IN UPPER PHASE (%)
DEGREE OF SUBSTITUTION (%)
b
CELLS IN UPPER PHASE (%)
DEGREE OF SUBSTITUTION (%)
c
CELLS IN UPPER PHASE (%)
DEGREE OF SUBSTITUTION (%)

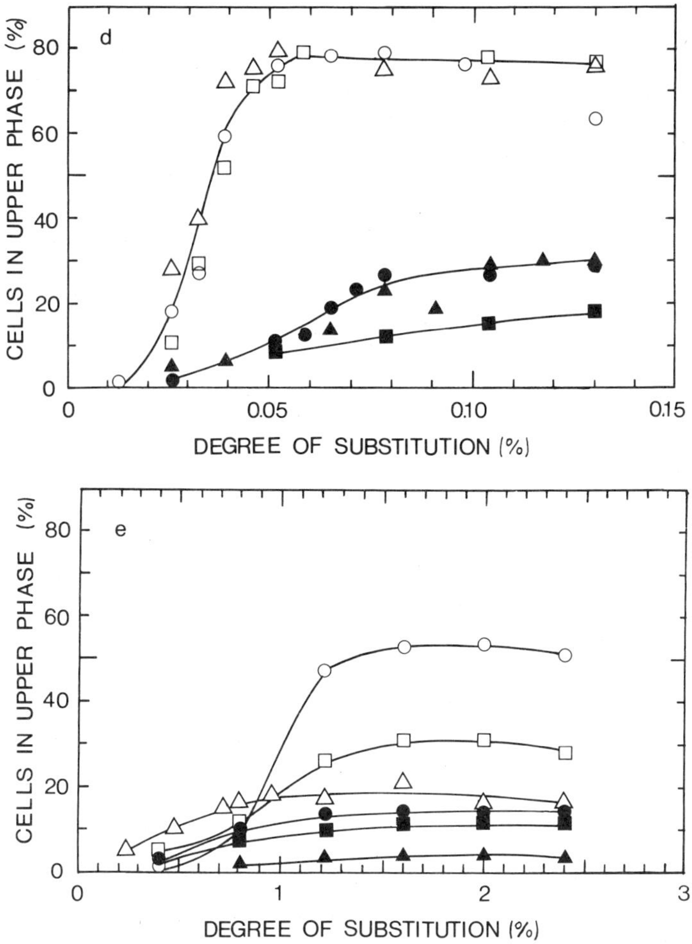

FIGURE 8.23. Hydrophobic affinity partition of erythrocytes. The percentage of erythrocytes in the upper phase is a function of the degree of substitution of PEG with esterified hydrophobic groups (% PEG replaced by esterified PEG). The erythrocytes are: ○, dog; △, rat; □, guinea pig; ●, sheep; ■, rabbit; ▲, human. PEG was esterified with (*a*) palmitic acid, (*b*) oleic acid, (*c*) linoleic acid, (*d*) linolenic acid, and (*e*) deoxy cholic acid (Eriksson et al., 1976).

and deoxycholic acid were esterified with polyethylene glycol via its hydroxyl end groups. The esters partition like polyethylene glycol, that is, they are concentrated in the upper phase of the dextran–polyethylene glycol system. These ligand–polymers have a strong influence on the partition of red blood cells. The effect of various PEG esters on red cells from different species is shown in Figure 8.23.

An extremely small degree of substitution is needed to transfer the cells from

TABLE 8.3 Phosphatidyl Choline and Sphingomyelin Content of the Erythrocyte Membrane from Different Species[a]

	Phosphatidyl Choline, %	Sphingomyelin, %
Dog	46.9	10.8
Rat	47.5	12.8
Guinea pig	41.1	11.1
Rabbit	33.9	19.0
Human	28.9	26.9
Sheep	0	51.0

[a]Compare with Figure 8.21. (Eriksson *et al*, 1976).

the lower phase into the upper phase. (In these experiments a low-molecular-weight dextran, Dextran T40, was used in combination with NaCl to keep the cells in the lower phase and also to prevent aggregation.) In the case of PEG palmitate, 0.005% of substitution is enough to transfer 50% of the dog red cells from the lower phase to the upper phase. This must be due to the binding of the PEG palmitate to the cell membrane. This ability to bind to the erythrocyte membrane is species-specific.

The effectiveness in transferring the cells into the upper phase decreases with increasing unsaturation. PEG palmitate is about 1.8 times more efficient than PEG oleate and 6 times more efficient than PEG linoleate or PEG linolenate. PEG deoxycholate is 2500 times less effective than PEG palmitate.

The erythrocytes fall into two groups independent of the acid esterified with PEG—one with relatively higher affinity for the PEG esters (rat, dog, and guinea pig), and one with relatively lower affinity (rabbit, sheep, and human). It is interesting to compare this with the phospholipid composition of these cells (Table 8.3). The group of cells with the higher affinity for the PEG esters has a relatively higher content of sphingomyelin than the other group. Since these phospholipids constitute the outer layer of the red cell membrane bilayer, these studies suggest that it is the outer layer which binds the PEG ester and that a higher content of phosphatidylcholine causes a stronger binding of these esters.

The effect of chain length of the fatty acid in the PEG ester would be expected to be the same as demonstrated for chloroplasts and for proteins, that is, a longer hydrocarbon chain is more effective as an affinity ligand (Chapter 4).

LYMPHOCYTES AND LEUKOCYTES

The partition of these cells differs according to the source of the cells. Lymphocytes and granulocytes from horse have been separated by countercurrent distribution (Walter, 1982). Rat leucocytes have been separated into different subpopulations (Fig. 8.24) which were characterized with respect to different cell activities (Malmström et al., 1978; Nelson et al., 1978). Also, human blood lymphocytes have been separated into different subpopulations having characteristic surface markers and functional abilities (Malmström et al., 1980a, b; Walter et al., 1979). Natural killer and K cells can be enriched by countercurrent distribution of human peripheral blood lymphocytes (Malmström et al., 1980b; Walter et al., 1981). The electrophoretic mobility of different subpopulations was also measured (Walter et al., 1980a). For some subpopulations there was a

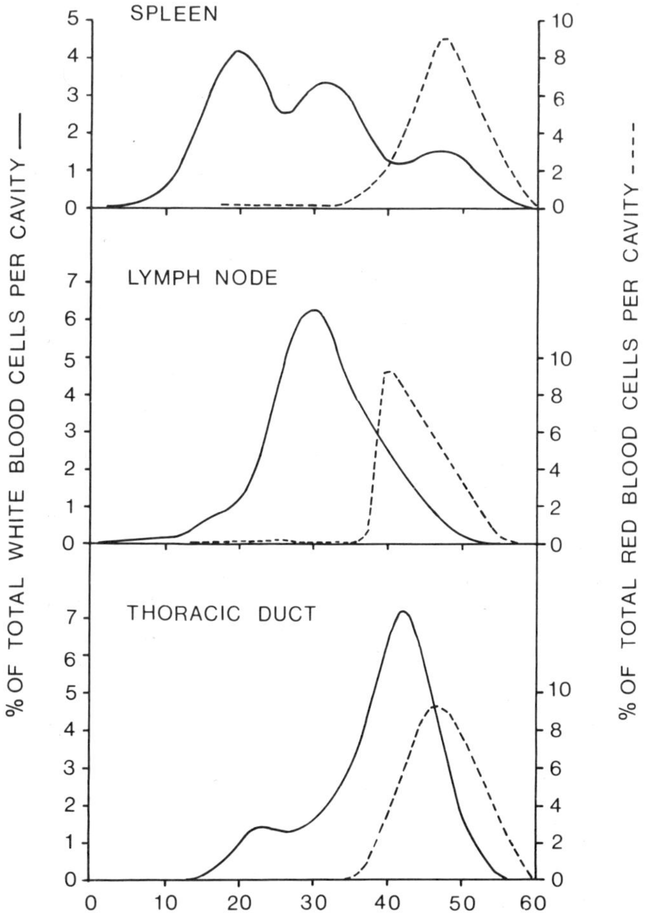

FIGURE 8.24*a*. Countercurrent distribution of leucocytes (——) and erythrocytes (– – –) from different organs of the rat (Malmström et al., 1978).

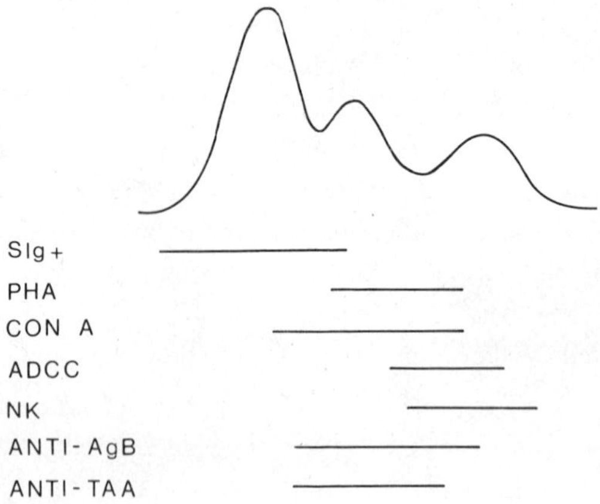

FIGURE 8.24*b*. Location of different markers and cell activities in fractions obtained by CCD of rat spleen cells.

correlation between the partition ratio and electrophoretic mobility. For example, the bulk of the B-lymphocytes had the lowest electrophoretic mobility and the lowest partition ratio, while T-lymphocytes had a higher partition coefficient and mobility. However, most of the F_c-receptor-bearing cells had the highest partition ratio but an intermediate mobility. Thus, other factors than charge determine the partition of these cells and it can therefore be misleading to draw conclusions on the net charge of particle from the position in a countercurrent distribution diagram. F_c receptor cells had such a high partition they should be purified by a simple batch procedure (Walter et al., 1980).

Monocytes from human blood have been separated into subpopulations having different phagocytizing ability; cells with higher partition ratio were more active in phagocytosis (Walter et al., 1980b).

TISSUE CULTURE CELLS

He La and mouse mast cells in different stages of growth have been analyzed by countercurrent distribution (Pinaev et al., 1976). The metaphase cells had a characteristic distribution pattern different from that of other cells in the same population (Fig. 8.25). Experiments with suspension cultures indicate a distinct change in the partition during exponential and stationary phases of growth. The freshly transferred cells were found to the right in the distribution diagram. With

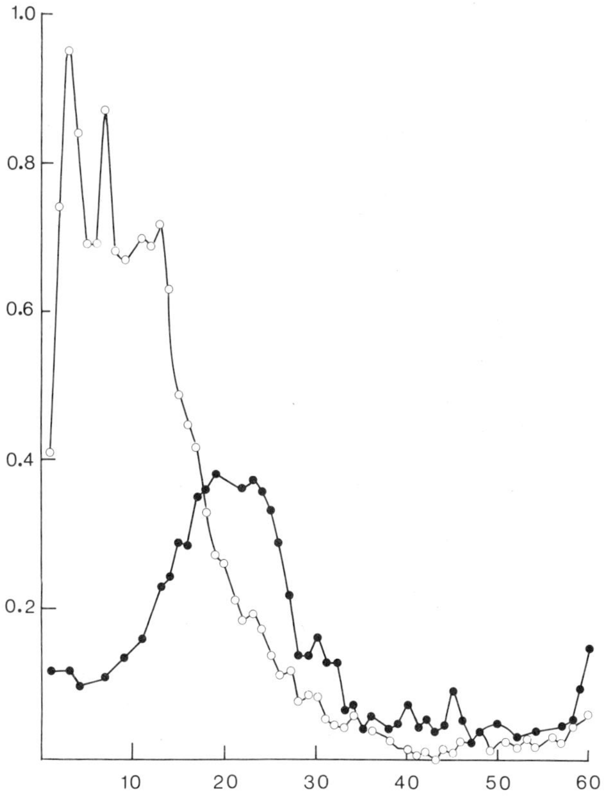

FIGURE 8.25. Countercurrent distributions of metaphase cells (●) and total cell population (○) of HeLa cells (Pinaev et al., 1976).

age the cell population moves to the left, that is, the older cells prefer the bottom phase.

Cell monolayer and suspension cultures of mouse fibroblasts gave different countercurrent distribution patterns. The monolayer cells had more affinity for the lower phase than the suspension cultured cells, which preferred the upper phase of the dextran–polyethylene glycol system (Petrov et al., 1981). The stage of the culture growth also influenced the partition. With increasing age the partition coefficient decreased. This effect was even more pronounced in the monolayer cells (Petrov et al., 1982). The effect of trypsin, EDTA, and DMSO, and suboptimal temperatures on partition was also studied by countercurrent distribution. The different treatments increased the number of cells with low partition coefficient (Petrov and Andreyeva, 1982).

CANCER CELLS

Malignant cell variants have been separated from the majority of cells of low malignant potential by countercurrent distribution of a culture of a murine lymphosarcoma cell line (Miner et al., 1981).

Hybrids between nonmalignant mouse 3T3 cells and highly malignant rat JFI cells were compared with parent cells by countercurrent distribution. The hybrid behaved more like the 3T3 cells, indicating that the cell surface of the hybrids was more like the nonmalignant surface (Freedlanskaya et al., 1983).

Partition of cancer cells was compared with normal cells of different cell systems (fibroblasts and hepatocytes from rat and mouse). In all cases there were large significant differences in the partition between normal cells and their malignant counterparts. Treatment of normal cells with carcinogens causes a shift in the partition of the cells indicating an alteration of the cell surface (Andreeva et al., 1982).

Cell surface glycosylation and number of surface sialic acid residues correlated with partition for lymphocytes (Kessel, 1981). The effects of porphyrines (Kessel, 1977), bromodeoxyuridine (Evans et al., 1977), and cholesterol enrichment (Walter, 1982) have also been studied.

REFERENCES

Albertsson, P.-Å. (1958). *Biochim. Biophys. Acta.* **27,** 378–395.

Albertsson, P.-Å. (1965). *Anal. Biochem.,* **11,** 121.

Albertsson, P.-Å. (1970). *Science Tools,* **17**(3), 56.

Albertsson, P.-Å. (1971). *Partition of Cell Particles and Macromolecules,* 1st ed., Almqvist & Wiksell, Stockholm, and Wiley, New York.

Albertsson, P.-Å., and Baird, G.D. (1962). *Exptl. Cell Res.,* **28,** 296.

Andreeva, E.A., Belisheva, N.K., Freedlanskaya, I.I., Pinaev, G.P., and Blomquist, G. (1982). *Chemosphere,* **11,** 377–381.

Baird, G.D., Albertsson, P.-Å., and Hofsten, B.V. (1962). *Nature,* **192,** 236.

Ballard, C.M., Dickinson, J.P., and Smith, J.J. (1979a). *Biochim. Biophys. Acta,* **582,** 89–101.

Ballard, C.M., Roberts, M.H.W., and Dickinson, J.P. (1979b). *Biochim. Biophys. Acta,* **582,** 102–103.

Blomquist, G.K., Ström, G.B., and Söderström, B. (1984). *Appl. Environm. Microbiol.,* **47,** 1316–1318.

Brunette, D.M., McCulloch, E.A., and Till, J.E. (1968). *Cell Tissue, Kinet.,* **1,** 319.

Burczyk, J. (1975). *Exptl. Cell Res.,* **90,** 211–222.

Colleen, S., Herrström, P., Wieslander, Å., and Mårdh, P.A. (1980). *Scand. J. Infect. Dis., Suppl.* **24,** 165–172.

Delafield, F.P., Somerville, H.J., and Rittenberg, S.C. (1968). *J. Bacteriol.*, **96,** 713–720.

Edebo, L., Lindström, F., Sköldstam, L., Stendahl, O., and Tagesson, C. (1975). *Imm. Comm.*, **4,** 587–601.

Edebo, L., Magnusson, K.E., and Stendahl, O. (1983). *Act. Path. Microbiol. Immunol. Scand.*, **B91,** 101–106.

Eriksson, E., Albertsson, P.-Å. and Johansson, G. (1976). *Mol. Cell. Biochem.*, **10,** 123.

Evans, J., Distefano, P., Case, K.R., and Bosman, H.B. (1977). *FEBS Lett.*, **78,** 109–112.

Fisher, D., and Walter, H. (1984). *Biochim. Biophys. Acta,* **801,** 106–110.

Freedlanskaya, I.I., Polanskaya, G.G., Blomquist, E., Tatulyan, S.A., and Pinaev, G.P. (1983). *Cytologia,* **25,** 600.

Gerson, D.F. (1980). *Biochim. Biophys. Acta,* **602,** 269–280.

Gerson, D.F., and Akit, J. (1980). *Biochim. Biophys. Acta,* **602,** 281–284.

Gersten, D.M., and Bosman, H.B. (1974). *Exptl. Cell Res.*, **88,** 225–230.

Hofsten, B.V. (1965). *Exptl. Cell. Res.*, **41,** 117.

Johansson, G. (1974). *Mol. Cell Biochem.*, **4,** 169–180.

Kessel, D. (1976). *Biochem. Pharm.*, **25,** 483–485.

Kessel, D. (1977). *Biochemistry,* **16,** 3443–3449.

Kessel, D. (1981). *Biochim. Biophys. Acta,* **678,** 24–249.

Kihlström, E., and Magnusson, K.E. (1980). *Cell Biophysics,* **2,** 177–189.

Kihlström, E., and Magnusson, K.-E. (1983). *Acta. Path. Microbiol. Immunol. Scand.*, **B91,** 113–119.

Leive, L., Culliname, L.M., Ito, Y., and Bramblett, G.T. (1984). *J. Liq. Chromatogr.*, **7,** 403–418.

Levy, E.M., Zanki, S., and Walter, H. (1981). *Eur. J. Immunol.*, **11,** 952–955.

Magnusson, K.E., and Johansson, G. (1977). *FEMS Microbiol. Lett.*, **2,** 225–228.

Magnusson, K.E., and Stjernström, J. (1982). *Immunology,* **45,** 239–248.

Magnusson, K.E., Stendahl, O., Tagesson, C., Edebo, L., and Johansson, G. (1977). *Acta. Path. Microbiol. Scand.*, **B85,** 212–218.

Magnusson, K.E., Stendahl, O., Stjernström, J., and Edebo, L. (1979a). *Immunology,* **36,** 439–447.

Magnusson, K.E., Kihlström, E., Norqvist, A., Davies, J., and Normark, S. (1979b). *Infect. Immunity,* **26,** 402–407.

Magnusson, K.E., Kihlström, E., Norlander, L., Norqvist, A., Davies, J., and Normark, S. (1979c). *Infect. Immunity,* **26,** 397–401.

Magnusson, K.E., Davies, J., Grundström, T., Kihlström, E., and Normark, S. (1980). *Scand. J. Infect. Dis. Suppl.*, **24,** 131–134.

Malmström, P., Nelson, K., Jönsson, Å., Sjögren, H.-O., Walter, H., and Albertsson, P.-Å. (1978). *Cell. Immunol.*, **37,** 409–421.

Malmström, P., Jönsson, Å., Hallberg, T., and Sjögren, H.-O. (1980a). *Cell. Immunol.*, **53,** 39–50.

Malmström, P., Jönsson, Å., and Sjögren, H.-O. (1980b). *Cell. Immunol.*, **53,** 51–64.

Michalski, J., Razandi, M., McCoombs, C.C., and Walter, H. (1983). *Clin. Imm. Immunopath.*, **29,** 15–28.

Miner, K.M., Walter, H., and Nicholson, G.L. (1981). *Biochemistry,* **20,** 6244–6250.

Miörner, H., Myhre, E., Björk, L., and Kronvall, G. (1980). *Inf. Immunity,* **29,** 879–885.

Miörner, H., Albertsson, P.-Å., and Kronvall, G. (1982). *Infect. Immunity,* **36,** 227–234.

Miörner, H., Johansson, G., and Kronvall, G. (1983). *Inf. Immunity,* **39,** 336–343.

Nakasawa, H., Yamaguchi, A., Kawaguchi, H., and Orii, H. (1979). *Biochim. Biophys. Acta,* **586,** 425–431.

Nelson, K., Malmström, P., Jönsson, Å., and Sjögren, H.O. (1978). *Cell. Immunol.,* **37,** 422–431.

Ohman, L., Norman, B., and Stendahl, O. (1981). *Infect. Immunity,* **32,** 951–955.

Pestka, S., Walter, H., and Wayne, L.C. (1977). *Antimicr. Agents. Chemotherapy,* **11,** 978–983.

Petrov, Y.P., and Andreyeva, E.V. (1982). *Cytologia,* **24,** 736.

Petrov, Y.P., Andreyeva, E.V., and Pinaev, G.P. (1981). *Cytologia,* **23,** 1192.

Petrov, Y.P., Andreyeva, E.V., and Pinaev, G.P. (1982). *Cytologia,* **24,** 591.

Pinaev, G., Hoorn, B., and Albertsson, P.-Å. (1976). *Exptl. Cell. Res.,* **98,** 127–135.

Raymond, F.D., and Fisher, D. (1980). *Biochim. Biophys. Acta,* **596,** 445–458.

Sharpe, P.T., and Watts, D.J. (1984). *FEBS. Lett.,* **168,** 89.

Sharpe, P.T., Treffry, T.E., and Watts, D.J., (1982). *J. Embryol. Exp. Morph.,* **67,** 181–193.

Sharpe, P.T., McDonald, B.R., Gallagher, J.A., Treffry, T.E., and Russel, R.G.G. (1984). *Bioscience Reports,* **4,** 415–419.

Schürch, S., Gerson, D.F., and McIver, D.J.L. (1981). *Biochim. Biophys. Acta,* **640,** 557–571.

Söderlund, G., and Kihlström, E. (1982). *Infect. Immunity,* **36,** 893–899.

Stendahl, O., Tagesson, C., and Edebo, M. (1973a). *Infect. Immunity,* **8,** 36–41.

Stendahl, O., Magnusson, K.-E., Tagesson, C., Cunningham, R., and Edebo, L. (1973b). *Infect. Immunity,* **7,** 573–577.

Stendahl, O., Tagesson, C., and Edebo, L. (1974). *Infect. Immunity,* **10,** 316–319.

Stendahl, O., Norman, B., and Edebo, L. (1979). *Acta Path. Microbiol. Scand.,* **B87,** 85–91.

Walter, H., (1982). In *Cell Separation: Methods and Selected Applications,* Vol. 1, Academic, New York, pp. 261–299.

Walter, H., and Krob, E.J. (1975). *Exptl. Cell. Res.,* **91,** 6–14.

Walter, H., and Krob, E.J. (1983). *Cell Biophysics,* **5,** 205–219.

Walter, H., and Krob, E.J. (1984). *Biochem. Biophys. Res. Comm.,* **120,** 250–255.

Walter, H., and Nagaya, H. (1975). *Cell. Immunol.,* **19,** 158–161.

Walter, H., and Selby, F.W. (1966). *Biochim. Biophys. Acta,* **112,** 146–153.

Walter, H., Selby, F.W., and Brake, J.M. (1964). *Biochem. Biophys. Res. Comm.,* **15,** 497–501.

Walter, H., Winge, R., and Selby, F. (1965). *Biochim. Biophys. Acta,* **109,** 293–301.

Walter, H., Eriksson, G., Taube, Ö., and Albertsson, P.-Å. (1971) *Exptl. Cell Res.,* **64,** 486–490.

Walter, H., Krob, E.J., Ascher, G.S., and Seaman, G.V.F. (1973a). *Exptl. Cell Res.,* **82,** 15.

Walter, H., Eriksson, G., Taube, Ö., and Albertsson, P.-Å. (1973b). *Exptl. Cell. Res.,* **77,** 361–366.

Walter, H., Webber, T.J., Michalski, J.P., McCombs, C.C., Moncla, B.J., Krob, E., and Graham, L.L. (1979). *J. Immunol.* **123,** 1687–1695.

Walter, H., Graham, L.L., Krob, E.J., and Hill, M. (1980a). *Biochim. Biophys. Acta,* **602,** 309–322.

Walter, H., Tamblyn, C.H., Levy, E.M., Brooks, D., and Seaman, G.F.V. (1980b). *Biochim. Biophys. Acta,* **598,** 193–199.

Walter, H., Tamblyn, C.H., Krob, E.J., and Seaman, G.V.F. (1983). *Biochim. Biophys. Acta,* **734,** 368–372.

Wayne, L.G., and Walter, H. (1974). *Antimicr. Agents Chemotherapy,* **5,** 203–209.

Weiser, E., Yeh, C. K., Lin, W., and Mazur, A. (1976). *J. Biol. Chem.,* **251,** 5703–5710.

Xin, J.H., Magnusson, K.-E., Stendahl, O., and Edebo, L. (1983). *J. Gen. Microbiol.,* **129,** 3075–3084.

9 BIOTECHNICAL APPLICATIONS

One of the attractive features of liquid–liquid extraction is that it can easily be scaled up. This is because the partition coefficient for soluble substances in most cases is independent of the total volume of the phase system, the volume ratio, and the concentration of the partitioned substance over a very large concentration range.

The partition behavior in large-scale phase partition can therefore be predicted with great accuracy from laboratory experiments with small test tubes. Furthermore, for aqueous two-phase systems, mixing and rapid equilibration are easily accomplished and require relatively little energy. Aqueous two-phase partition therefore offers many advantages for large-scale separation of biological material, both from a technical and an economical point of view. Three main biotechnical applications will be dealt with here: concentration, purification, and bioconversion.

CONCENTRATION

Many biotechnically interesting substances are available in dilute solutions or suspensions. It may be liquid cultures of bacteria, algae, or virus, or an enzyme extract. For their isolation and purification one has to start with large volumes. An essential step is, therefore, the concentration of the material into a small volume.

A two-phase system can carry out such a concentration provided it can be constructed in such a way that most of the desired substance is transferred to a phase with a small volume compared to the original solution. Since impurities may be less concentrated or not concentrated at all, a concomitant purification may also be achieved.

Concentration may be achieved by a one-step procedure or by multistep procedures.

The One-Step Procedure

The principle for the concentration in one step is that solutions of two polymers are added to the particle suspension so that a small volume phase is obtained containing most of the particles (Fig. 9.1). Suppose that the volume of the original particle suspension is V_0 mL and that the concentration of its particles is C_0 particles/mL. The two polymer solutions with a total volume of v are then added to the particle suspension to get the desired phase system. After mixing and phase separation, a bottom phase with a volume of V_b is formed; the volume of the top phase is then

$$V_t = V_0 + v - V_b \tag{1}$$

The particles distribute with a partition coefficient K, and no adsorption occurs at the interface. The concentration in the top phase is C_t and in the bottom phase C_b. The number of particles in the top phase, together with the number present in the bottom phase, should equal the number of particles in the original solution:

$$V_tC_t + V_bC_b = V_0C_0 \tag{2}$$

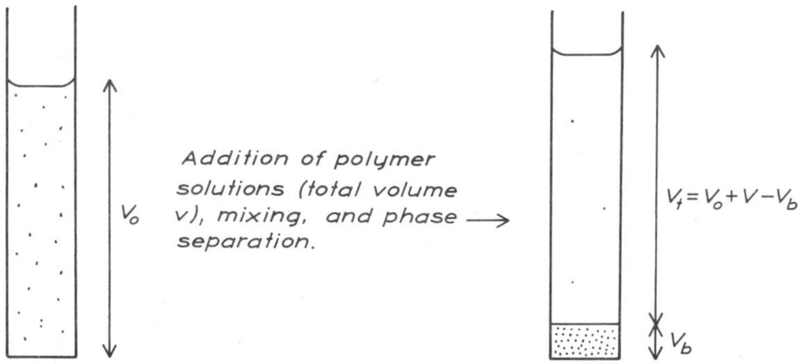

FIGURE 9.1. The principle for concentration of particles by phase partition. Solutions of polymers are added to the particle suspension, which has a volume of V_0, in such proportions that a small phase is formed into which the particles are concentrated. The points represent the particles.

We are mainly interested in knowing two things. The first is the concentration effect of the system; this is the concentration of particles in the particle-rich phase, in this case the bottom phase, compared with the concentration of particles in the original suspension. The second is the yield of virus particles in the virus-rich phase.

The Concentrating Effect. As a measure of the concentrating effect of a system, we use the concentration factor α which is defined as

$$\alpha = \frac{C_b}{C_0} \tag{3}$$

This may be calculated from Eqs. (1) and (2) and is

$$\alpha = V_0 \Big/ V_b\left(1 + \frac{V_t}{V_b} K\right) \tag{4}$$

To obtain large α values, that is, a good concentrating effect, V_b should be small compared to V_0, K should be small, and V_t should not be too large compared with V_0 (v should be as small as possible).

From Eq. (5) we also learn that if the ratio of the volumes of the two phases in a given system is kept constant, the concentrating factor will be larger the smaller the partition coefficient K of the particles, until a value of $\alpha = V_0/V_b$ is approached when K approaches zero.

If, for a system, V_0, v, and K are kept constant, but V_b is decreased, then α will increase and approach a value of $1/K \times (V_0/V_t)$ as V_b approaches zero.

The Yield of Concentrated Particles. The yield y of the concentrated particles is expressed as percentage of the total amount of particles, that is,

$$y = 100\,\frac{C_b V_b}{C_0 V_0} \tag{5a}$$

or

$$y = 100\alpha\,\frac{V_b}{V_0} \tag{5b}$$

TABLE 9.1 Yield y and Concentration Factor α of a System[a]

V_b, mL	y, %	α
100	99	9.9
10	90	90
1	48	480

[a]Calculated from Eqs. (4) and (5c) with $V_0 = 1000$ mL, $v = 100$ mL, $K = 0.001$, for different V_b values.

or

$$y = 100 \Big/ 1 + \frac{V_t}{V_b} K \tag{5c}$$

From these equations we learn that when the volume ratio is constant, the yield becomes greater the smaller the partition coefficient K.

However, for constant K, V_0, and v, the yield becomes less the smaller V_b [Eq. (5c)]. For a system where K is given, one has to compromise between the increasing concentrating effect [Eq. (4)] and a decreasing yield [Eq. (5c)] when V_b is made smaller; see also Table 9.1, where the yield and the concentrating factor have been calculated for systems with different bottom phase volumes.

TABLE 9.2 Minimum Volume V_b of the Bottom Phase and Concentration Factor α for 90% Concentration[a]

K	V_b, mL	α
0.1	521	1.7
0.01	91	10
0.005	47.4	19
0.001	9.8	92
0.0005	4.9	183
0.0001	0.99	910

[a]These are the values that allow a concentration of 90% of the virus particles in the bottom phase. Calculated from Eqs. (4) and (5c) for different K values; $V_0 = 1000$ mL; $v = 100$ mL; $y = 90\%$.

If one requires a yield above a certain value, for example 90%, there will then be for each K value a lower limit to the volume of the bottom phase which has the maximum concentrating effect. These values have been calculated for a system with $V_0 = 1000$ mL and $v = 100$ mL, and are given in Table 9.2.

These calculations apply to systems in which the particles are concentrated into the bottom phase. Relations that apply to systems in which the virus particles are concentrated into the top phase may be obtained by exchanging V_b and V_t and by replacing K with $1/K$ in the above equations.

Selection of a Phase System. The composition of a phase system suitable for concentration should be represented by a point on the phase diagram which is far enough from the critical point to represent a stable system and to allow a one-sided distribution without adsorption at the interface. To get a small phase, the point representing the mixture should be near the binodial, for example, point A of Figure 9.2. In addition, the time of phase separation should be short and it should be possible to add the polymers as concentrated solutions so that v (see the equations above) can be made small.

To find a system which fulfills these requirements, one may proceed as follows. If the phase diagram of a system is known, one first determines the K value for the particles in a series of mixtures represented by points increasingly removed from the critical point, for example, points 1–5 in Figure 9.3. A system which gives a one-sided distribution and which is not too far away from the critical point is then selected. Suppose, for example, that points 3, 4, and 5 give a fully one-sided distribution of the particles. Of these systems the one represented by

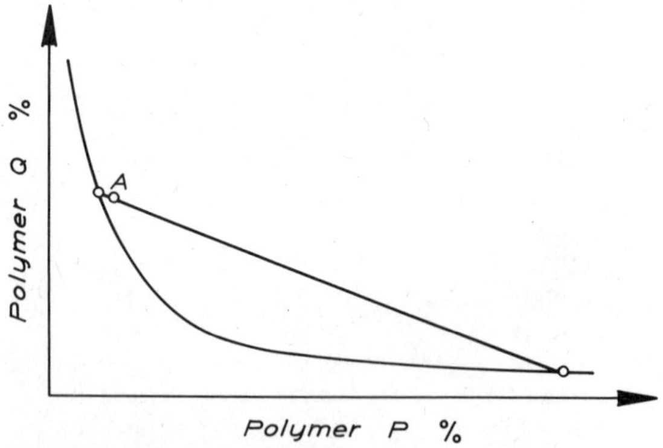

FIGURE 9.2. To obtain a phase system with one phase small compared to the other, the total composition of the system should be represented by a point lying close to the binodial, such as point A of this figure. In this case, the polymer-P-rich phase will be the smaller.

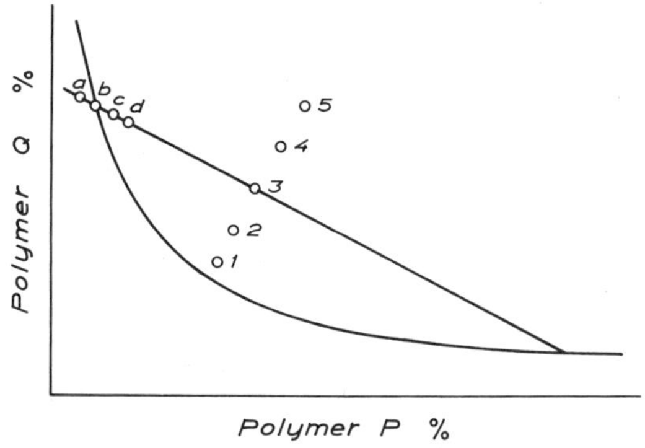

FIGURE 9.3. To obtain a system such as point A of Figure 9.2, which also allows a concentration of virus particles, one may proceed as follows: the partition of the particles is first determined in systems represented by the points 1, 2, 3, 4, and 5. If, for example, system 3 gives a fully one-sided distribution, a number of systems a, b, c, d on the tie line going through point 3 are tested to find out which gives a suitable small phase in which the particles are concentrated.

point 3 is far enough from the critical point to be a stable system but not too far from it to have phases which are too viscous. A system with a small phase is therefore selected from a point lying on the tie line going through point 3. For example, a number of systems represented by points a, b, c, and d (Fig. 9.3) are set up and the one with a suitably small phase is selected. Some of these points may lie outside the binodial because, for systems with polydisperse polymers, there is not a sharp change from a one-phase system to a two-phase system such as indicated by the binodial (see Chapter 2, Influence of Polydispersity of the Polymers).

The Time of Phase Separation. The settling time for a small bottom phase in a large top phase is relatively long. For practical purpose and on a large scale it is therefore advantageous to speed up the separation time by centrifugation, for example, by using a continuously working centrifugal separator.

Multistep Procedure

By this procedure a substance is first concentrated into a small, for example, bottom phase of a given phase system. The bottom phase is collected and to it is added a much smaller new top phase. At the same time conditions are changed so that the substance now prefers the upper phase and as a result will be concentrated in the upper phase (Fig. 9.4). In principle, this alternate partition into upper and lower phase with concomitant concentration can be repeated several

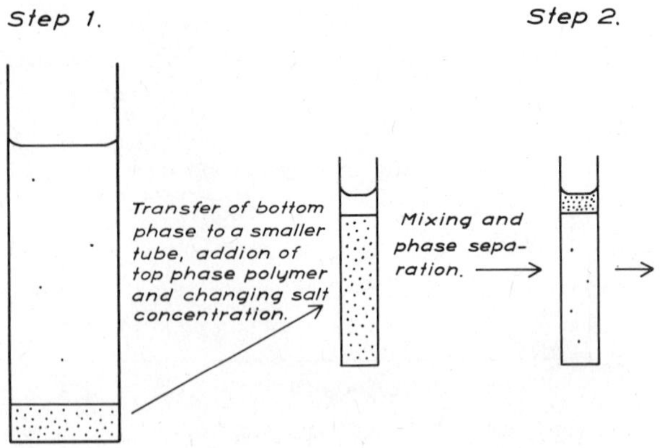

FIGURE 9.4. Concentration of particles alternately into the bottom and top phases.

times. In order to shift the partition coefficient so that the substance is concentrated into either the upper or lower phase one can exploit one or several of the different factors which determine the partition coefficient, such as pH, ionic composition, type and molecular weight of polymers, charged polymers, and biospecific ligands coupled to polymers. Some examples will be given.

1. Many proteins, for example, almost all serum proteins, have a very low partition coefficient in a phase system of low-molecular-weight dextran and PEG with NaCl as the dominating salt. Thus, more than 99% of the total plasma proteins partition into the lower phase of a system containing 10% w/w Dextran 40 and 7% w/w PEG 6000 in 0.1 *M* KSCN. By using 1% Dextran 40 and 10% PEG 6000 the lower phase will be small and the proteins will be concentrated into a small volume. If this volume is collected, a desired protein can now be concentrated into an upper phase by the addition of a small upper phase containing a PEG derivative which will increase the partition coefficient so that the protein favors the upper phase. For example, palmitoyl–PEG will selectively extract serum albumin into the upper phase.

2. In a dextran–PEG system containing negatively charged PEG (S-PEG; see Chapter 4) at low ionic strength, negatively charged proteins or particles will favor the dextran phase. If this phase is made small, concentration into this phase is obtained. After this has been collected the proteins or particles can be concentrated in a small, new upper phase containing positively charged (PEG) (TMA-PEG; see Chapter 4).

As an example, a model concentration experiment with Chlorella cells is described

(Albertsson, 1974). A suspension of Chlorella cells is first concentrated about 100 times by partition in a system composed of 0.2% w/w Dextran 500, 4% w/w PEG 6000, and 3% w/w S-PEG 6000 in water. The total volume is 1.000 mL and the lower phase volume is 8 mL. Six milliliters of the lower phase is collected, and to it is added 2 mL of a new top phase composed of 4% PEG 6000 and 3% TMA-PEG 6000 in water. Thereby, all the Chlorella cells concentrate into the upper phase which is now positive because of the presence of TMA-PEG. The upper phase is collected. To this an equal volume (2 mL) of a negative top phase composed of 4% PEG 6000 and 3% S-PEG in water is added. The two charged PEG's thereby neutralize each other.

If an additional negative upper phase is added, for example 2 mL of 4% PEG 6000, 3% S-PEG 6000 in water, and a small volume of a 20% w/w Dextran 500 solution, a phase system with a small lower phase will form. Since the upper phase will now be negative, the cells will concentrate into the lower phase. In principle, this procedure could be repeated an unlimited number of times. Since each cycle concentrates the particles about 100 times, two cycles would accomplish a concentration of the order of 10,000 times. This procedure certainly would be worth testing for concentration of virus from large volumes of dilute suspensions.

Virus Concentrations

Several viruses have been concentrated by phase partition. See Albertsson (1967, 1971) for details and further references. Concentration can be used either as a first step in purification of a virus or to increase the titer to allow detection for example in samples from waste water (Grindrod and Chiver, 1970).

The dextran–PEG system has been found to be mild towards virus; the polymers have been reported not to inhibit virus growth in cell cultures (Grindrod and Chiver, 1970). In this respect dextran is to be preferred to dextran sulfate, which inhibits certain viruses.

LARGE SCALE PURIFICATION OF ENZYMES

Application of aqueous two-phase systems for large scale enzyme purification (Table 9.3) has recently been developed, notably by Kula and co-workers (Kroner et al., 1982a, b; Kula et al., 1978, 1982a, b; Kopperschläger and Johansson, 1982; Johansson, 1985). They have studied the partition of a number of enzymes

TABLE 9.3 Enzymes or Proteins Which Have Been Purified on a Large Scale

Enzyme or Protein	Reference
Aspartase	Kula et al. (1982a)
Chlorophyll a/b-protein (LHCP)	Albertsson and Andersson (1981)
Formate dehydrogenase	Kula et al. (1982a)
β-galactosidase	Veide et al. (1983)
Fumarase	Kula et al. (1982a)
Glucose isomerase	Kula et al. (1982a)
Interferon	Kula et al. (1982a)
β-glucosidase	Kula et al. (1982a)
Phosphofructokinase	Kopperschläger and Johansson (1982)
Phospholipase	Albertsson (1973)

in either salt–PEG or dextran–PEG phase systems. Salt concentration, molecular weight of polymers, and also charged PEG and biospecific ligand PEG were used to optimize partition so that the desired enzyme was extracted with high yield into one of the phases, leaving most of the impurities in the other phase. Various techniques were studied to accomplish fast separation of the phases. The settling time for a phase system which is heavily loaded with cell homogenate is usually long because of the presence of DNA and cell debris which increase the viscosity. Satisfactory separation can be obtained, however, by using commercial continuous centrifugal separators. Calculations were also made which showed that large-scale purification by partition is far more economical than conventional methods using precipitation and chromatography (see, for example, Table 9.4) (Kula et al., 1982a; Kroner et al., 1984).

Kopperschläger and Johansson (1982) have described a method for purification of phosphofructokinase from yeast cells by affinity partition employing cibachrome blue–PEG as the affinity polymer. They demonstrated that extract from as much as 1 kg of yeast could be incorporated into 250 mL of phase system and that the time of purification was reduced to 1 day from the 4 days needed by standard procedures involving chromatography steps. The polymers stabilized the enzyme activity. Johansson (1985) has purified glucose 6-phosphate dehydrogenase by a similar procedure.

Membrane proteins can also be isolated on a large scale. Phospholipase A_2 was isolated from large quantities of *E. coli* by a dextran–PEG system including a detergent, Triton X-100, for solubilization of the enzyme (Albertsson, 1973). A three-phase partition step was used in order to separate the detergent from the

TABLE 9.4 Comparison of Technical and Economic Performance[a]

	Partition	Conventional
Total cell mass	50	5
Initial units	460×10^3	31×10^3
Purity, U/mg	2.2	2.2
Yield, %	70	51
Net time, hr	18	121
Performance factor, U/kg hr	356	26
Cost index, DM/unit	7×10^{-3}	374×10^{-3}

[a]The comparison is of preparations of the enzyme formate dehydrogenase by an aqueous-partition method or a conventional precipitation–chromatographic method (from Kula et al., 1982a).

enzymes. Monoamin oxidase from beef liver has also been purified by a three-phase system including Triton X-100 (Salach, 1978). The light harvesting chlorophyll a/b protein from chloroplast can also be isolated on a large scale by a detergent-containing phase system (Albertsson and Andersson, 1981).

Partition by aqueous two-phase systems has the following advantages for large scale enzyme purification:

1. Scale-up can easily and reliably be predicted from small laboratory experiments.
2. Rapid mass transfer and equilibrium is reached by relatively little input of energy in the form of mechanical mixing.
3. It can be developed as a continuous process.
4. The polymers stabilize the enzymes.
5. Separation can be made selective and rapid.
6. Because of the rapid separation it may be carried out at room temperature instead of in the cold, thus at lower investment costs.
7. It is more economical than other purification methods.

BIOCONVERSION

Aqueous two-phase systems can be used for extractive bioconversion or enzyme reactors. By confining an enzyme or a cell suspension in one phase together with substrate while the product partitions into the other phase the product can be

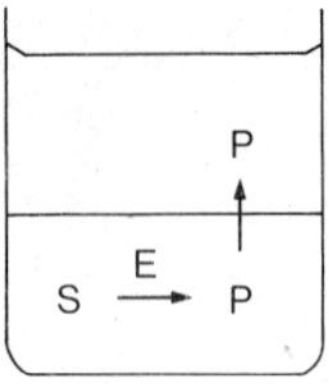

FIGURE 9.5. Principle of bioconversion in a two-phase system. A catalyst E (an enzyme or a cell) converts the substrate S to P. Both E and S have low partition coefficients and are confined to the lower phase while the product is extracted into the upper phase.

removed by repetitive partition steps or by a continuous extraction procedure. The principle is shown in Figure 9.5.

Extraction can be used both for removing the product from the enzyme and the substrate for further purification and also for the purpose of increasing the velocity of the reaction if the enzyme is inhibited by the product, which is often the case.

By using a mixer–settler the process can be carried out continuously (Fig. 9.6). Substrate in the extracting phase, in this case the top phase, is fed into the mixer. The mixture of the two phases is removed from the settler together with the product. The top phase together with the product leaves the settler. The other phase containing enzyme or cells together with substrate is pumped back from the settler to the mixing vessel. Ideally, no substrate or catalyst should leave the system with the extracting phase. In practice, certain losses are observed, depending on the partition coefficient of the substrate and the catalyst.

Puziss and Hedén (1965) used a dextran–polyethylene glycol phase system as growth medium for *Clostridium tetani* and for concomitant toxin production. By choosing a favorable phase volume ratio (top/bottom phase = 15/1) the greater part of the toxin was in the top phase while the cells were in the bottom

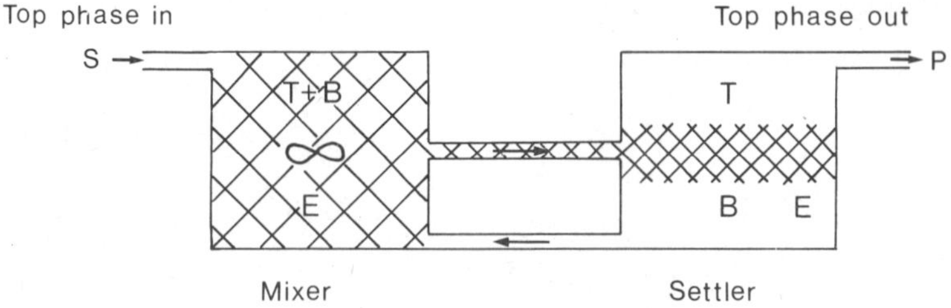

FIGURE 9.6. Continuous bioconversion using a mixer–settler. S, substrate; P, product; E, enzyme or cell; T, top phase; B, bottom phase.

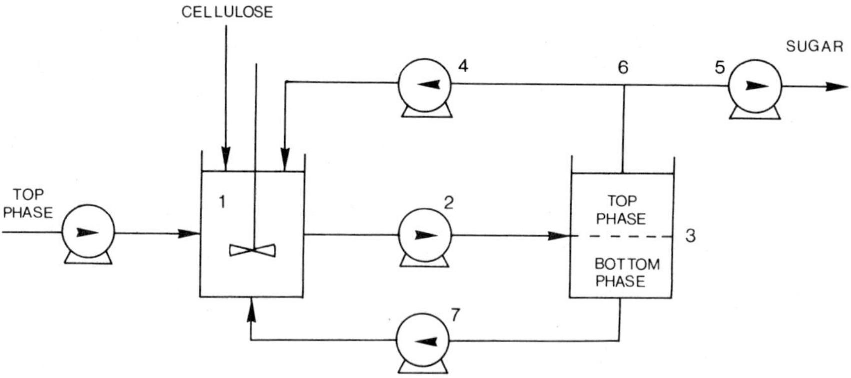

FIGURE 9.7. Experimental set up for semicontinuous hydrolysis of cellulose. Cellulose together with top phase enters the mixer (1). The mixture is pumped (2) to the settler (3). Top phase can either be recirculated to the mixer (4) or removed from the system together with product glucose (5). Glucose can also be separated from the top phase polymer by ultrafiltration at (6). The bottom phase of the settler is pumped (7) back to the mixer (Tjerneld et al., 1985b).

phase or at the interface. In another study it was found that higher yields of streptococcus cells were obtained when grown in a similar phase system compared to a conventional culture medium (Hedén and Holmström, 1962).

Alcoholic fermentation of glucose by yeast in a dextran–polyethylene glycol phase system was studied by Kühn (1980). Fermentation occurred in the stirred phase system. After stopping the stirring the phases were separated, and the upper phase collected. Its ethanol was distilled off and the alcohol-free upper phase was then added to the yeast containing lower phase in the fermentation flask. After adding a new batch of glucose the fermentation started again. Several of such cycles were carried out. The author suggests several advantages of the procedure.

The bioconversion of cellulose to glucose by a combined action of cellulase and α-glucosidase in a dextran–polyethylene glycol phase system is a good example of how an insoluble substrate can be continuously degraded by an enzyme-containing phase system (Tjerneld et al., 1985b). Cellulose particles are confined to the lower phase together with most of the cellulase and α-glucosidase, while the product, glucose, partitions evenly between the phases. By using a mixer–settler, the upper phase can be continuously recovered. The glucose can be removed from it by dialysis and the upper phase, now depleted of glucose, is then remixed with lower phase in the mixing chamber. Figure 9.7 shows schematically how the continuous operation can be carried out and Figure 9.8 shows the production of glucose from cellulose.

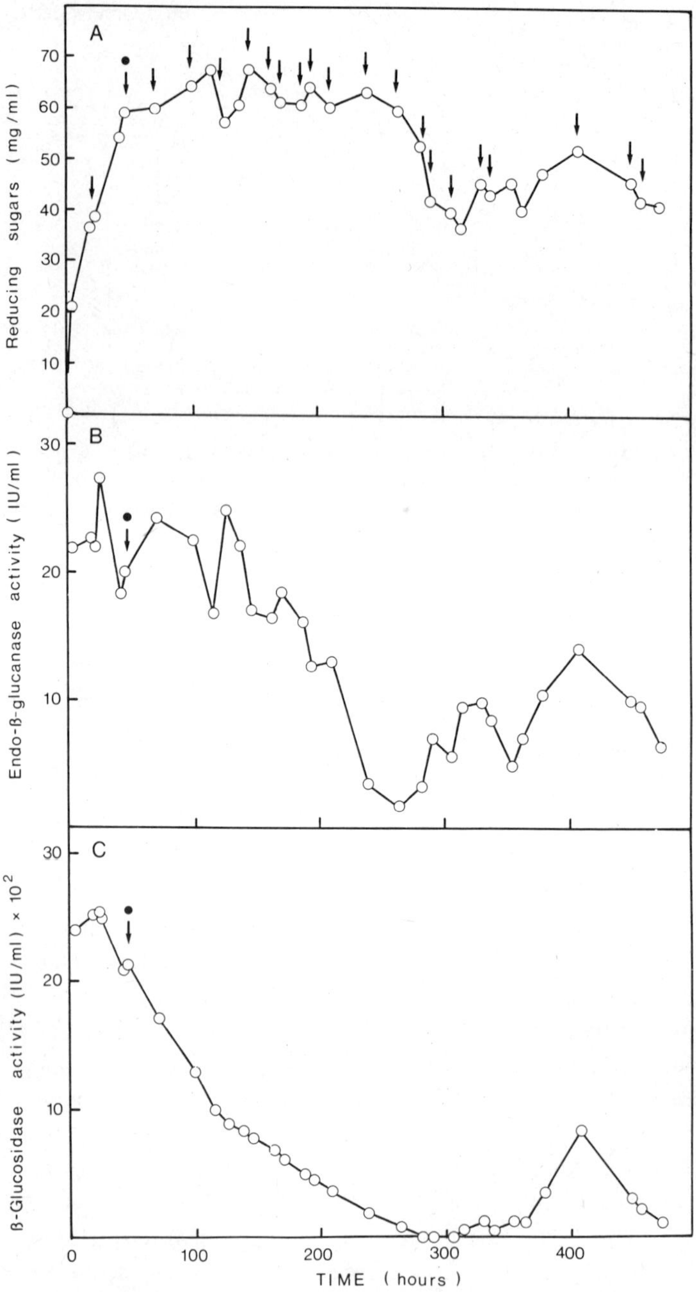

FIGURE 9.8. Semicontinuous hydrolysis of cellulose by the set-up shown in Figure 9.7. Phase system 5% w/w crude dextran, 3% w/w PEG 20,000, 50 m*M* sodium acetate pH 4.8. Initial cellulose concentration 7.5%. Enzyme concentration 2 FPU/mL. Flow rate 5 mL/hr. Total volume, of the system 800 mL. ●, start of withdrawal of top phase; ↓, addition of cellulose; (A) Concentration of reducing sugars in the top phase; (B) endo-β-glucanase activity in the top phase; (C) β-glucosidase activity in the top phase (Tjerneld et al., 1985b).

TABLE 9.5 Examples of Bioconversion in Aqueous Two-Phase Systems

Process	Reference
Toxin production of *Clostridium tetani*	Puziss and Hedén (1965)
Production of glucose 6-Phosphate from glucose	Yamazaki and Suzuki (1978)
Glucose fermentation	Hahn-Hägerdal et al. (1981, 1982); Kühn (1980)
Cellulose saccharification	Hahn-Hägerdal et al. (1981, 1982); Tjerneld et al. (1985b)
Butanol, acetone, and butyric acid formation by *Clostridium acetobutylicum*	Mattiasson et al. (1982)
Hydrolysis of starch by amylase	Wennersten et al. (1983)
Cellulase production by *Trichoderma reesei*	Persson et al. (1985)
Deacylation of benzylpenicillin	Andersson et al. (1983, 1984)
Regeneration of ATP by chromatophores	Smeds et al. (1983)

For this process to work effectively it is important that the enzymes are concentrated in the lower phase, that is, that they have a low partition coefficient. To achieve this all the factors which determine the partition (described in Chapter 4) were considered and systematically studied for cellulase and α-glucosidase. Their confinement to the lower phase was facilitated by adsorption of some of the enzymes to the cellulose particles (Tjerneld et al., 1985b).

The combination of the degradation of cellulose to glucose with its alcoholic fermentation has also been studied (Hahn-Hägerdal et al., 1981).

Other applications of bioconversion in two-phase systems are found in Table 9.5.

REFERENCES

Albertsson, P.-Å. (1960). *Partition of Cell Particles and Macromolecules,* 1st ed., Almqvist & Wiksell, Stockholm, and Wiley, New York.

Albertsson, P.-Å. (1967). In *Methods in Virology,* Vol. 2, K. Maramorosch and H. Koprowski, Eds., Academic, New York, pp. 303–321.

Albertsson, P.-Å. (1971). In *Methods in Microbiology,* Vol. 58, J.R. Norris and D.W. Ribbons, Eds., Academic, New York, pp. 385–423.

Albertsson, P.-Å. (1973). *Biochemistry,* **12,** 2525.

Albertsson, P.-Å. (1974). In *Virus Survival in Water and Wastewater Systems,* J.F. Malnia and B.P. Sogik, Eds., Water Res. Synop. No. 7, Univ. of Texas, Austin, pp. 16–18.

Albertsson, P.-Å. and Andersson, B. (1981). *J. Chromatogr.*, **215,** 131–141.

Andersson, E., Mattiasson, B., and Hahn-Hägerdal, B. (1983). *Acta. Chem. Scand.*, **B37,** 749–750.

Andersson, E., Mattiasson, B., and Hahn-Hägerdal, B. (1984). *Enzym. Microbiol. Technol.*, **6,** 301–306.

Grindrod, J., and Chiver, D.O. (1970). *Arch. Gesamte Virusforschung,* **31,** 365–372.

Hahn-Hägerdal, B., Mattiasson, B., and Albertsson, P.-Å. (1981). *Biotechnol. Lett.*, **3,** 53–58.

Hahn-Hägerdal, B., Mattiasson, B., Andersson, E., and Albertsson, P.-Å. (1982). *J. Chem. Tech. Biotechnol.*, **32,** 157–161.

Hedén, C.G., and Holmström, H.B. (1962). *Abstr. 8th Int. Congr. Microbiology,* Montreal.

Kopperschläger, G. and Johansson, G. (1982). *Analyt. Biochem.*, **124,** 117–124.

Kroner, K.H., Schutte, H., Stach, W., and Kula, M.-R., (1982a). *J. Chem. Tech. Biotechn.*, **32,** 130.

Kroner, K.H., Hustedt, H., and Kula, M.-R. (1982b). *Biotechnol. Bioeng.*, **24,** 1015–1045.

Kroner, K.H., Hustedt, H., and Kula, M.-R. (1984). *Process Biochemistry,* **19,** 170–179.

Kula, M.-R., Johansson, G., and Buckmann, A.F. (1978). *Biochem. Soc. Transact.*, **7,** 1–5.

Kula, M.-R., Kroner, K.H., and Hustedt, H. (1982a). *Adv. Biochem. Eng.*, **24,** 73–118.

Kula, M.-R., Kroner, K.H., Hustedt, H., and Schutte, H. (1982b). *Enzyme Eng.*, **6,** 69.

Kühn, J. (1980). *Biotechnol. Bioeng.*, **22,** 2393–2398.

Mattiasson, B., Suominen, M., Andersson, E., Häggström, L., Albertsson, P.-Å., and Hahn-Hägerdal, B. (1982). *Enzyme Eng.*, **6,** 153–155.

Persson, I., Tjerneld, F., and Hahn-Hägerdal, B. (1984). *Enzyme. Microb. Techn.* **6,** 415–418.

Puziss, M., and Hedén, C.G. (1965). *Biotechnol. Bioeng.* **7,** 355–366.

Salach, J.J. (1978). *Methods Enzymol.*, **53,** 495.

Smeds, A.-L., Veide, A., and Enfors, S.-O. (1983). Enzyme Microb. Techn., **5,** 33–36.

Tjerneld, F., Persson, I., Albertsson, P.-Å., and Hahn-Hägerdal, B. (1985a). *Biotechn. Bioeng.* **27,** 1036–1043.

Tjerneld, F., Persson, I., Albertsson, P.-Å., and Hahn-Hägerdal, B. (1985b). *Biotechn. Bioeng.* **27,** 1044–1050.

Veide, A., Smeds, A.L., and Enfors, S.-O. (1983). *Biotech. Bioeng.*, **25,** 1789–1800.

Veide, A., Lindbäck, T., and Enfors, S.-O. (1984). *Enzyme and Microb. Techn.*, **6,** 325–330.

Wennersten, R., Tjerneld, F., Larsson, M., and Mattiasson, B. (1983). *Proc. International Solvent Extraction Conf.* Denver, p. 505–506.

Yamazaki, Y. and Suzuki, H. (1979). *Rep. Fermentation Res. Inst.*, No. 52.

10 BINDING STUDIES

The study of interactions between different biomolecules is of increasing importance for our understanding of the function of these molecules. All molecules of the cell interact more or less strongly with other molecules. Some interactions are strong and specific, giving rise to stable complexes such as oligomeric enzymes. Other interactions are weaker, and the complexes formed are so unstable that they dissociate when the cells are broken and the content diluted.

In this chapter I describe how phase partition can be used to detect and quantify interactions of, for example, the following types: protein–small ligand, nucleic acid–small ligand, protein–protein, protein–nucleic acid, and protein–lipid.

The principle of this method is as follows. Two substances are partitioned separately and together in a two-phase system. If there is no interaction, the two substances partition independently of each other; the presence of one substance does not influence the other. If there is interaction, the presence of one substance perturbs the partition of the other. Suppose the molecules A and B interact and form a complex AB. If the apparent partition of A at constant concentration is studied as a function of added B a curve such as that in Figure 10.1 may be obtained. The apparent partition coefficient of A will shift from the true partition coefficient of A to the partition coefficient of AB as the concentration of B is increased. A similar curve is obtained for B if its partition is studied as a function of added A. The changes in partition can be used for calculation of dissociation constants.

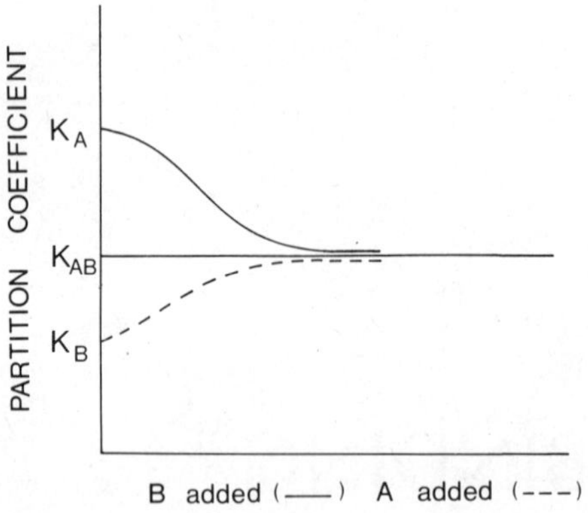

FIGURE 10.1. Binding between two molecules A and B can be detected by phase partition. The partition coefficients for A and B are K_A and K_B. If B is added to the system and a complex AB is formed, the apparent partition of A will change (solid curved line) and approach that of the partition coefficient K_{AB} of the complex. If A is added to B, its apparent partition coefficient will change from K_B to K_{AB} (dashed line).

THEORY

The following symbols are used to denote concentrations, partition coefficients, and dissociation constants:

$[A^0]_t$ = total concentration of A in top phase

$[B^0]_t$ = total concentration of B in top phase

$[A]_t$ = concentration of free A in top phase

$[B]_t$ = concentration of free B in top phase

$[AB]_t$ = concentration of complex AB in top phase; by replacing the subscript t with b, corresponding symbols for the bottom phase are obtained

K_A, K_B, K_{AB} = partition coefficients for A, B, and AB

K_t, K_b = dissociation constants in the top and bottom phase

Interactions between Two Molecules A and B

Two cases will be considered.

Interaction between Two Molecules A and B Having the Partition Coefficients K_A and K_B. In this case we have the equilibria shown in Fig. 10.2. It is assumed that A and B form a 1:1 complex. There are two dissociation equi-

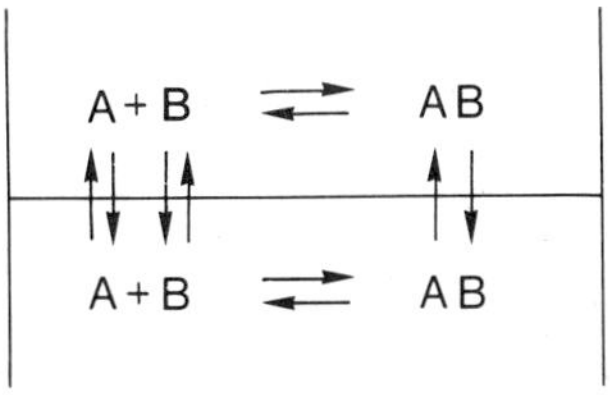

FIGURE 10.2. Equilibria for two interacting species A and B in a two-phase system.

libria, one for each phase, and three partition equilibria. If the total concentrations of A and B can be determined in each phase, the equilibrium constants can be calculated. In this manner interactions between molecules can be detected and studied quantitatively.

The following equations can be written:

$$K_A = \frac{[A]_t}{[A]_b} \tag{1}$$

$$K_B = \frac{[B]_t}{[B]_b} \tag{2}$$

$$K_{AB} = \frac{[AB]_t}{[AB]_b} \tag{3}$$

$$[A]_t + [AB]_t = [A^0]_t \tag{4}$$

$$[B]_t + [AB]_t = [B^0]_t \tag{5}$$

$$[A]_b + [AB]_b = [A^0]_b \tag{6}$$

$$[B]_b + [AB]_b = [B^0]_b \tag{7}$$

$$K_t = \frac{[A]_t[B]_t}{[AB]_t} \tag{8}$$

$$K_b = \frac{[A]_b[B]_b}{[AB]_b} \tag{9}$$

If we take a sample from the top phase and dilute it so that the complex AB dissociates, and if we can assay A and B separately, for example, by an enzymatic, immunological, or radioactive assay, then we can determine the total concentration of A and B in the upper phase. In the same way, the total concentration of A and B in the bottom phase can be determined. Thus $[A^0]_t$, $[B^0]_t$, $[A^0]_b$, and $[B^0]_b$ will be known.

Both K_A and K_B can be determined by measuring the partition coefficients of the proteins separately. The remaining nine unknowns can be solved by means of the above equations. We obtain the following relationships for the dissociation constants and the partition coefficient of the complex:

$$K_t = \frac{\{[A^0]_t - [B^0]_t - K_B([A^0]_b - [B^0]_b)\}\{[A^0]_t - [B^0]_t - K_A([A^0]_b - [B^0]_b)\}}{(K_B - K_A)[(1/K_A)[A^0]_t - (1/K_B)[B^0]_t - [A^0]_b + [B^0]_b]} \qquad (10)$$

$$K_b = \frac{\{[A^0]_t - [B^0]_t - K_B([A^0]_b - [B^0]_b)\}\{[A^0]_t - [B^0]_t - K_A([A^0]_b - [B^0]_b)\}}{(K_A - K_B)(K_A[A^0]_b - K_B[B^0]_b - [A^0]_t + [B^0]_t)} \qquad (11)$$

$$K_{AB} = K_A K_B \frac{[B^0]_t/K_B - [A^0]_t/K_A + [A^0]_b - [B^0]_b}{K_A[A^0]_b - K_B[B^0]_b - [A^0]_t + [B^0]_t}$$

The following relation also holds:

$$\frac{K_t}{K_b} = \frac{K_A K_B}{K_{AB}} \qquad (12)$$

Hence the dissociation constants and the partition coefficient of the complex can be determined by one partition experiment only.

It is assumed that a 1:1 complex between A and B is formed. Also, the partition coefficients K_A and K_B must be different. If they are only slightly different, the method is not very accurate, and if they are identical, the calculation cannot be used.

The partition coefficient of the complex, K_{AB}, can also be determined if an excess of A over B is added to the system (Fig. 10.1). If the excess is so large that all B is in the complex, its partition coefficient can be determined. Alternatively, an excess of B over A can be used. The following expression for the dissociation constant in the bottom phase can then be written:

$$K_b = \frac{([A^0]_b - \varphi)([B^0]_b - \varphi)}{\varphi} \qquad (13)$$

where

$$\varphi = \frac{[A^0]_t - K_A[A^0]_b}{K_{AB} - K_A} \qquad (14)$$

The dissociation constant in the top phase is obtained from (13). In this case K_A and K_B may be similar, provided that they are different from K_{AB}.

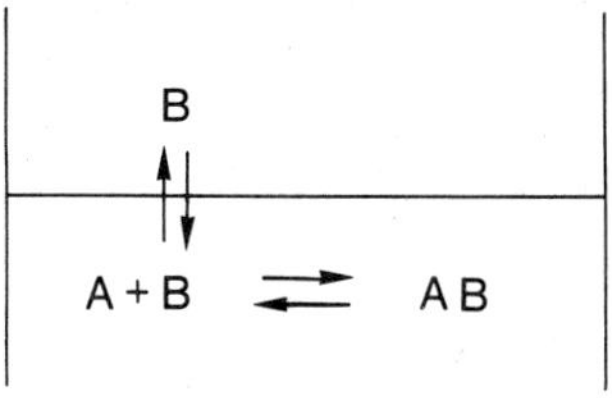

FIGURE 10.3. The same as Figure 10.2, but the partition coefficients of A and AB are so low that their concentration in the upper phase can be neglected.

The Partition of A is One-Sided, That Is, the Partition Coefficient of A Is Either 0 or ∞. In this case we have the situation as shown in Figure 10.3.

The partition coefficient of B, K_B, is determined by partitioning B alone. In the equilibrium-partition experiment we determine the concentration of B in the upper phase, $[B]_t$, and we can thereby calculate the concentration of free B in the lower phase, $[B]_b$, since

$$[B]_b = \frac{[B]_t}{K_B} \tag{15}$$

If we know the total amounts of A and B added to the system, and if we know the volumes of the phases, we can calculate both $[A]_b$ and $[AB]_b$ and therefore also the dissociation constant. We may also employ the Scatchard plot, that is, we plot bound/free versus bound in order to obtain both the dissociation constant and the number of binding sites on A.

Association of Two Identical Molecules

The association of two molecules of A to the dimer A_2 can also be detected and studied quantitatively (Middaugh and Lawson, 1980) by partition. In this case we can use (10) and (11) and set B equal to A and K_{AB} equal to K_{A_2}, the partition coefficient of the dimer. This can be determined by extrapolation of the partition coefficient at a high concentration when the dimers dominate.

The following expression can then be written for the dissociation constant of the A_2 complex in the lower phase:

$$K_b = 2\frac{([A^0]_t - [A^0]_b K_{A_2})^2}{(K_{A_2} - K_A)([A^0]_t - K_A[A^0]_b)} \tag{16}$$

and for the upper phase,

$$K_t = \frac{(K_A)^2}{K_{A_2}} K_b \tag{17}$$

Treatment of Data

According to (10) and (11), K_t and K_b can be calculated from one partition experiment only. However, more reliable values of the dissociation constants are obtained if the partition is carried out at different concentrations of A and B. The dissociation constant so obtained should be independent of concentration for a 1:1 complex. However, one has to be sure that the concentrations of A and B are chosen such that part of A or B is free and the rest bound. Also one should check that K_A and K_B are independent of concentration in the concentration range used. A convenient approach is to measure the partition coefficients of A and B alone and mixed in a series of tubes with increasing concentrations of A and B, and then plot K_t and K_b as a function of concentration, as seen in Figure 10.6.

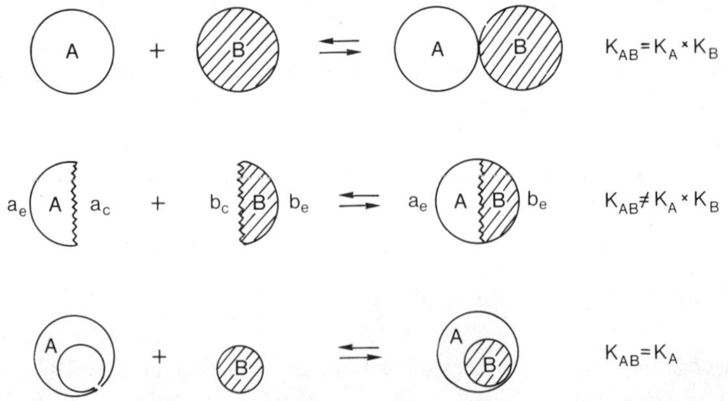

FIGURE 10.4. Three different models for complex formation between A and B. (*Upper*) Molecules A and B have a very small surface of contact. In this case the partition coefficient K_{AB} should be the product of the partition coefficients of A and B (see text). This model might represent some protein–protein interactions. (*Middle*) Molecules A and B have a very large surface of contact. In this case one would expect the partition coefficient of the complex to differ from the product of the partition coefficients of A and B. An example of this model might be interactions between subunits of proteins, for example hemoglobin. An extreme example of this case is the complex between two single-stranded nucleic acids. (*Lower*) Molecule A encloses B in the complex. In this case the partition coefficient of the complex would be the same as that of A, since A and AB expose the same surfaces to the medium. An example of this case is that of a detergent micelle enclosing a hydrophobic protein, or a ligand buried in a crevice of an enzyme (Albertsson, 1983).

Partition of two different substances A and B can be used to detect if there is interaction between them. The dissociation constant can be determined easily in the case of a 1:1 complex. In the special case when one of the substances has a very low or a very high K value, the number of binding sites can also be determined from the Scatchard plot.

However, one may get an indication of the surface of contact between the two molecules A and B. This information can be obtained by comparing the partition coefficient of the complex, K_{AB}, with K_A and K_B (Fig. 10.4).

We use the following argument. First, we assume that the free energy of transfer ΔG_A of a particle A between one phase and the other is the difference between the surface free energy G^s of the particle in the two phases:

$$RT \ln K_A = \Delta G^s{}_A = G^s{}_{A,bot} - G^s{}_{A,top} \tag{18}$$

In the same way for particle B,

$$RT \ln K_B = \Delta G^s{}_B = G^s{}_{B,bot} - G^s{}_{A,top} \tag{19}$$

Further, we assume that the free energy of transfer of the complex is the difference between the surface free energy of the complex in the lower and the upper phase.

$$RT \ln K_{AB} = \Delta G^s{}_{AB} = G^s_{AB,bot} - G^s{}_{AB,top} \tag{20}$$

If there is a small contact area between the two interacting particles, then the surface of the complex is the sum of the surfaces of the two particles. Therefore

$$\Delta G^s{}_{AB} = \Delta G^s{}_A + \Delta G^s{}_B \tag{21}$$

and

$$RT \ln K_{AB} = RT \ln K_A + RT \ln K_B \tag{22}$$

or

$$K_{AB} = K_A K_B \tag{23}$$

If, on the other hand, the contact surface between the two interacting particles is large, Eq. (21) no longer holds. In this case, surfaces of each particle disappear from contact with the surrounding phase. We call the contact surfaces of particles A and B a_c and b_c, respectively, and the remaining surfaces that are exposed to the surrounding liquid a_e and b_e (Fig. 10.4). The partition coefficient of particle A is determined by the surfaces a_c and a_e such that

$$RT \ln K_A = \Delta G^s_{a_c} + \Delta G^s_{a_e} \tag{24}$$

where $\Delta G^s_{a_c}$ and $\Delta G^s_{a_e}$ are the surface free energies of transfer of the surfaces a_c and a_e. In the same way the partition coefficient of particle B depends on its surfaces such that

$$RT \ln K_B = \Delta G^s_{b_c} + \Delta G^s_{b_e} \tag{25}$$

where $\Delta G^s_{b_c}$ and $\Delta G^s_{b_e}$ are the surface free energies of transfer of the surfaces b_c and b_e.

The partition coefficient of the complex depends on its surfaces as follows:

$$RT \ln K_{AB} = \Delta G^s_{a_e} + \Delta G^s_{b_e} \tag{26}$$

The following relation between the partition coefficient of the complex AB and the partition coefficients of A and B is obtained:

$$\ln K_{AB} = \ln K_A + \ln K_B - \frac{\Delta G^s_{a_c} + \Delta G^s_{b_c}}{RT} \tag{27}$$

The last term is zero when

1. Both $\Delta G^s_{a_c}$ and $\Delta G^s_{b_c}$ are zero. This is the case when the two contact surfaces are very small (see above).
2. The terms $G^s_{a_c}$ and $G^s_{b_c}$ have equal values but are of opposite sign. For large contact-surface areas this would be unlikely.

We can therefore conclude that if

$$K_{AB} = K_A K_B \qquad \text{or} \qquad K_t = K_b \tag{28}$$

it is a strong indication that the contact-surface area is very small, perhaps involving only a few amino acids in the case of protein. And we also conclude that if K_{AB} is very different from $K_A K_B$ (or K_t very different from K_b), the contact-surface area is large, or there is a general conformational change so that the exposed surface of the complex is different from the exposed area of the two interacting particles.

EXAMPLES (See Table 10.1)

1. DNA–Methyl Green

In the dextran–PEG system containing sodium chloride or sodium acetate, DNA has a very low K value, whereas the dye methyl green has a K value close to 1. In a system of 5% w/w Dextran 500 and 4% w/w PEG 6000, the binding of methyl green and other triphenylmethane dyes to DNA was determined at different concentrations of sodium chloride (Norden *et al*, 1978). Figure 10.5 shows a Scatchard plot from such an experiment. The plot is typical for binding of small ligands to DNA. The linear part of the plot indicates binding to noninteracting binding sites with $K_d = 1.8 \times 10^6\ M^{-1}$, and the curved line indicates either binding to sites with weaker binding or negative cooperativity.

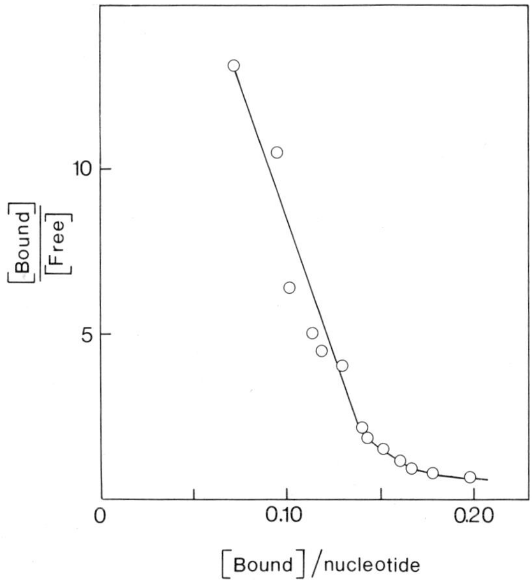

FIGURE 10.5. Scatchard plot of the binding of methyl green to DNA (Nordén et al., 1978).

2. Lipase–Colipase

Pancreatic lipase needs for its lipolytic activity in the presence of bile salts a protein cofactor colipase. Lipase binds colipase in a 1 : 1 complex. This binding was studied by partition in a dextran–PEG phase system: 7% w/w Dextran 500, 4.4% w/w PEG 6000, 150 m*M* NaCl, 5 m*M* tris-malate) (Patton *et al*, 1978). In this system lipase has a partition coefficient of 0.5, and that of colipase is 1.4. The partition coefficient of the complex is 0.7. The dissociation constant was determined by assaying the content of the two proteins in both phases and applying Eqs. (10) and (11). The dissociation constant was found

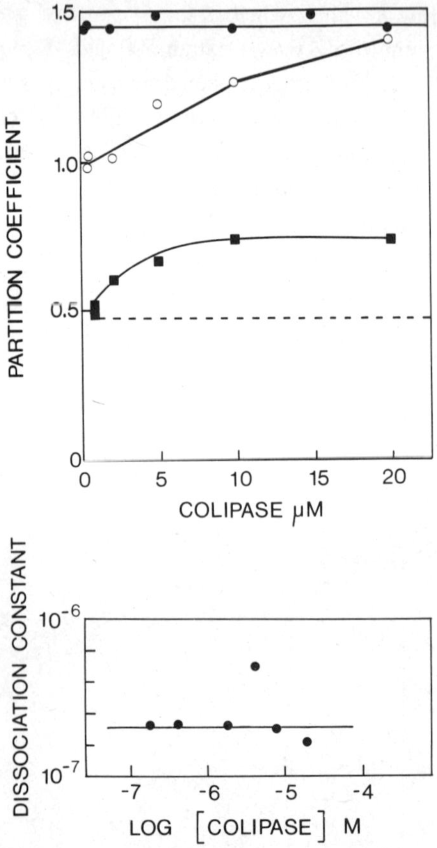

FIGURE 10.6. (*Upper*) ■, Partition coefficient of lipase (2.0×10^{-7} *M*) as a function of added colipase. At higher colipase concentrations most of the lipase is in the form of a complex that has a partition coefficient of 0.7; ○, colipase in the presence of 2×10^{-7} *M* lipase; ●, colipase alone; ---, lipase alone. (*Lower*) Plot of dissociation constant as a function of colipase concentration (Patton et al., 1978).

to be rather similar in the two phases (4.4 and 4.8 × 10^{-7} *M* in the top and the bottom phases, respectively). The dissociation constant is constant over a wide range of lipase and colipase concentrations, suggesting that a 1:1 complex is indeed formed (Fig. 10.6). Since the partition coefficient of the lipase–colipase complex (0.7) is the product of the partition coefficient of lipase (0.5) and that of colipase (1.4), it can be concluded that the contact surface between the two proteins is small, as discussed above. In this case the dissociation constants in the two phases, K_t and K_b, are the same within experimental error (4.4 and 4.8 × 10^{-7} *M*).

The lipase–colipase system is very favorable for application of phase partition since its partition coefficients are fairly different.

3. *Cytochrome* c*–Cytochrome Oxidase and Cytochrome P450–Cytochrome* b_5

A phase system of 7% w/w Dextran 500, 4.4% w/w PEG 6000, 0.5% Tween 80, 0.5% digitonin, 0.1% sodium cholate, and potassium phosphate buffer pH 7.4 at 25°C was used to study the interaction between cytochrome *c* and cytochrome oxidase (Petersen, 1978). This phase system contains detergents;

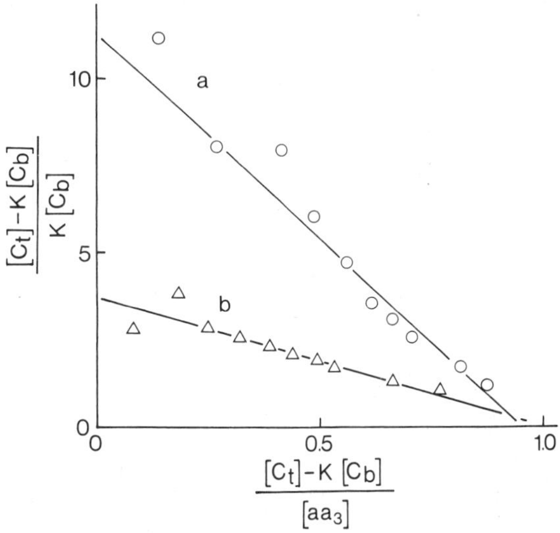

FIGURE 10.7. Binding of cytochrome *c* to cytochrome oxidase. Bound cytochrome divided by free cytochrome *c* is plotted against bound cytochrome *c* divided by the concentration of cytochrome oxidase. (a) Oxidized, (b) reduced cytochrome *c*. The plot demonstrates a 1:1 complex in both cases but a stronger binding for the oxidized form of cytochrome *c* (Petersen, 1978).

Tween 80 partitions into the upper phase. Cytochrome oxidase, an intrinsic membrane protein solubilized by this detergent, also partitions into the upper phase ($K = 20$); whereas, the cytochrome *c* partitions with a K of 0.28 (Petersen, 1978). Figure 10.7 shows a Scatchard plot of the interaction between cytochrome *c* and cytochrome oxidase.

This example is of general interest since it demonstrates interaction between a membrane protein and a soluble protein. Since membrane proteins solubilized by detergents such as Tween and Triton give high K values (Chapter 5), this system should be applicable to other membrane proteins, too.

Interaction of purified microsomal cytochrome P450 with cytochrome b_5 was studied using a dextran–PEG phase system (Chiang, 1981). Purified cytochrome P450 was almost exclusively partitioned to the bottom phase ($K = 0.06$), whereas cytochrome b_5 preferred the top phase ($K = 2.5$), possibly due to bound detergents. When P450 and b_5 were mixed in various molar ratios the apparent partition coefficients of both cytochromes varied depending on the amount of the other cytochrome present. When P450 was in large excess the b_5 partitioned into the bottom phase showing that the complex formed preferred this phase. In this case the partition coefficient of the complex would be roughly the product of the partition coefficients of the two enzymes (0.06×2.5). The data were interpretated as formation of a 1:1 complex. The interaction between cytochrome P450 and P450 reductase was also studied in the same phase system. Trypsin treatment, which removes a polypeptide from b_5 or the reductase, prevented the interaction with P450, showing that the removed polypeptide is important for the enzyme–enzyme interaction (Chiang, 1981).

4. Dissociation of Hemoglobin

Upon dilution, hemoglobin dissociates into half molecules. If hemoglobin is partitioned in the dextran–PEG system at different concentrations of hemoglobin, its partition coefficient changes in the same concentration range as that in which dissociation occurs (see Figure 10.8). By extrapolation of the partition coefficient at high and low concentrations, the partition coefficients of the undissociated hemoglobin (tetramer) and the dimers can be determined (Middaugh and Lawson, 1980). From such data the association constant can be calculated. In order to obtain the association constant in water, experiments were carried out with systems of different polymer concentrations. By extrapolation of the association constants so obtained for the upper and the lower phase at different polymer concentrations to zero polymer concentration, the association constant in water was obtained (Fig. 10.9).

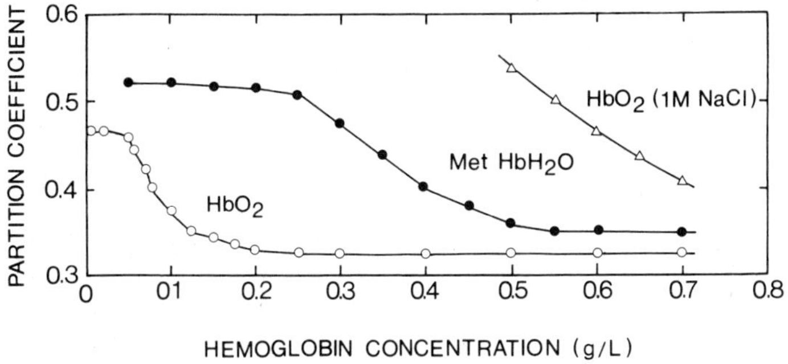

FIGURE 10.8. Effect of protein concentration on the partition coefficients of human oxyhemoglobin and methemoglobin (and of oxyhemoglobin in 0.1 *M* NaCl) in a two-phase system of the following composition: 8% Dextran 40, 4% PEG 6000, 0.02 *M* potassium phosphate, 0.1 *M* potassium sulfate, 0.1 *M* NaCl, pH 7.2, 23°C (Middaugh and Lawson, 1980).

The partition coefficient of the dimer of oxyhemoglobin is 0.47 (Fig. 10.8), whereas that of the tetramer is 0.33, deviating from 0.22, which is the square of the partition coefficient of the dimer. As expected from Eq. (17), the dissociation constants in the two phases also differ (Fig. 10.9). This may be due to the relatively large contact surface between two dimers in the hemoglobin molecule (Fig. 10.4). (Methemoglobin, however, displays rather similar dissociation constants in the upper and the lower phase for a given phase system.)

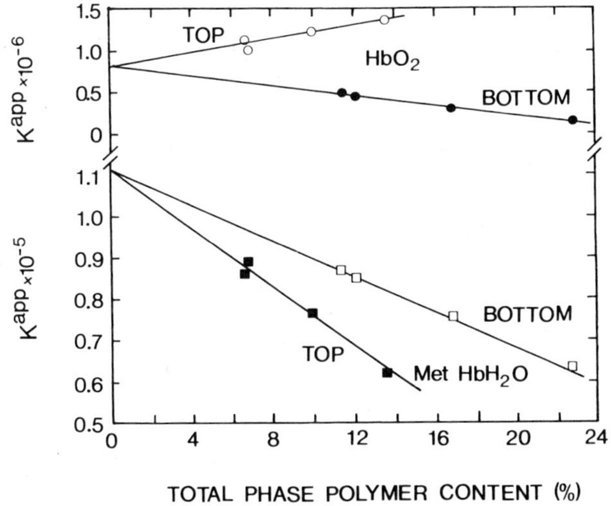

FIGURE 10.9. Extrapolation of the tetramer–dimer association constants of oxyhemoglobin and methemoglobin to zero polymer concentrations (Middaugh and Lawson, 1980).

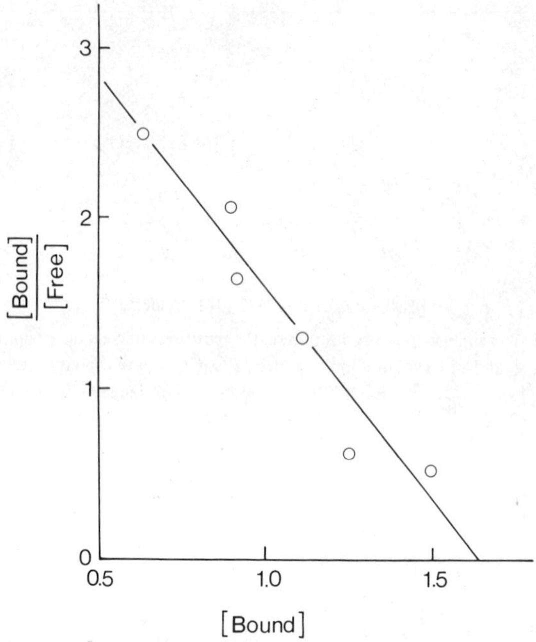

FIGURE 10.10. Scatchard plot of the binding of $tRNA^{-Leu}$ to leucyl–tRNA synthetase (concentration of synthetase 2.1 μ*M*) (Hustedt et al., 1977).

5. *RNA–Leucyl-tRNA Synthetase*

Hustedt and Kula (1977) made a systematic study of the partition of $tRNA^{Leu}$ and leucyl-tRNA synthetase in different dextran–PEG systems. By varying the molecular weight of the PEG and the ionic composition, they constructed a phase system in which the synthetase is concentrated mainly in the upper phase ($K = >10$), whereas the tRNA has a K near 1. With this system they could study the binding between the tRNA and the enzyme. Figure 10.10 shows a Scatchard plot for the binding of $tRNA^{Leu}$ to leucyl–tRNA synthetase. It gives a K_t of $2.6 \times 10^{-6} M^{-1}$, and the intercept at the abscissa demonstrates a 1:1 complex.

6. *Protein–Lipid Droplets*

The binding of colipase (a protein cofactor for lipase) to emulsified triglycerides was studied in a dextran–PEG phase system; 7% Dextran 500, 4.4% PEG 6000, 150 m*M* NaCl, 5 m*M* Tris-malate (pH 7.4), and 4 m*M* bile salts

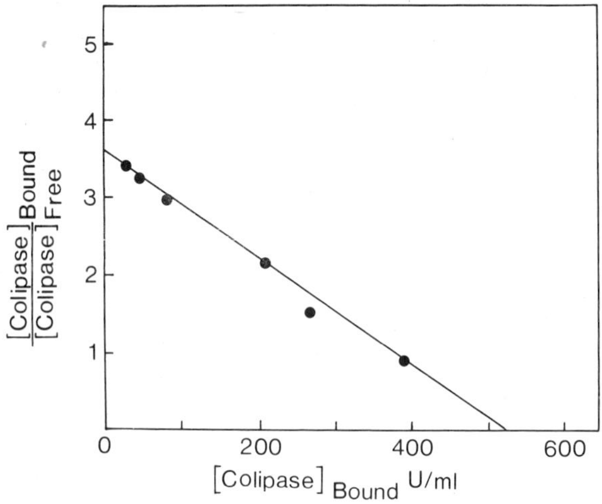

FIGURE 10.11. Scatchard plot of the binding of colipase to a suspension of lipid droplets (Erlanson-Albertsson, 1980a).

(Erlanson-Albertsson, 1980a). The triglycerides partition quantitatively into the lower phase, whereas colipase when partitioned alone has a partition coefficient of 2. The concentration of colipase was measured in the upper phase as a function of added colipase, with a constant amount of triglyceride in the phase system. The procedure described under B above was used, and the data were plotted according to Scatchard. In this way the dissociation constant of the binding of colipase to the triglyceride droplets was determined, as was the number of binding sites available at the lipid surface. A Scatchard plot of the binding of colipase to tributyrin is shown in Figure 10.11.

This example is of particular interest because it involves binding between a protein and a surface and suggests that interactions between protein and membrane surfaces could be studied by the phase-partition technique.

7. *Nucleic Acid–Nucleic Acid*

Single-stranded nucleic acids partition quite differently from double-stranded nucleic acids. Partition can therefore be used to separate the two forms and also to study association–dissociation of nucleic acids (Albertsson, 1965; Mak et al., 1976). Figure 10.12 shows partition of the single-stranded polyribonucleotides polycytidylic acid (poly C) and polyinosinic acid (poly I) when

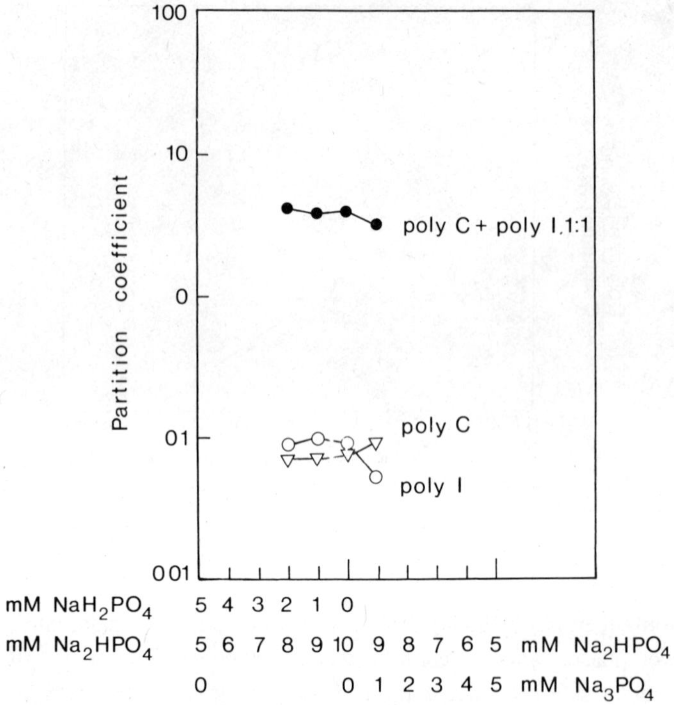

FIGURE 10.12. Partition of polycytidylic acid and polyinosinic acid separately and mixed, when they form a complex (Albertsson, 1965).

they are partitioned separately and together. When mixed, these two single-stranded molecules form a 1:1 double-stranded complex. As seen in Figure 10.12, the complex has a much higher partition coefficient than either of the two single-stranded forms.

8. *Analytical Applications*

When a specific binding molecule, such as an antibody, a lectin, or a specific binding protein, is included in polymer two-phase systems, it can be used for competitive-binding assays for routine analysis. For example, steroids can be determined using a steroid-binding protein in one of the phases (Södergård et al., 1982). Lectins can be used for the assay of sugars (Mattiasson and Ling, 1980), and antibodies for the assay of digoxin (Mattiasson, 1980) and bacteria (Mattiasson, et al., 1982).

COUNTERCURRENT DISTRIBUTION

The theory and the examples given above involve partition in a single tube. The separation of two compounds can be improved considerably by repeating the partition procedure several times, for example, by countercurrent distribution. This can also be used for studying interactions between molecules (Backman,

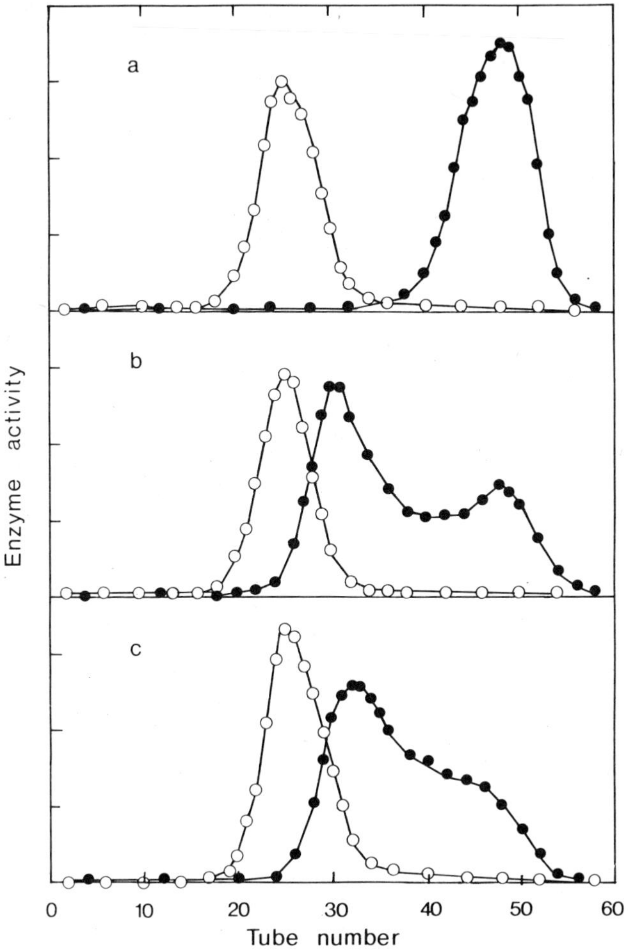

FIGURE 10.13. Demonstration of interaction between two enzymes, aspartate aminotransferase (○) and malate dehydrogenase (●), by countercurrent distribution. The two enzymes were first run separately (*a*) and then together, (*b*) and (*c*), at a ratios of aminotransferase to malate dehydrogenase of 12 and 24, respectively (Backman and Johansson, 1976).

TABLE 10.1 Applications of Interactions between Biomolecules Studied by Phase Partition

Interacting Molecules	Reference
Macromolecule–Small Ligand	
Aspartate transcarbamylase–cytidine triphosphate	Gray and Chamberlain (1971)
Formyltetrahydrofolate synthetase–ATP, ADP	Curthoys and Rabinowitz (1971)
Steroid-binding protein–5α-dihydrotestosterone	Shanbhag et al. (1973)
Phe-tRNA synthetase–phenylalanine	Fasiolo et al. (1974)
Arg-tRNA synthetase–arginine, ATP	Parfait and Grosjean (1972)
Ribosomes–antibiotics	Pestka et al. (1976); LeGoffic et al. (1980)
Ileu-tRNA synthetase–isoleucine	Hustedt et al. (1977)
DNA–methyl green	Nordén et al. (1978)
Lectin–sugar	Mattiasson and Ling (1980)
DNA–glucocorticoids steroids	Akhrem et al. (1978)
DNA–estradiol	Alberga et al. (1976)
Serum albumin–ritamycin	Assandri and Moro (1977)
Serum albumin–tryptophan	Backman (1980)
Serum albumin–cibachrome blue	Ling et al. (1982)
Plasma protein–testosterone, estradiol	Södergård et al. (1982)
Serum albumin–triiodothyronine	Mattiasson and Eriksson (1982)
Plasma protein–ioglycamide, steroids, metrizamide	Wirell (1978, 1982a, b); Wirell et al. (1982)
Protein–Protein	
Hemocyanin subunit–hemocyanin subunit	Albertsson (1971)
Hemoglobin–lysozyme	Hartman et al. (1974)
Aspartate transaminase–malate dehydrogenase	Backman and Johansson (1976)
Lactoferrin–carboxylesterhydrolase	Erlanson-Albertsson et al. (1985)
Lipase–colipase	Patton et al. (1978); Sternby and Erlanson-Albertsson (1982)
Cytochrome *c*–cytochrome oxidase	Petersen (1978)
Cytochrome P 450–cytochrome b_5	Chiang (1981)
Cytochrome P 450–cytochrome P 450 reductase	Chiang (1981)
Retinol-binding protein–prealbumin	Fex et al. (1979)
Albumin–prealbumin	Birkenmeier et al. (1984)
Calmodulin–spectrin	Berglund et al. (1984)

TABLE 10.1 (*Continued*)

Interacting Molecules	Reference
Hemoglobin–carboanhydrase	Silverman et al. (1979); Backman (1981)
Cytochrome *c*–phosvitin	Petersen and Cox (1980)
Hemoglobin dimer–hemoglobin dimer	Walter and Sasakawa (1971); Middaugh and Lawson (1980)
Glytolytic enzymes–actin	Westrin and Backman (1983)
Thrombin–antithrombin	Petersen and Jörgensen (1983)
Protein–Nucleic Acid	
Aminoacyl–tRNA synthetase–tRNA	Hustedt et al. (1977)
DNA–Qb RNA polymerase	Silverman (1973)
DNA–rhofactor	Goldberg and Hurwitz (1972)
Nucleic Acid–Nucleic Acid	
Poly A–Poly U	Albertsson (1965); Mak et al. (1976)
Poly A–Poly I	Albertsson (1965); Mak et al. (1976)
DNA–RNA	Mak et al. (1976)
DNA–glucocorticoid receptors	Andreasen et al. (1981)
Antigen–Antibody	
Phycoerythrin–antibody	Albertsson and Philipson (1960)
Serum albumin–antibody	Albertsson and Philipson (1960)
Polio virus–antibody	Philipson et al. (1966); Shanbhag et al. (1973)
b_2-Microglobulin–antibody	Ling and Mattiasson (1983)
Peptide hormone–antibody	Desbuquois and Aurbach (1972)
Digoxin–antibody	Mattiasson (1980)
Somatomammotropin–antibody	Urios et al. (1982)
Bacteria–antibody	Ling et al. (1982); Mattiasson et al. (1981); Miörner et al. (1980); Stendahl et al. (1974, 1977)
Protein–Lipid	
Colipase–lipid droplets	Erlanson-Albertsson (1980a)
Colipase–lecithin	Erlanson-Albertsson (1980b)
Protein–Polysaccharide	
Protein–heparin	Petersen and Cox (1980)
Cell–Cell	
Leucocytes–liposomes	Dahlgren et al. (1977)
Erythrocytes–lymphocytes	Walter et al. (1978)

1980, 1981, 1982; Backman and Johansson, 1976; Backman and Shanbhag, 1979). By comparing the countercurrent distribution diagrams of two molecules when they are run separately or together, one can draw conclusions about their interaction. The advantage of this technique is that countercurrent distribution allows a very accurate determination of shifts in partition and should be very sensitive. The analysis of the countercurrent diagrams to get quantitative data involves complicated calculations which can be carried out conveniently by modern computers. For a detailed account of the use of countercurrent distribution with aqueous polymer two-phase systems, the reader is referred to the references given above.

Figure 10.13 shows a good example of how countercurrent distribution can be used to study interaction between two enzymes: malate dehydrogenase and aspartate transaminase. These two enzymes catalyze two consecutive metabolic steps. Each enzyme has one cytoplasmic and one mitochondrial isoenzymic form. Using countercurrent distribution, Backman and Johansson (1976) demonstrated a physical interaction between the cytoplasmic forms of malate dehydrogenase and aspartate transaminase and also between the mitochondrial forms of the two enzymes. However, no interaction between the heterotopical enzymes was found, that is, between cytoplasmic malate dehydrogenase and mitochondrial aspartate transaminase or between mitochondrial malate dehydrogenase and cytoplasmic aspartate transaminase.

Thus each enzyme seems to recognize its appropriate neighbor enzyme. Therefore, in addition to catalytic and regulatory sites, enzymes also must have recognition or social sites that interact with neighboring enzymes *in vivo*. Countercurrent distribution has also been used to demonstrate interaction between hemoglobin and carbonic anhydrase (Backman, 1981; Silverman et al., 1979).

GENERAL COMMENTS

Both phases in the phase systems described here are aqueous. They are mild to biological material and can dissolve many proteins and other biopolymers. Partition in these systems can therefore be used for studying interaction between several molecules of biochemical interest. Since equilibration is reached rapidly—it is a matter of only seconds during shaking of the phase system—the method is fast and advantageous for studying interaction between labile components. It should be stressed that the dissociation constants obtained by this method using the milieus of the phases are not necessarily the same as the constants obtained with water as solvent. In those cases in which we can compare the dissociation constant obtained by this method with the constant obtained by other methods, there does not seem to be much difference, indicating that the polymers do not influence the interaction to a high degree. But this should depend greatly on the interacting system. One can imagine cases in which the polymers would either inhibit or promote interaction. For example, if one of the polymers binds strongly to A or B, it might interfere with the binding between A and B, or if A or B has a very low solubility in the phases, but not in the complex AB, its formation will be promoted.

A particular advantage of the partition technique is that it can be applied to interactions between two macromolecules, such as protein–protein or protein–nucleic acid interactions, and also between a macromolecule and a particle surface, as exemplified by the colipase–lipid droplet interactions. It should be possible to use the technique in studying binding of proteins to membrane surfaces of, for example, cell organelles and membrane vesicles. Also, phase partition could be used to detect and quantify interactions between cells, cell organelles, and membrane vesicles. In fact, interaction between liposomes and leukocytes (Dahlgren et al., 1977) and between erythrocytes and lymphocytes (Walter et al., 1978) have been studied by partition and countercurrent distribution.

The number of interactions studied so far is too small to allow a comparison between the experimental values and the examples of different types of interaction shown in Figure 10.4. For the interaction of both lipase–colipase and prealbumin–retinol-binding protein, the complex has a partition coefficient that is a product of the partition coefficients of the two interacting proteins. Therefore, these would be examples of the type of complex shown at the top of Figure 10.4, that is, a relatively small contact surface.

An example of the type of interaction shown in the middle of Figure 10.4 is given by hemoglobin, where A and B are equal and represent the dimer that associates to the tetramer. An extreme example of this type of interaction is the formation of a double-stranded nucleic acid from two single strands (Fig. 10.12). In this case the surface properties of the complex and those of the reactants are completely different. The partition coefficient of the single-stranded nucleic acids is 0.1 or lower, whereas that of the double-stranded nucleic acid is 10 or higher, that is, the partition coefficient of the complex is very different from the product of the partition coefficients of the two reactants.

When a membrane protein is solubilized by a detergent, it is more or less included in the micelles of the detergent. Its partition is very similar to that of the micelle, and this case could be an illustration of the type of interaction shown at the bottom of Figure 10.4. This type of interaction also can be exemplified by the binding of small ligand, for example, a cofactor or an inhibitor of an enzyme, to a protein when the ligand is buried in a deep crevice of the protein.

How to Find a System Suitable for Interaction Studies

It is important to know how a phase system should be selected in order to increase the possiblity of detecting an interaction and to increase the accuracy of the determination of the dissociation constant. Generally, a larger difference in partition between the interacting species and the complex should facilitate a quantitative study. When K_{AB} is approximately equal to $K_A K_B$, one should not use a phase system in which K_A and K_B are both close to 1, because then K_{AB} will also be close to 1. Rather one should try to find a system in which at least one of the interacting species has a K different from 1. For example, if K_A is 1 and K_B is 0.1, then K_{AB} will be 0.1, and there will be a large difference in partition between free A, and A bound to B. By trying different methods to adjust the partition of biomolecules, as outlined in Chapter 4, one can increase the accuracy of the method considerably. Thus the following factors can be employed: ionic composition, pH, molecular weight of phase polymer, charged groups or ligands coupled to the polymers. I particularly want to point to the possibility of using different molecular weight polymers to adjust the partition. By using a low-molecular-weight fraction of dextran together with PEG, one gets a phase system in which large proteins and nucleic acids have very low partition coefficients. These systems can generally be used for interaction studies between a small molecule ($K = 1$) and a large molecule ($K = 0.1$). Such phase systems should also be of use for studies of interactions between different proteins, since the difference between the partition of the free protein and that of the complexes would be very large. Consider, for example, the case when $K_A = 0.1$ and $K_B = 0.01$. Then K_{AB} is 0.001, and a measurement of A in the upper phase of such a phase system with and without B is a sensitive way of measuring binding between A and B. Phase systems can also be selected after simplex optimization (Backman and Shanbhag, 1984).

REFERENCES

Akhrem, A.A., Barai, V.N., Zinchenko, A.I., Martsev, S.P., and Chaschin, V.L. (1978). *Biochemistry (USSR)*, **43,** 933–938.

Alberga, A., Ferrez, M., and Baulieu, E.-E. (1976). *Febs. Lett.*, **61,** 223–226.

Albertsson, P.-Å. (1965). *Biochim. Biophys. Acta,* **103,** 1–12.

Albertsson, P.-Å. (1971). *Partition of Cell Particles* and *Macromolecules,* 2nd ed., Almqvist and Wiksell, Stockholm, and Wiley, New York.

Albertsson, P.-Å. (1978). *J. Chromat.,* **159,** 111–122.

Albertsson, P.-Å. (1983). *Meth. Biochem. Anal.,* **29,** 1–24.

Albertsson, P.-Å. and Philipson, L. (1960). *Nature,* **185,** 38–40.

Andreasen, P.A. and Gehring, U. (1981). *Eur. J. Biochem.,* **120,** 443–449.

Assandri, A., and Moro, L. (1977). *J. Chromat.,* **135,** 37–48.

Backman, L. (1980). *J. Chromat.,* **196,** 207–216.

Backman, L. (1981). *Eur. J. Biochem.,* **120,** 257–261.

Backman, L. (1982). *J. Chromat.,* **237,** 185–198.

Backman, L., and Johansson, G. (1976). *FEBS Lett.,* **65,** 39–43.

Backman, L., and Shanbhag, V.P. (1979). *J. Chromat.,* **171,** 1–13.

Backman, L., and Shanbhag, V.P. (1984). *Anal. Biochem.,* **138,** 372–379.

Berglund, Å., Backman, L., and Shanbhag, V.P. (1984). *FEBS Lett.,* **172,** 109–113.

Birkenmeier, G., Tschechonien, B., and Kopperschläger, G. (1984). *FEBS Lett.,* **174,** 162–166.

Chiang, J.Y.L. (1981). *Arch. Biochem. Biophys.,* **211,** 662–673.

Curthoys, N.P., and Rabinowitz, J.C. (1971). *J. Biol. Chem.,* **246,** 6942–6952.

Dahlgren, C., Kihlström, E., Magnusson, K.-E., Stendahl, O., and Tagesson, C. (1977). *Exptl. Cell. Res.,* **108,** 175–184.

Desbuquois, B., and Aurbach, G.D. (1972). *Biochem. J.,* **126,** 717–726.

Erlanson-Albertsson, C. (1980a). *Biochim. Biophys. Acta,* **617,** 371–382.

Erlanson-Albertsson, C. (1980b). *FEBS Lett.,* **117,** 295–298.

Erlanson-Albertsson, C., Sternby, B., and Johannesson, (1985). *Biochim. Biophys. Acta* **829,** 282–287.

Fasiolo, F., Remy, P., Pouyet, J., and Ebel, J.-P. (1974). *Eur. J. Biochem.,* **50,** 227–236.

Fex, G., Albertsson, P.Å., and Hansson, B. (1979). *Eur. J. Biochem.,* **99,** 353–360.

Goldberg, A.R., and Hurwitz, J. (1972). *J. Biol. Chem.,* **247,** 5637–5645.

Gray, C.W., and Chamberlain, M.J. (1971). *Anal. Biochem.,* **41,** 83–104.

Hartman, A., Johansson, G., and Albertsson, P.-Å. (1974). *Eur. J. Biochem.,* **46,** 75–81.

Hustedt, H., and Kula, M.-R. (1977). *Eur. J. Biochem.,* **74,** 191–198.

Hustedt, H., Flossdorf, J., and Kula, M.-R. (1977). *Eur. J. Biochem.,* **74,** 199–202.

Kegeles, G. (1973). *Methods Enzymol.,* **27,** 456–464.

LeGoffic, F., Moreau, N. Langrene, S., and Pasquier, A. (1980). *Anal. Biochem.,* **107,** 417–423.

Ling, T.G.I., and Mattiasson, B. (1982). *J. Chromat.,* **252,** 159–166.

Ling, T.G.I., and Mattiasson, B. (1983). *J. Immunol. Meth.,* **59,** 327–337.

Ling, T.G.I., Ramstorp, M., and Mattiasson, B. (1982). *Anal. Biochem.,* **122,** 26–32.

Mak, S., Öberg, B., Johansson, K., and Philipson, L. (1976). *Biochemistry,* **15,** 5754–5761.

Mattiasson, B. (1980). *J. Immunol. Meth.,* **35,** 137–146.

Mattiasson, B. and Eriksson, H. (1982). *Clin. Chem.,* **28,** 680–683.

Mattiasson, B., and Ling, T.G.I. (1980). *J. Immunol. Meth.,* **38,** 217–223.

Mattiasson, B., Ling, T.G.I., and Ramstorp, M. (1981). *J. Immunol. Meth.,* **41,** 105–114.

Mattiasson, B., Ramstorp, M., and Ling, T.G.I. (1982). *Adv. Appl. Microbiol.*, **28,** 117–147.

Mattiasson, B., Eriksson, H., and Nilsson, J. (1983). *Clin. Chim. Acta,* **127,** 301–304.

Middaugh, C.R., and Lawson, E.Q. (1980). *Anal. Biochem.*, **105,** 364–368.

Miörner, H., Myhre, E., Björk, L., and Kronvall, G. (1980). *Infect. Immunity,* **29,** 879–885.

Nordén, B., Tjerneld, F., and Palm, E. (1978). *Biophys. Chem.*, **8,** 1–15.

Parfait, R., and Grosjean, H. (1972). *Eur. J. Biochem.*, **30,** 242–249.

Patton, J.S., Albertsson, P.-Å., Erlanson, C., and Borgström, B. (1978). *J. Biol. Chem.*, **253,** 4195–4202.

Pestka, S., Weiss, D., and Vince, R. (1976). *Anal. Biochem.*, **71,** 137–142.

Petersen, L.C. (1978). *FEBS Lett.*, **94,** 105–108.

Petersen, L.C., and Cox, P. (1980). *Biochem. J.*, **192,** 687–693.

Petersen, L.C., and Jörgensen, M. (1983). *Biochem. J.*, **211,** 91–97.

Philipson, L. (1966). *Virology,* **28,** 35–46.

Philipson, L., Killander, J., and Albertsson, P.-Å. (1966). *Virology,* **28,** 22–34.

Shanbhag, V.P., Södergård, R., Carstensen, H., and Albertsson, P.-Å. (1973). *J. Steroid. Biochem.*, **4,** 537–550.

Silverman, P.M. (1973). *Arch. Biochem. Biophys.*, **157,** 234–242.

Silverman, D.N., Backman, L., and Tu, C. (1979). *J. Biol. Chem.*, **254,** 2588–2591.

Södergård, R., Bäckström, T., Shanbhag, V., and Carstensen, H. (1982). *J. Steroid. Biochem.*, **16,** 801–810.

Stendahl, O., Tagesson, C., and Edebo, L. (1974). *Infect. Immunity,* **10,** 316–319.

Stendahl, O., Tagesson, K.-E., and Edebo, L. (1977). *Immunology,* **32,** 11–18.

Sternby, B., and Erlanson-Albertsson, C. (1982). *Biochim. Biophys. Acta,* **711,** 193–195.

Urios, P., Rajkowski, K.M., Engler, R., and Cittanova, N. (1982). *Anal. Biochem.*, **119,** 253–260.

Walter, H., and Sasakawa, S. (1971). *Biochemistry,* **10,** 108–113.

Walter, H., Krob, E.J., and Moncla, B.J. (1978). *Exptl. Cell Res.*, **115,** 379–385.

Westrin, H., and Backman, L. (1983). *Eur. J. Biochem.*, **136,** 407–411.

Winlund, C.C., and Chamberlain, M.J. (1970). *Biochem. Biophys. Res. Commun.*, **40,** 43.

Wirell, S. (1978). *Acta Radiol. Diagn.*, **19,** 289–296.

Wirell, S. (1982a). *Acta Radiol. Diagn.*, **23,** 497–502.

Wirell, S. (1982b). *Acta Radiol. Diagn.*, **23,** 239–243.

Wirell, S., Södergård, R., and Selstam, G. (1982). *Acta Radiol. Diagn.*, **23,** 567–572.

11 PARTITION IN 3- AND 4-PHASE SYSTEMS

When three different polymers are mixed above certain concentrations a three-phase system may be obtained. If four polymers are used a four-phase system may be formed, and so forth. Aqueous phase systems with as many as 18 liquid phases have been obtained (Chapter 2, Fig. 2.2). Furthermore, one may add a solid phase to a liquid phase system and thereby increase the number of phases.

Suppose a phase system contains n phases. Soluble substances, such as proteins, will partition between the bulk phases, and one partition step will result in n fractions. Particles, such as cell organelles, will distribute between the bulk phases and the different interfaces. For a phase system containing n phases there are $n - 1$ interfaces and the material can thus distribute among $2n - 1$ layers. Polyphase systems have the advantage that a mixture can almost instantly be resolved into several fractions by one single partition step.

Dextran–Ficoll–PEG–water forms a three-phase system above certain concentrations. (See the phase diagram, Chapter 12, Figure 12.24*c*.) From bottom to top the three phases are enriched in dextran, Ficoll, and PEG, respectively. When the polymers are in the proportion 1 : 1 : 1 by weight, the system consists of a single phase at 22°C if the concentration of each polymer is less than 4.1% w/w. Three liquid phases are obtained if the concentration of each polymer exceeds 6.9% w/w. Between these concentration limits two-phase systems are obtained. If PEG is replaced by charged PEG, such as trimethylamino PEG (TMA-PEG) higher concentrations are needed for the formation of three phases (Hartman *et al*, 1974). The polymer composition of the three phases of a system which includes TMA-PEG and which has been used for partition studies is shown in Table 11.1. PEG predominates in the top phase which contains 17% PEG; the

TABLE 11.1 Composition of the Phases in a Three-Phase System[a]

Phase	Dextran, % w/w	Ficoll, % w/w	PEG + TMA-PEG, % w/w	Water, % w/w
Top	0	4	17	79
Middle	2	22	2	74
Bottom	26	3	0	71

[a] 8% w/w Dextran 500, 8% w/w Ficoll 400, 4% w/w PEG 6000, 4% w/w trimethylamino polyethylene glycol (TMA-PEG 6000), and 5 m*M*/Kg K_3PO_4 buffer, pH 6.8 at 22°C (Hartman *et al*, 1974).

middle phase contains 2% and the bottom phase less than 0.5%. TMA-PEG distributes roughly as does PEG. This unequal distribution of TMA-PEG gives rise to electrical potentials, both between the top and the middle, and between the middle and the bottom phase. Both potentials act in the same direction; the top phase being more positive than the middle phase, which in turn is more positive than the bottom phase. The potential between the top phase and the bottom phase is the sum of the two other potentials.

Partition of a protein in this system can be described by either of the three different partition coefficients

$$K_{\mathrm{M/B}} = \frac{\text{concentration in middle phase}}{\text{concentration in bottom phase}} \tag{1}$$

$$K_{\mathrm{T/M}} = \frac{\text{concentration in top phase}}{\text{concentration in middle phase}} \tag{2}$$

$$K_{\mathrm{T/B}} = \frac{\text{concentration in top phase}}{\text{concentration in bottom phase}} \tag{3}$$

or by the concentration in each phase, or by the percentage distribution between the phases.

Each of the partition coefficients depends on the same factors as described for two-phase systems, that is, type and molecular weight of polymer, ionic composition, charge, and hydrophobic properties of the protein, and so forth. By using different ions or charged PEG and ligand-PEG, electrochemical, hydrophobic, and biospecific affinity affects can be introduced.

A theoretical treatment of partition in a three-phase system has been published

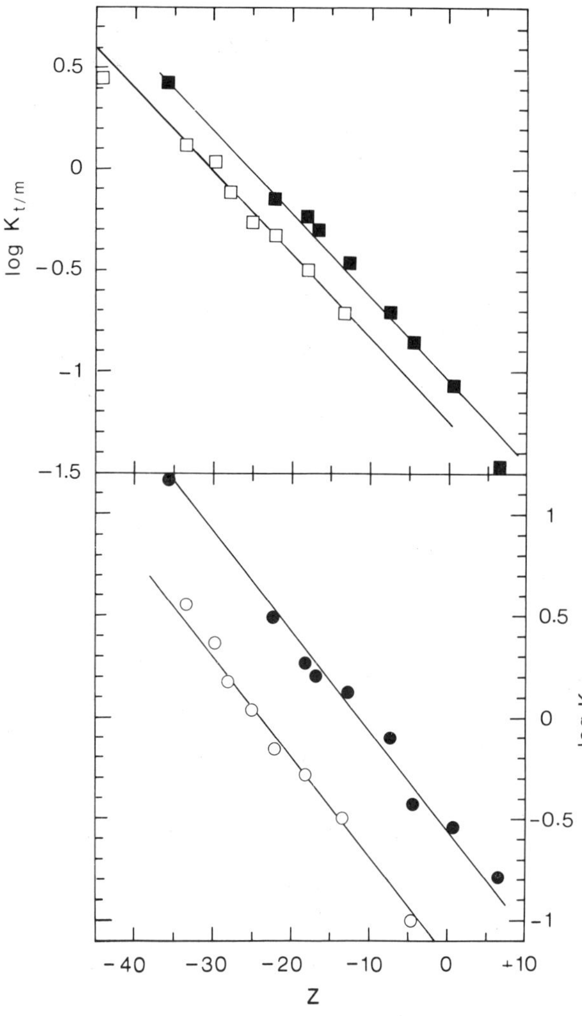

FIGURE 11.1 Two CO-hemoglobins (filled symbols, human; open symbols, porcine) were partitioned in a three-phase system [dextran–Ficoll–(PEG + TMA-PEG)] at different pH. Log partition coefficient for top–middle or middle–bottom phases was plotted against net charge of protein calculated from data in Figure 11.3. Compare Eqs. (4) and (5) (Johansson and Hartman, 1976).

(Johansson and Hartman, 1976). For each partition coefficient Eq. (3) of Chapter 4 should hold, that is,

$$\ln K_{M/B} = \ln K^{0}{}_{M/B} + \frac{ZF}{RT}(U_B - U_M) \tag{4}$$

$$\ln K_{T/M} = \ln K^{0}{}_{T/M} + \frac{ZF}{RT}(U_M - U_T) \tag{5}$$

$$\ln K_{T/B} = \ln K^{0}{}_{T/B} + \frac{ZF}{RT}(U_B - U_T) \tag{6}$$

Equations (4) and (5) have been tested with CO-hemoglobin (Fig. 11.1). Excellent agreement between the experiment results and the equations was obtained.

THE EXTRACTION PROFILE

The extraction profile of a single protein is obtained if the fractions of the protein present in the top, middle, and bottom phase are plotted against a parameter which is used to change the partition, for example, pH when TMA-PEG is included in the phase system. TMA-PEG makes the top phase more positive than the middle phase, which in turn is more positive than the lower phase. Consequently, when a protein is partitioned in a three-phase system containing TMA-PEG, a protein is usually found in the bottom phase at low pH values. With increasing pH, the affinity of the protein for the other two phases is successively enhanced, first for the middle phase, and then for the top phase.

An extraction profile, such as is shown in Figure 11.2*a*, is obtained. The fraction in the bottom phase will fall while that of the top phase will rise. The middle-phase curve, however, goes through a maximum (pH 5). The profile of the three curves and the position of the maximum of the middle curve depends on the K^0 values in the three phases [Eqs. (4)–(6)], the net charge of the protein, and the potential differences between the phases. Another protein may give a profile such as that shown in Figure 11.2*b*. Because of lower $K^{0}{}_{T/M}$ and $K^{0}{}_{M/B}$ values and/or different charge properties this protein is extracted at much higher pH, with a middle-phase maximum at pH 9.

If the two proteins in equal proportions are partitioned together in the same phase system, an extraction profile shown in Figure 11.2*c* is obtained. The top-phase and the bottom-phase curves (when a nondiscriminating assay is used) have plateaus, while there are two maxima in the middle-phase curve.

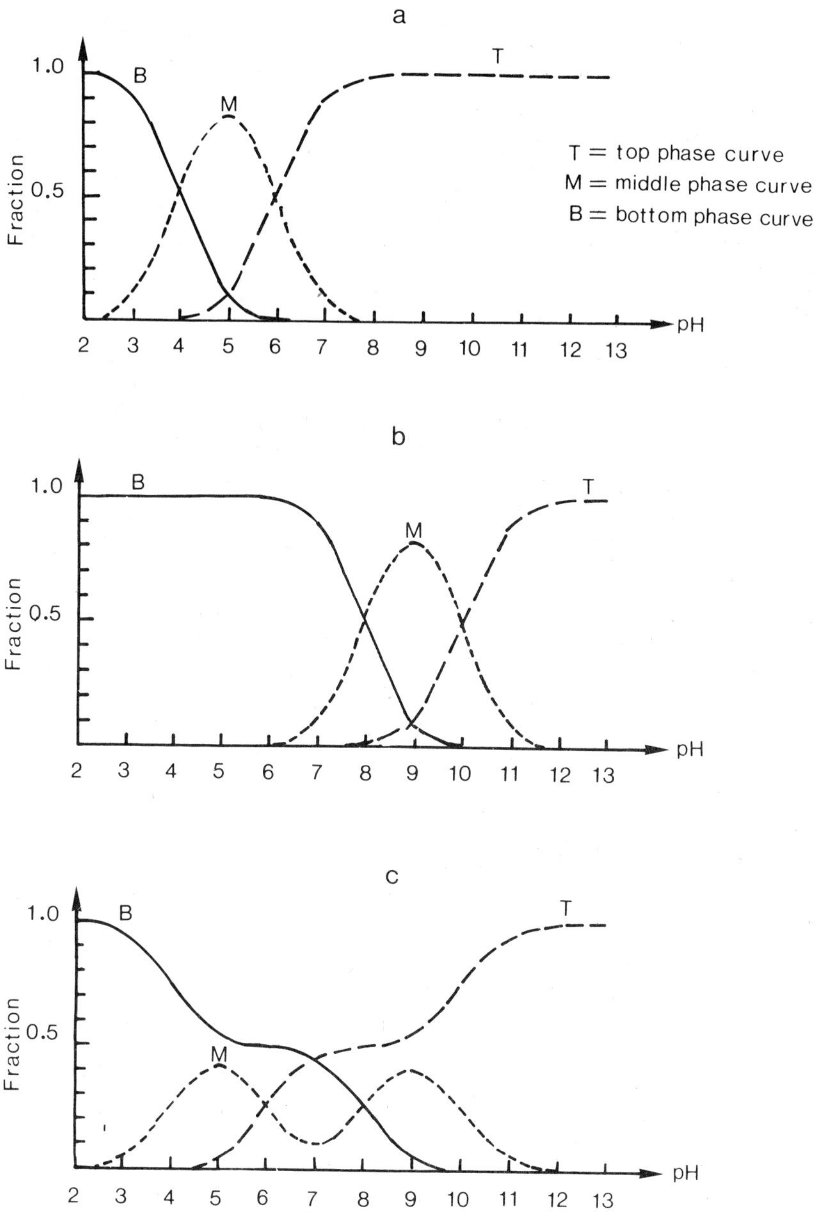

FIGURE 11.2 Extraction profiles for proteins in a three-phase system: (*a*) protein a alone, (*b*) protein b alone, (*c*) mixture of a and b (Hartman, 1976b).

In the same way a mixture of three components can display three maxima in the middle-phase curve, a mixture of four components displays four maxima and so forth.

PROTEINS

Figure 11.3 shows partition of two CO-hemoglobins in a three-phase system. The diagram shows the percentage distribution between the three phases of each protein at different pH. The three-phase system used consists of dextran–Ficoll–PEG and the positively charged trimethylamino-PEG.

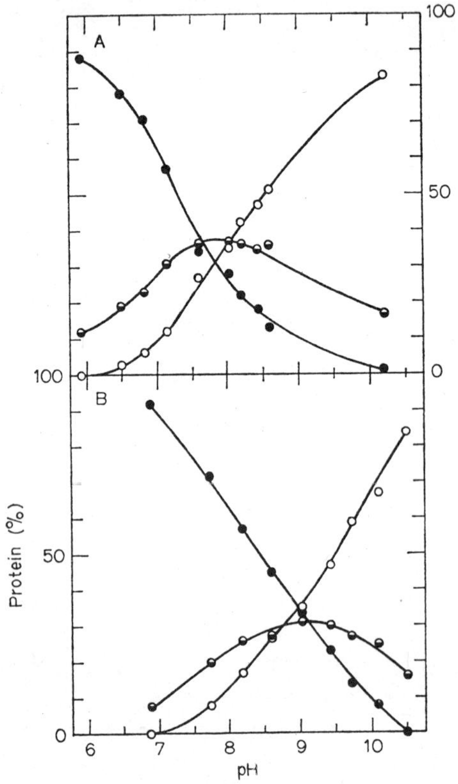

FIGURE 11.3 Partition of human CO-hemoglobin (A) and porcine CO-hemoglobin (B) in a three-phase system; ○, top phase; ◒, middle phase; ●, bottom phase. Composition: 8% w/w Dextran 500, 4% Ficoll 400, 4% PEG 6000, 4% TMA-PEG, 5 m*M* potassium phosphate (Hartman et al., 1974).

At its isoelectric point the partition of a protein is not influenced by the interfacial potential. Hemoglobin at its isoelectric point (pH 7) is found mainly in the bottom phase (Fig. 11.3). As pH is increased and the net negative charge of protein increases, more is found first in the middle phase and then in the top phase. At pH below the isoelectric point the tendency to favor the bottom phase is increased. Depending on the electrical potential, the net charge of the protein, and other noncharge factors, different shapes of the partition curves are obtained.

In Figure 11.4 the partition of some glycolytic enzymes is shown. Hexokinase partitions mainly between the middle and the top phase. Pyruvate kinase is extracted in two steps by the top phase when pH is increased. Also, the enzyme is extracted by two steps by the middle phase as shown by the two distinct peaks of the middle phase curve. This indicates the presence of two forms of pyruvate

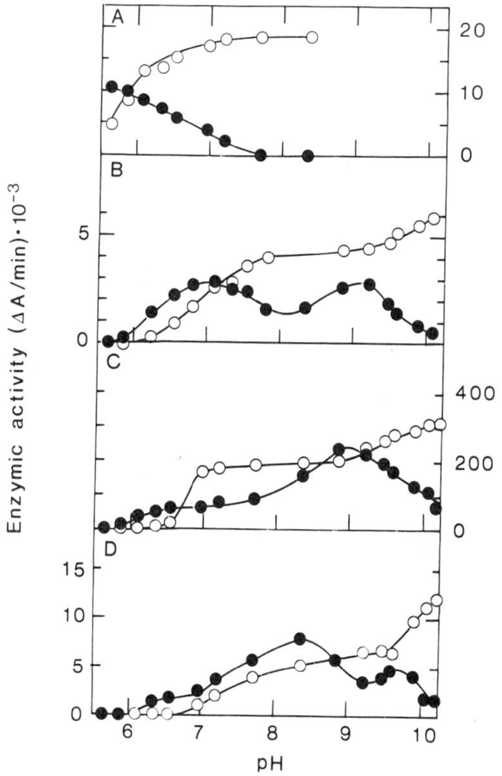

FIGURE 11.4 Partition of glycolytic enzymes from baker's yeast in a three-phase system. (A) hexokinase, (B) pyruvate kinase, (C) 3-phosphoglycerate kinase, (D) enolase. ○, top phase; ●, middle phase. (Hartman et al., 1974).

TABLE 11.2 Distribution of the Three Isoenzymes of Enolase when Lysate of Baker's Yeast was Partitioned in the Three-Phase System of Table 11.1 at pH 8.05[a]

	Enzyme Activity (% of Total Activity in the System) in		
Isoenzyme	Top Phase	Middle Phase	Bottom Phase
I	—	1	44
II	18	19	1
III	11	6	—
Total	29	26	45

[a]The distribution in each phase was determined by isoelectric focusing. Isoenzyme I has I_P = 7.2; isoenzyme II has I_P = 6.7; and isoenzyme III has I_P = 5.7 (Hartman et al., 1974).

kinase. Thus, at pH 8 one form is in the top phase and the other is mainly in the bottom phase. 3-Phosphoglycerate kinase is also extracted in two steps, one between 6.5 and 7 and other between 9 and 10. Enolase, which consists of three isoenzymes, shows a more complex partition pattern. The diagram indicates the presence of three forms with different affinities for the phases. Around pH 8 one form of the enzyme should be concentrated in the bottom phase, another in the middle phase, and a third in the top phase. This was also demonstrated by the experiment shown in Table 11.2; 98% form I was found in the bottom phase, 50% form II in the middle phase, and 65% form III in the top phase. A considerable separation is, therefore, obtained by one partition step.

In the examples given above a charged PEG was used to create a large potential difference between the phases, and change in pH was used to change the net charge of the protein and thereby its partition. However, one also can use hydrophobic or biospecific ligands attached to PEG in order to extract proteins or particles from the lower phase into the middle or the top phase.

PARTICLES

When particles are partitioned in a three-phase system they can adsorb to the two interfaces in addition to being suspended in the three bulk phases. Figures 11.5–11.8 show four examples of behavior of different particles in three-phase systems. In the first (Fig. 11.5), a microorganism *Sarcina lutea* is moved from the bottom phase, via, first the middle phase, and then the interface between the

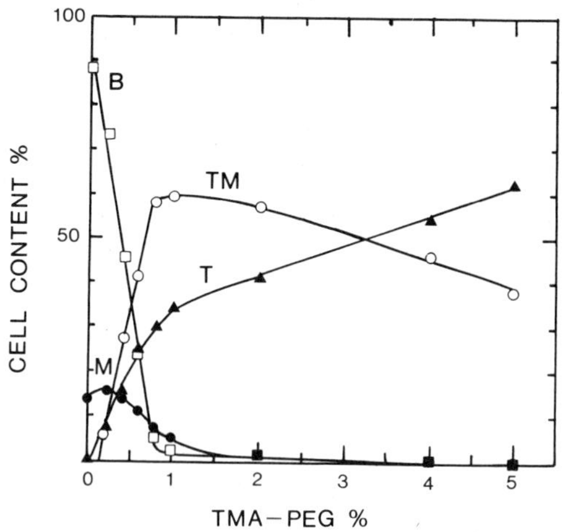

FIGURE 11.5 Partition of *Sarcina lutea* in a three-phase system of 7% w/w Dextran 500, 12% w/w Ficoll, 12% w/w PEG 4000, 5 m*M* potassium phosphate, pH 6.8, with increasing amount of PEG replaced by TMA-PEG (For phase composition, see Table 11.3). B, bottom phase; T, top phase; M, middle phase; TM, interface between top and middle phase (Hartman, 1976a).

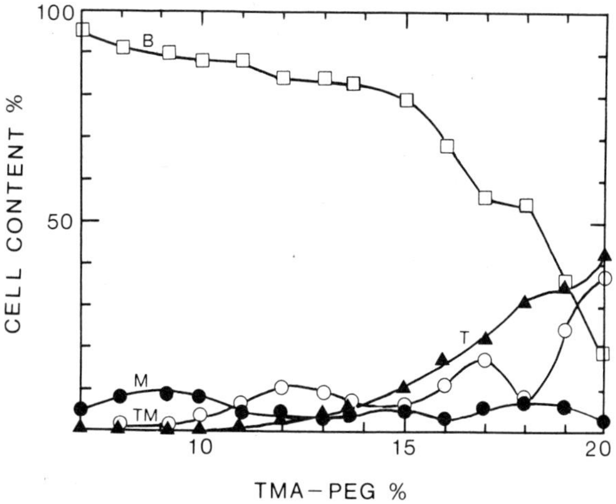

FIGURE 11.6 Partition of baker's yeast cells in a three-phase system with increasing amounts of PEG replaced by TMA-PEG. Phase system composition and symbols as in Figure 11.5. Note the three peaks in the M and TM curves, which indicate the presence of three subpopulations of cells.

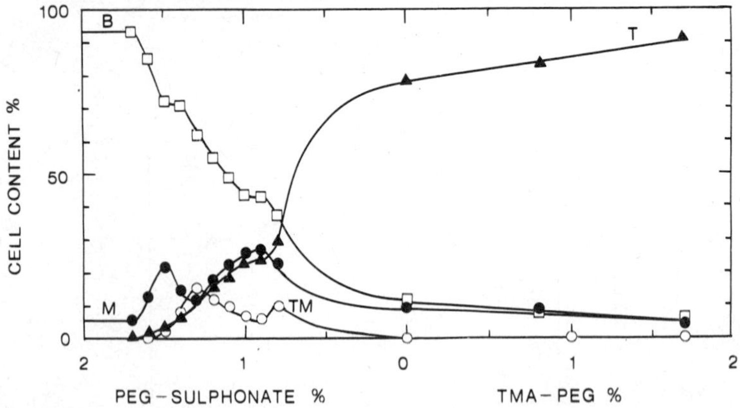

FIGURE 11.7 Partition of *Chlorella pyrenoidosa* in a three-phase system as a function of negatively charged PEG (S-PEG) or positively charged PEG (TMA-PEG). Phase system composition and symbols as in Figure 11.5.

top and middle phase to the top phase upon increasing the concentration of positively charged PEG in the phase system. The second example shows the partition of baker's yeast cells (Fig. 11.6). In this case the middle phase curve and the curve for the interface between the top and the middle phase both have three maxima, indicating that the yeast cells consisted of three subpopulations.

The third example (Fig. 11.7) shows the partition of a unicellular algae *Chlorella pyrenoidosa* in a three-phase system as a function of either a negatively charged PEG which pushes the cells down into the bottom phase, or a positively

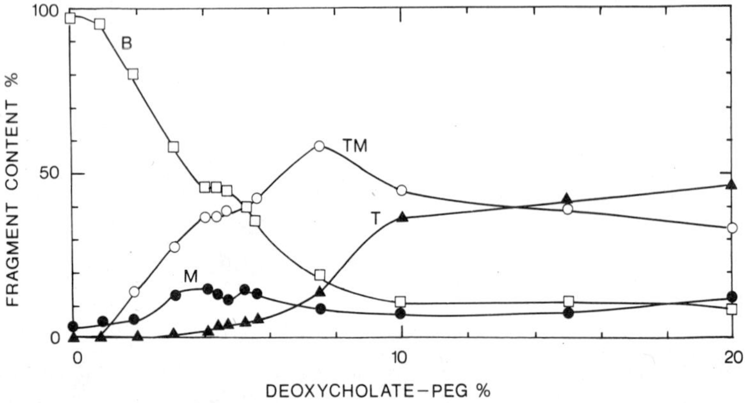

FIGURE 11.8 Partition of thylakoid vesicles in a three-phase system as a function of deoxycholate-PEG. Composition and symbols as in Figure 11.5.

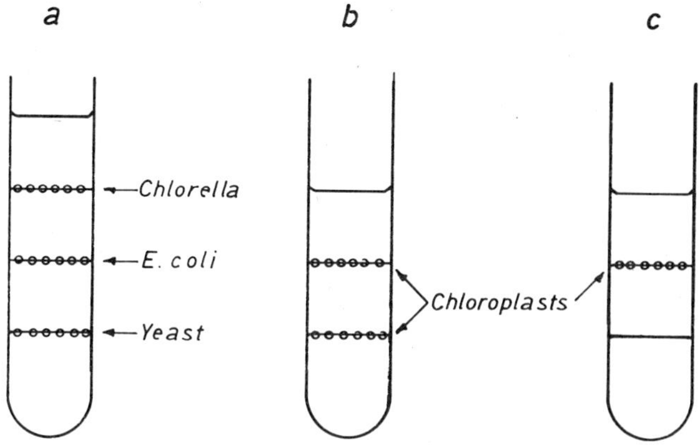

FIGURE 11.9 (*Left*) Collection of different microorganisms in a four-phase system (5% Dextran 500, 6% hydroxypropyldextran, 10.5% Ficoll, 5.5% PEG 6000 with 0.01 *M* Li phosphate, pH 7. (*Middle*) Chloroplasts in a three-phase system of 7% Dextran 500, 7% Ficoll, 6% PEG 6000 with 0.01 *M* Li phosphate, pH 7.9. (*Right*) The same as in the middle except that 1% PEG was replaced by 2% TMA-PEG.

charged PEG which attracts the cells into the top phase. The cells also are found in the middle phase and at the interface between the middle phase and the top phase. The partition profiles indicate heterogeneity of the cell population; note the bimodial distribution of the M-line and the step-wise change in the B and T curves. This probably reflects the different surface properties of cells at different stages in the growth cycle, as demonstrated by countercurrent distribution of these cells (Chapter 8, Fig. 8.12).

The fourth example (Fig. 11.8) shows the partition of thylakoid fragments in a three-phase system where a hydrophobic ligand, deoxycholate, bound to PEG, was used to change the partition.

TABLE 11.3 Composition of the Phases in a Three-Phase System[a]

Phase	PEG + TMA-PEG, % w/w	Ficoll, % w/w	Dextran, % w/w	Water, % w/w
Top	22	3	0	75
Middle	7	25	1	67
Bottom	0	3	37	60

[a]7% w/w Dextran 500, 12% w/w Ficoll 400, 12% w/w total polyethylene glycol (90% w/w unsubstituted PEG 4000 + 10% w/w TMA-PEG 4000), and 0.5 mM/Kg K_3PO_4 buffer pH 6.8 at 22°C (Hartman, 1976).

Figure 11.9 shows the partition in a four-phase system. Different microorganisms collect at different interfaces.

DETERGENT CONTAINING PHASES

Three- and four-phase systems have been used for purification of a membrane-bound enzyme phospholipase Al (Albertsson, 1973). The enzyme, solubilized by the detergent Triton X-100, was enriched in the top phase of a dextran–Ficoll–PEG three-phase system, while the majority of the proteins and nucleic acids were found in the bottom phase or at the two interfaces (Fig. 11.10). The detergent was removed by adding a phase of polypropylene glycol to the combined top (PEG) and middle (Ficoll) phases. The detergent went to the polypropylene glycol phase while the protein, now insoluble, was found at the interface between the Ficoll and PEG phases. This procedure could be adapted for large-scale purification of the enzyme.

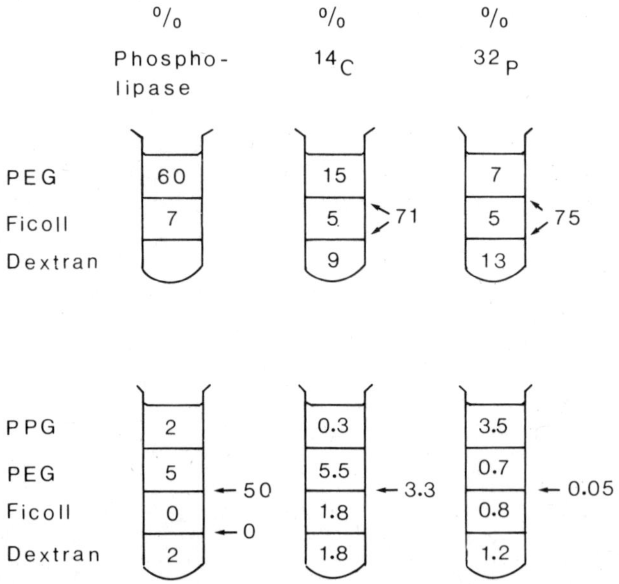

FIGURE 11.10a Distribution of phospholipase Al activity and ^{14}C and ^{32}P counts of lysates from *E. coli* labeled with ^{14}C-leucine and ^{32}P-phosphate. The lysates were first partitioned in the three-phase system (upper row). The two upper phases (Ficoll and PEG phases) were then withdrawn and combined with a fresh dextran phase and a PPG phase to give the four-phase system (lower row). All figures are percentage of total input. The three-phase polymer composition was 6.67% Dextran 500, 6% Ficoll, and 5.33% PEG 6000 (Albertsson, 1973).

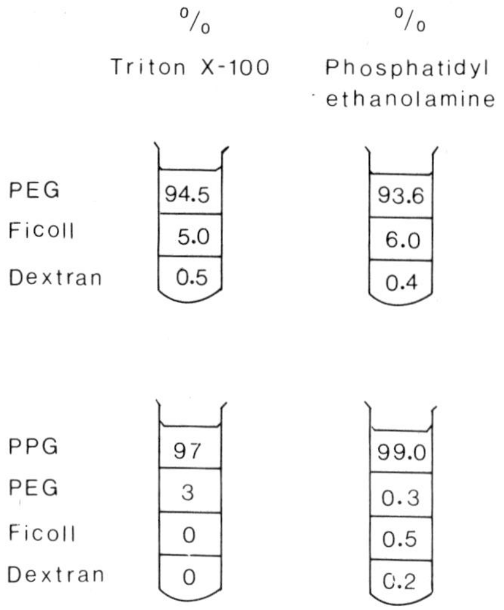

FIGURE 11.10*b* Distribution of Triton X-100 and phosphatidylethanolamine. Phase system as in (*a*).

SOLID–LIQUID POLYPHASE SYSTEMS

If a solid phase, such as hydroxylapatite, is added to a liquid-phase system, the solid phase may adsorb a partitioned substance and withdraw it from the liquid phases. The selectivity of adsorption is then superimposed on the selective phase distribution. This may lead to an increased overall separation effect.

Of particular interest is the distribution of particles and DNA in the hydroxylapatite–dextran–PEG system (Fig. 11.11). The concentration of buffer is so low that the particles are adsorbed on the hydroxylapatite if only a buffer–hydroxylapatite system is used. Also, in a system of hydroxylapatite–bottom-phase or hydroxylapatite–top phase the particles are adsorbed on the solid phase. However, in the complete system containing the two liquid phases and hydroxylapatite, the particles go to the upper phase as they do when partitioned between the two liquid phases alone. The top phase and the hydroxylapatite therefore compete for the particles; when the top phase is in direct contact with hydroxylapatite the latter wins, but when the bottom phase is also present the particles prefer the upper phase. It seems that the presence of the liquid–liquid interface is necessary for the detachment of the particles from the solid phase into the upper phase. (The same phenomenon holds for DNA when partitioned in a similar way.)

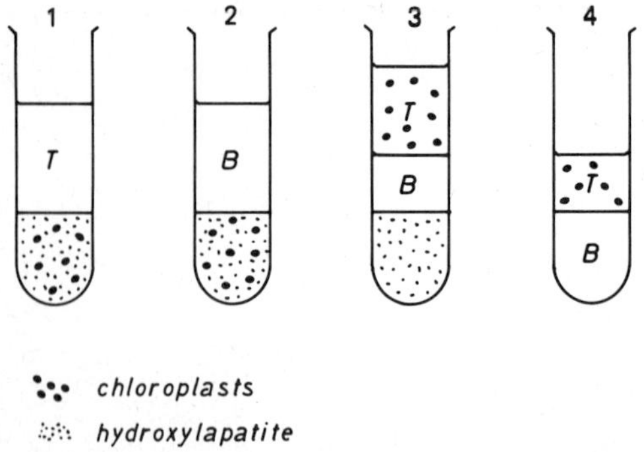

FIGURE 11.11 Partition paradox, showing competition between adsorption of chloroplasts to hydroxylapatite and collection in the top phase. (1) Top phase and hydroxylapatite: chloroplasts are strongly adsorbed to the solid phase. (2) Bottom phase and hydroxylapatite: chloroplasts are strongly adsorbed to the solid phase. (3) When (1) and (2) are mixed, chloroplasts are detached from the solid phase and collect in the liquid top phase as they normally do in the liquid–liquid system (4). Phase system: 5% Dextran 500, 4% PEG 6000, in all tubes with 0.01 *M* Li phosphate.

The hydroxylapatite–dextran–PEG system might therefore be useful in removing various proteins or other substances that are adsorbed on the hydroxylapatite from particle suspensions or DNA which partitions into the upper phase.

Recently a phase system of dextran and PEG including a gel phase has been used for separation purposes as described above (Hedman and Gustavsson, 1984).

REFERENCES

Albertsson, P.-Å. (1973). *Biochemistry*, **12**, 2525–2530.

Hartman, A. (1976a). *Acta Chem. Scand.*, **B30**, 585–594.

Hartman, A. (1976b). Thesis, Umeå University, Umeå, Sweden.

Hartman, A., Johansson, G., and Albertsson, P.-Å. (1974). *Europ. J. Biochem.*, **46**, 75–81.

Hedman, P.O. and Gustavsson, J.-G. (1984). *Anal. Biochem.*, **138**, 411.

Johansson, G. and Hartman, A. (1976). *Europ. J. Biochem.*, **63**, 1–8.

12 PHASE DIAGRAMS

In this chapter, phase diagrams of various aqueous polymer systems are given together with some data on the polymers used. It includes almost all phase diagrams which have been determined in this laboratory for many years. Some of the phase systems have never been used for partition but are nevertheless included because they might be useful for partition of certain material.

THE POLYMERS AND POLYMER SOLUTIONS

In general, it is a good rule always to standardize the procedure for making a polymer solution, since the property of such solution may sometimes depend on its previous history.

Dextran (D). All the dextran fractions used were supplied by Pharmacia, Uppsala, Sweden. The chemistry of dextran and its application is described by Murphy and Whistler (1973), where many references may be found. The dextran used here is a branched polyglucose built up by the bacterium *Leuconostoc mesenteroides,* strain B 512 when it grows on appropriate solutions of sucrose. After isolation and partial acid hydrolysis, different fractions are obtained by differential alcohol precipitation, and subsequently purified. Each fraction is characterized by the manufacturer, using the limiting viscosity number $[\eta]$, number average molecular weight $\overline{M}_n$, and weight-average molecular weight $\overline{M}_w$. The fractions which have been used here are given in Table 12.1. The greater the quotient $\overline{M}_w/\overline{M}_n$, the more polydisperse is the fraction. The physicochemical properties of these various fractions have been investigated by Granath (1958).

The glucose units of the main chain in these dextran molecules are connected by α—1→6 linkages. The branches are connected with this chain through α—

TABLE 12.1 Data of Various Dextran Fractions[a]

Abbreviation	Limiting Viscosity Number,[b] ml/g	Number Average Molecular Weight, M_n	Weight Average Molecular Weight, M_w
D 5	4.5	2,300	3,400
D 17	16.8	23,000	~30,000
D 19 or Dextran 40	19	20,000	42,000
D 24	24	40,500	
D 37	37	83,000	179,000
D 48 or Dextran 500	48	180,000	460,000
D 68	68	280,000	2,200,000
D 70	70	73,000	—

[a]Data according to the manufacturer.
[b]The same as intrinsic viscosity, which is usually expressed in dL/g.

1→3 linkages. For the dextran used here about 95% of the linkages are α—1→6 linkages, the rest being α—1→3 linkages. (Lindberg and Svensson, 1968; Sloan et al., 1954). It has a specific optical rotation of $[\alpha]_D^{25} = +199°$, if the $\overline{M}_n$ value exceeds about 10,000.

The dextran used contained 5–10% moisture. Solutions were originally prepared from dextran dried at 120° for 12 hr. It was found, however, that drying by heat occasionally caused a brown-yellow color change and an increase in absorption of ultraviolet light. Solutions were therefore prepared directly from the undried dextran and the concentration of the solution was determined by polarimetry. The procedure is as follows: The undried dextran is first wetted and mixed to a paste with a small amount of water. The rest of the water is then added and the dextran dissolved by stirring and slowly heating the mixture to boiling. The solution is then allowed to cool down with the flask covered by a watch glass. For the determination of the concentration, about 10 g of the solution are weighed into a 25-mL measuring flask, which is then filled to the mark with water. The optical rotation of this solution is then determined in a polarimeter using a tube 20 cm long. For stock solutions, 10–20% w/w is a suitable concentration range for the higher molecular weight dextrans and 20–30% w/w for the lower molecular weight dextrans.

Nowadays the dextran batches are supplied as a fine powder which dissolves easily even without heating. However, it may still be useful to heat the solution to boiling to speed up the dissolution and also partly sterilize the solution. For complete sterilization the dextran solutions may be autoclaved in the usual manner.

The dextran solutions have a small absorbance in the UV. A typical spectrum is shown in Figure 12.1.

Dextran solutions, except those of the low-molecular-weight fractions are stable if kept under sterile conditions. The low-molecular-weight fractions such as D5 and D17 undergo crystallization after some time and give rise to milky solutions. These however, may, be clarified by heating up to boiling.

Concentrated solutions of D5 do not become clear if dissolved as above, and must be clarified by filtration. Their concentration has been determined by dry-weight determination since the $[\alpha]_D^{25}$ value used for the other fractions, does not apply to D5.

Hydroxypropyldextran (HPD) was obtained from Pharmacia, Uppsala, Sweden. One sample, HPD-70, had been prepared from D70 and had a degree of substitution of approximately 1 per glucose unit. Three other samples, HPD-A, HPD-B and HPD-C, had been prepared from D500 and had a degree of substitution of 0.39, 0.76, and 1.5, respectively, per glucose unit. Four samples, HPD-a, HPD-b, HPD-c, and HPD-d also had been prepared from D500 and had a degree of substitution of 0.21, 0.42, 0.93, and 1.72. It may be dissolved directly in water and the concentration determined by dry-weight determination.

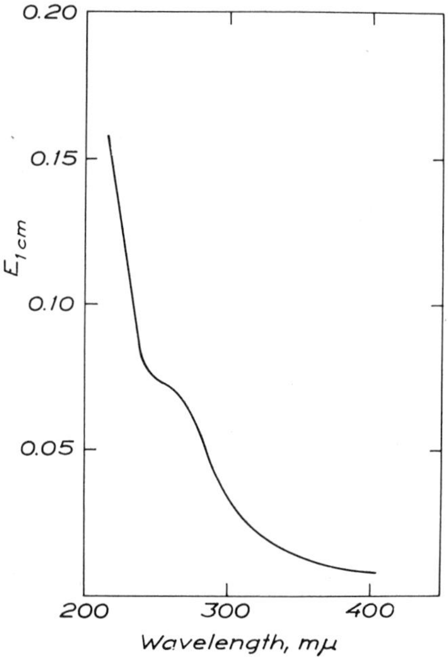

FIGURE 12.1. Absorption curve for a 10% w/w solution of dextran (D48).

Ficoll (F) was obtained from Pharmacia, Uppsala, Sweden. It is a synthetic polymer made by polymerization of sucrose. It is a nonionic polymer with $M_w \sim 400{,}000$. The concentration of the solutions was determined by polarimetry, $[\alpha]_D^{25} = 54.9$.

Carboxymethyldextran, sodium salt (NaCMD), was obtained from Pharmacia, Uppsala, Sweden. The fraction which was used had been prepared from a dextran fraction with a limiting viscosity number of 68 mL/g. Its equivalent weight was 297. The product was dissolved in water, filtered, dialyzed against water, freeze dried, and kept over P_2O_5.

Dextran sulfate, sodium salt (NaDS), was supplied by Pharmacia, Uppsala, Sweden. Two samples were used which had been prepared from dextran fractions with the limiting viscosity numbers of 50 and 68 mL/g. They will be referred to as NaDS50 and NaDS68. The sulfur content was 16.8%. It was dried and kept over P_2O_5 and used without further purification. The specific optical rotation for dextran sulfate varies slightly with the salt concentration. $[\alpha]_D^{21}$ was found to be +100° in water, +99.6° in 0.15 *M* NaCl, +99.5° in 0.3 *M* NaCl, and +98.9° in 1 *M* NaCl.

Diethylaminoethyl dextran, chloride form (DEAE dextran · HCl), was obtained from Pharmacia, Uppsala, Sweden. It had been prepared from a dextran fraction with a limiting viscosity number of 50 mL/g. Its nitrogen content was 3.5%.

Methylcellulose (MC) was obtained from the Dow Chemical Company, U.S.A. Three different products were used: Methocel 4000, Methocel 400, and Methocel 10, all U.S.P. grades. They will be referred to as MC4000, MC400, and MC10. The figures indicate viscosity in centipoises of a 2% solution. Data on the properties of Methocel may be obtained from Methocel Handbook, published by the Dow Chemical Company. Information about the chemistry of cellulose ethers in general may be obtained from Whistler (1973). From data on the degree of polymerization as a function of the viscosity of a 2% solution, MC4000 may be calculated to have a mean molecular weight of about 140 000, MC400 about 80 000, and MC10 about 30 000. These values are, however, somewhat uncertain and the fractions are rather polydisperse. 1–2, 5, and 10% w/w are suitable concentrations for stock solutions of MC4000, 400, and 10, respectively. The methylcellulose was dried at 110°C for 24 hrs. One kilogram of a 1% w/w solution of MC4000 is prepared in the following way. Ten grams of dry methylcellulose are weighed into an Erlenmeyer flask and 300–500 mL hot (80–90°C) water added. The flask is closed and shaken vigorously for a few minutes in order to wet the powder. Cold water (500 mL) is then added and the flask is shaken and allowed to stand with occasional stirring until it reaches room temperature. The powder now swells and is slowly dissolved; but, note well, it should not be allowed to settle on the bottom of the flask. The flask is put on the balance and water added to bring it to the desired weight. The solution is then cooled down to 4°C and kept at this temperature before use. The concentration is checked by dry-weight determination at 110°C. The final cooling of the solution is necessary in order to get a clear solution. Since the solution properties, for example, viscosity, of cellulose ethers in general depend on the lowest temperature to which they have been subjected all solutions are cooled to the same temperature. An absorption curve of methylcellulose is shown in Figure 12.2.

Carboxymethylcellulose, sodium salt (NaCMC), was obtained from Uddeholms Ltd., Sweden. It was dissolved directly in water to a 2% w/w solution.

Ethyl hydroxyethylcellulose (EHEC) was supplied by Mo och Domsjö, Ltd., Örnsköldsvik, Sweden, with the trade name Modocoll. Its properties are described by Lindenfors and Jullander (1973). A sample designated Modocoll E600 was used. It was dissolved directly in water, cooled to 4°C, and before use it was clarified by centrifugation. This product is rather polydisperse with a weight-average molecular weight of about 200,000.

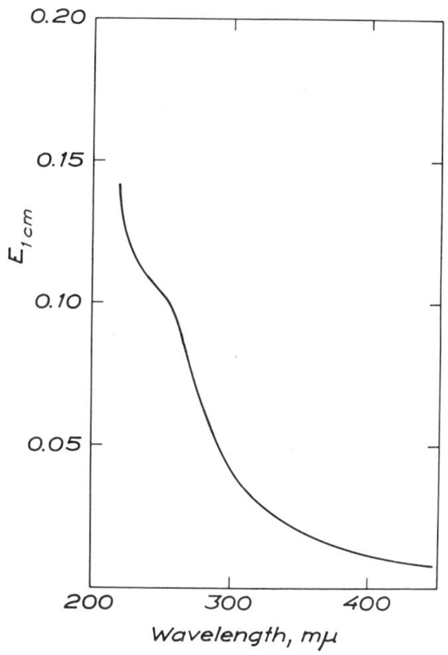

FIGURE 12.2. Absorption curve for a 1% w/w solution of methylcellulose (MC4000).

Polyvinyl alcohol (PVA) was obtained from Firma Wacker Chemie, GMBH, Munich, Germany. It is marketed under the trade name Polyviol. Polyviol 48/20 (PVA 48/20) and 28/20 (PVA 28/20 were the products used). A solution of polyvinylalcohol is prepared by first wetting the powder with hot water to a paste and then adding the rest of the water and keeping the mixture at 100°C until a solution is obtained. The concentration is determined by dry-weight determination.

Polyethylene glycol (PEG) was obtained in the form of Carbowax compounds produced by Union Carbide Chemicals, New York. The samples used are described in Table 12.2. Information on the properties of polyethylene glycol may be obtained from the manufacturer. Polyethylene glycol is dissolved directly in water and the concentration is determined by freeze drying. (Dry-weight determination by heating is unsatisfactory because part of the polyethylene glycol evaporates and during drying a cake is formed which may retain much water.) The Carbowax compounds contain small amounts of an impurity which has a peak of absorption in UV light at 290 mμ. This may be removed by precipitation of the polyethylene glycol from an acetone solution with ether. Such a purification of PEG6000 is done in the following way: Three hundred grams PEG6000 are dissolved by careful warming in 6 L acetone; 3 L ether are then added during stirring. The mixture is allowed to stand overnight. The precipitate is collected by filtration through a filter paper, washed with (2:1) acetone–ether mixture and dried in air. The UV spectrum of such a preparation is given in Figure 12.3.

Polypropylene glycol (PPG) was obtained from Union Carbide with the name Polypropylene glycol 425 and had an average molecular weight of 400–500.

Methoxy polyethylene glycol (MPEG) was obtained as Carbowax methoxy polyethylene glycol 550 from Union Carbide. It had an average molecular weight of 525–575.

TABLE 12.2 Samples of Polyethylene Glycols Which Have Been Used

Abbreviation	Commercial Name: Carbowax Polyethylene Glycol	Number Average Molecular Weight (11), M_w
PEG 20000	20 *M*	15,000–20,000
PEG 6000	6000 or 8000	6,000–7,500
PEG 4000	4000 or 3400	3,000–3,700
PEG 1540	1540	1,300–1,600
PEG 1000	1000	950–1,050
PEG 600	600	570–630
PEG 400	400	380–420
PEG 300	300	285–315

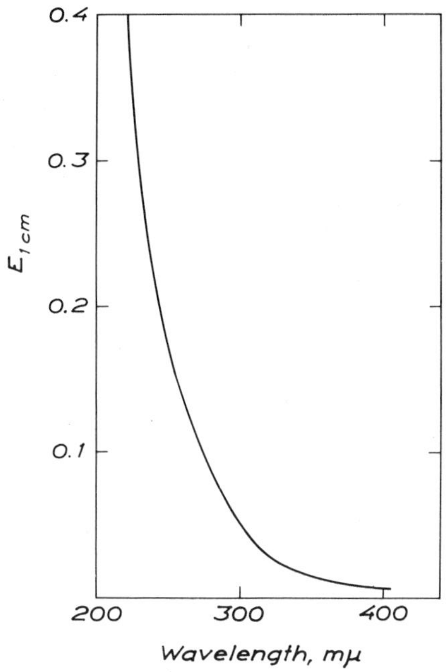

FIGURE 12.3. Absorption curve for a 10% w/w solution of purified polyethylene glycol (PEG 6000).

Ucon HB are random copolymers of ethylene oxide and propylene oxide, made by Union Carbide Chemicals, New York.

Pluronic block polymers of polyethylene glycol and polypropylene glycol were obtained from Wyandotte Chemicals Co., Michigan, USA. Pluronic F-98, F-88, F-77, and P-85 have been used. The first digit in the code gives the molecular weight of the polypropylene glycol, the second digit is one tenth of the percentage polyethylene glycol in the molecule.

Tergitol NP, produced by Union Carbide Chemicals, New York, are nonyl phenyl polyethylene glycol ethers. The polymers that have been used are NP-44, NP-40, and NP-35 containing 40, 20, and 15 mole ethylene oxide per mole.

ANALYSIS OF THE PHASES

For most of the phase systems the compositions of the different phases were calculated from determinations of the water content and the concentration of one of the polymers; the concentration of the other polymer is then obtained by subtraction. The total composition is always known since the phase systems are set up by mixing weighed amounts of water and polymer solutions with known concentrations. Thus the compositions of the total system, the top phase, and the bottom phase are usually determined in each experiment. The condition that these three compositions can be represented by three points on a straight line (see Fig. 2.5) in the phase diagram is thus a check on the accuracy of the analysis.

The phase systems are prepared in the following way. Suitable amounts of water and polymer solutions are weighed into a separating funnel. Because of their high viscosity, polymer solutions are measured by weight instead of volume. The funnel is equilibrated in a thermostat and the contents mixed by inverting the funnel about 50 times. The funnel is then allowed to stand for 24–48 hrs. for phase separation. Samples from the top phase are taken by a pipette and from the bottom phase through the outlet of the funnel. The samples are then analyzed as follows.

The Dextran–Polyethylene Glycol System

For dry-weight determination, samples of about 3–5 g from each phase were weighed into wide 50-mL weighing glasses or 100-mL Erlenmeyer flasks, diluted with 10–20 mL water, and freeze dried. After freeze drying, the samples were further dried over P_2O_5 to constant weight; this latter drying was necessary since dextran, which retains water rather firmly, was not always dried completely by freeze drying. For dextran determinations samples of about 10 g from each phase were weighed into 25–mL measuring flasks which were then filled with water. The dextran concentration was then determined using a polarimeter and tubes 20 cm long.

In some experiments the drying of the bottom phase was not satisfactory. In these cases the point representing the bottom phase composition was obtained from the intersection of the line joining the points representing the top phase and the total compositions and the line representing the dextran concentration of the bottom phase.

For some phase systems the dry-weight determination was replaced by refractive index measurements. By measuring both the refractive index and the optical rotation of a phase the concentration of PEG and dextran could be calculated.

The increase in refractive index is proportional to the concentration of polymer (in weight per volume) and when two polymers are mixed the increments are additive.

The dextran–polyvinyl alcohol system was analyzed in a similar way to the dextran-polyethylene glycol system.

The Dextran–Methylcellulose System

For dry-weight determinations, 10–15-g samples were kept at 110°C for 48 hrs. Dextran was determined polarimetrically on another sample. Methylcellulose exhibits a small optical rotation and is removed before dextran is determined. This was done by adding 5 g of 30% w/w $(NH_4)_2SO_4$ to a sample of about 10 g of the phase. This precipitated the methylcellulose, which could then be removed by filtration through a Pyrex glass filter, marked S.F. 1A1 or S.F. 1A2. (Filter paper was unsatisfactory.) The filtrate was collected in a 25-mL measuring flask. The precipitate was washed with 10% w/w $(NH_4)_2SO_4$ and the flask filled with water. The dextran concentration was then determined polarimetrically.

The dextran–ethylhydroxyethylcellulose system was analyzed in a similar way to the dextran-methylcellulose system.

The Dextran–Hydroxypropyldextran System

For dry-weight determinations samples of about 5 g were kept at 110°C for 24–48 hrs. Dextran was determined polarimetrically on another sample. Hydroxypropyldextran exhibits optical rotation and is removed before the dextran is determined. This was done by adding ammonium sulfate up to 25% w/w. The precipitate was removed by filtration through a glass filter (Pyrex, marked S.F. 1A1 or S.F. 1A2) or by centrifugation.

The Na Dextran Sulfate–Polyethylene Glycol–Sodium Chloride System

The dry weight was determined in the same way as for the dextran–polyethylene glycol system. The Na dextran sulfate was determined polarimetrically in a similar manner to dextran in the dextran–polyethylene glycol system. The phase system analyzed contained 0.3 *M* NaCl and the dextran sulfate has an $[\alpha]_D^{21}$ value of $+99.5°$ (see above). The sodium chloride content of each phase was obtained by passing a sample through a Dowex 50 column in acid form. The liberated acid was determined by titration using methyl blue-methyl red as indicator. Part of the liberated acid is dextran sulfuric acid and its equimolar content has to be subtracted. It was calculated from the percentage weight per weight content of the phase as determined polarimetrically and its equivalent weight; the latter was determined to be 209 by passing a known solution of dialyzed Na dextran sulfate through a Dowex 50 column. Thus the dry weight, the sodium chloride, and the Na dextran sulfate contents are known. The polyethylene glycol content is then obtained by subtraction.

The Na Dextran Sulfate–Methylcellulose–Sodium Chloride System

Dry weight was determined by freeze drying, since dextran sulfate cannot be heated for a long time. The Na dextran sulfate was determined polarimetrically after the methylcellulose had been precipitated and removed as described for the dextran–methylcellulose system. The Na dextran sulfate solution thus obtained contains 10% w/w ammonium sulfate and has an $[\alpha]_D^{21}$ value of $+100°$. Sodium chloride distributes almost equally between the two phases and its concentration in each phase was considered to be 0.15 and 0.3 *M*, respectively, for the two systems studied (see Figures 12.39 and 12.40). The methylcellulose concentration was then obtained by subtracting the sodium chloride and the Na dextran sulfate concentrations from the dry weight. In some experiments the drying of the bottom phase was unsatisfactory and in these cases the point representing the bottom phase was obtained from the intersection of the line joining the points representing the top phase and the total compositions and the line representing the Na dextran sulfate concentration of the bottom phase.

The Na dextran sulfate–polyvinyl alcohol–sodium chloride system was analyzed in the same way as the Na dextran sulfate-methylcellulose-sodium chloride system; the sodium chloride concentration was considered to be 0.3 *M* in both phases since it distributes almost equally between the phases (see Fig. 12.41).

The Potassium Phosphate–Polyethylene Glycol System

Two to six-gram samples were freeze dried as described for the dextran-polyethylene glycol system. The dried samples were then dissolved in water and passed through a column of the acid form of Dowex 50. The liberated phosphoric acid was then titrated with 0.1 *M* NaOH using bromcresol green as indicator. In this way the salt can be determined and the polyethylene glycol concentration is obtained by subtraction.

The Dextran–Ucon, Pluronic, Tergitol, or Ficoll and Ficoll–Polyethylene Glycol Systems

These systems were analyzed by measuring optical rotation and refractive index of the phases. For all mixtures it was found that the optical rotation and refractive index were each additive, and by measuring pure standard solutions the concentration of each polymer could be calculated.

PHASE DIAGRAMS

The results of the analysis of the selected systems listed in Table 12.3 are given in Figures. 12.4–12.52. As a convention, the concentration of the polymer which distributes in favor of the bottom phase is plotted as abscissa, and the concentration of the polymer which distributes in favor of the top phase is plotted as the ordinate. The composition of the phases and the total system are recorded under each phase diagram. The systems are listed in alphabetical order, this being the same as the order of their distances from the critical point. The potassium phosphate used was a mixture with the ratio 306.9 g K_2HPO_4 to 168.6 g KH_2PO_4.

TABLE 12.3 Analysis of Selected Systems

	Figures Showing Phase Diagram
Two-Phase Systems	
Dextran–polyethylene glycol–water	12.4–12.20
Dextran–Ucon–water	12.21
Dextran–Pluronic–water	12.22
Dextran–Tergitol–water	12.23
Dextran–Ficoll–water	12.24*a*
Ficoll–polyethylene glycol–water	12.24*b*
Dextran–hydroxypropyldextran–water	12.25–12.27
Hydroxypropyldextran–hydroxypropyldextran–water	12.27
Hydroxypropyldextran–polyethylene glycol–water	12.28
Dextran–methylcellulose–water	12.29–12.32
Dextran–polyvinyl alcohol–water	12.33–12.34
DEAE–dextran–polyethylene glycol–lithium sulfate–water	12.35
Na dextran sulfate–polyethylene glycol–sodium chloride–water	12.36–12.38
Na dextran sulfate–methylcellulose–sodium chloride–water	12.39–12.40
Na dextran sulfate–polyvinyl alcohol–sodium chloride–water	12.41
Na carboxymethyldextran–polyethylene glycol–sodium chloride–water	12.42
Polyvinyl alcohol–polyethylene glycol–water	12.42*b*, *c*
Polyvinyl pyrrolidone–polyethylene glycol–water	12.42*d*
Potassium phosphate–polyethylene glycol–water	12.43–12.49
Potassium phosphate–methoxypolyethylene glycol–water	12.50
Potassium phosphate–polypropylene glycol–water	12.50
Ammonium sulfate–polyethylene glycol–water	12.51
Magnesium sulfate–polyethylene glycol–water	12.52
Three-Phase systems	
Dextran–Ficoll–polyethylene glycol	12.24*c*

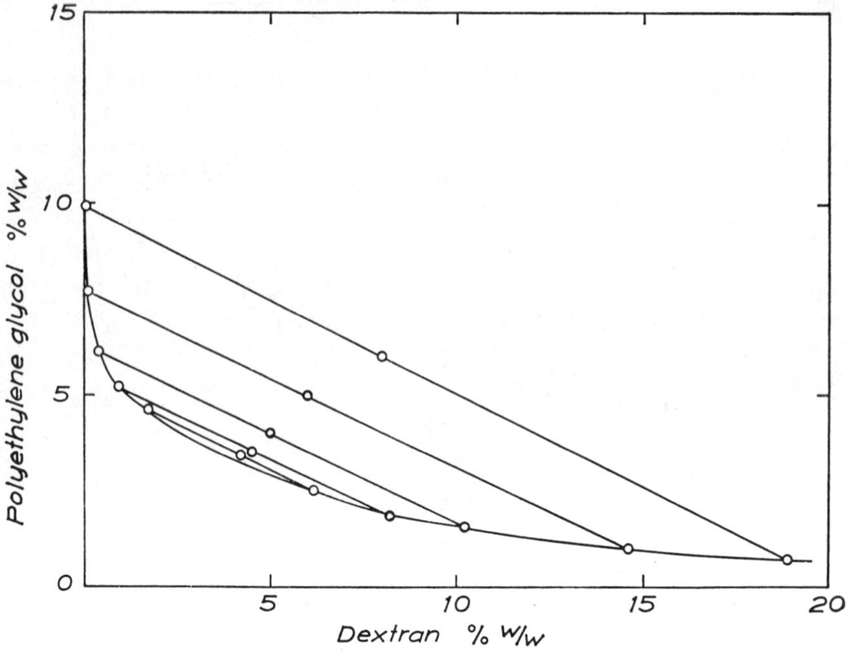

System	Total system: Dextran % w/w	Total system: Polyethylene glycol % w/w	Total system: H_2O % w/w	Bottom phase: Dextran % w/w	Bottom phase: Polyethylene glycol % w/w	Bottom phase: H_2O % w/w	Top phase: Dextran % w/w	Top phase: Polyethylene glycol % w/w	Top phase: H_2O % w/w
A	4.20	3.40	92.40	6.14	2.50	91.36	1.72	4.62	93.66
B	4.50	3.50	92.00	8.22	1.83	89.95	0.90	5.22	93.88
C	5.00	4.00	91.00	10.20	1.55	88.25	0.36	6.13	93.51
D	6.00	5.00	89.00	14.59	0.98	84.43	0.11	7.71	92.18
E	8.00	6.00	86.00	18.93	0.70	80.37	0.05	9.88	90.07

FIGURE 12.4. Phase diagram and phase compositions of the dextran–polyethylene glycol system D68–PEG 6000 at 20°C.

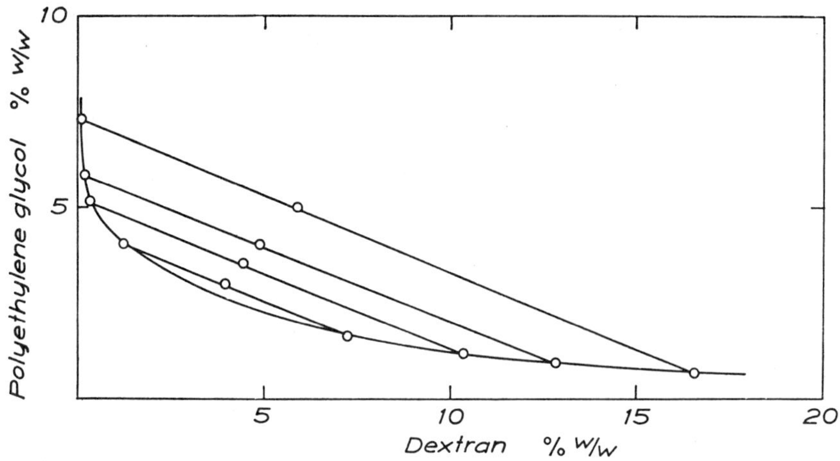

Sys-tem	Total system: Dextran % w/w	Total system: Poly-ethylene glycol % w/w	Total system: H_2O % w/w	Bottom phase: Dextran % w/w	Bottom phase: Poly-ethylene glycol % w/w	Bottom phase: H_2O % w/w	Top phase: Dextran % w/w	Top phase: Poly-ethylene glycol % w/w	Top phase: H_2O % w/w
A	3.92	3.00	93.08	7.23	1.64	91.13	1.20	4.02	94.78
B	4.41	3.51	92.08	10.37	1.19	88.44	0.26	5.14	94.60
C	4.89	4.00	91.11	12.82	0.91	86.27	0.16	5.83	94.01
D	5.87	5.00	89.13	16.52	0.67	82.81	0.07	7.29	92.64

FIGURE 12.5. Phase diagram and phase compositions of the dextran–polyethylene glycol system D68–PEG 6000 at 0°C.

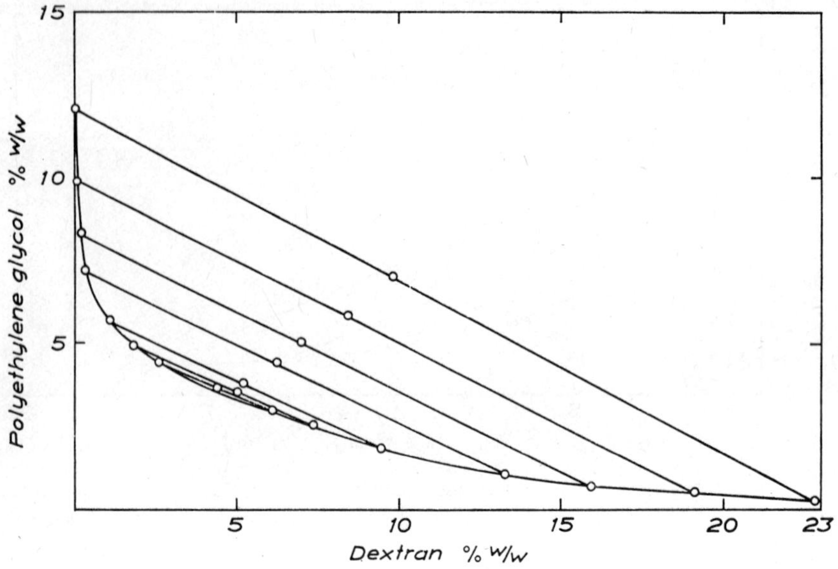

Sys-tem	Total system Dextran % w/w	Polyethylene glycol % w/w	H_2O % w/w	Bottom phase Dextran % w/w	Polyethylene glycol % w/w	H_2O % w/w	Top phase Dextran % w/w	Polyethylene glycol % w/w	H_2O % w/w
A	4.40	3.65	91.95	6.10	2.98	90.92	2.63	4.43	92.94
B	5.00	3.50	91.50	7.34	2.55	90.11	1.80	4.91	93.29
C	5.20	3.80	91.00	9.46	1.85	88.69	1.05	5.70	93.25
D	6.20	4.40	89.40	13.25	1.07	85.68	0.30	7.17	92.53
E	7.00	5.00	88.00	15.89	0.68	83.43	0.14	8.29	91.57
F	8.40	5.80	85.80	19.08	0.52	80.40	0.06	9.93	90.01
G	9.80	7.00	83.20	22.77	0.24	76.99	0.05	12.03	87.92

FIGURE 12.6. Phase diagram and phase compositions of the dextran–polyethylene glycol system D48–PEG 6000 at 20°C.

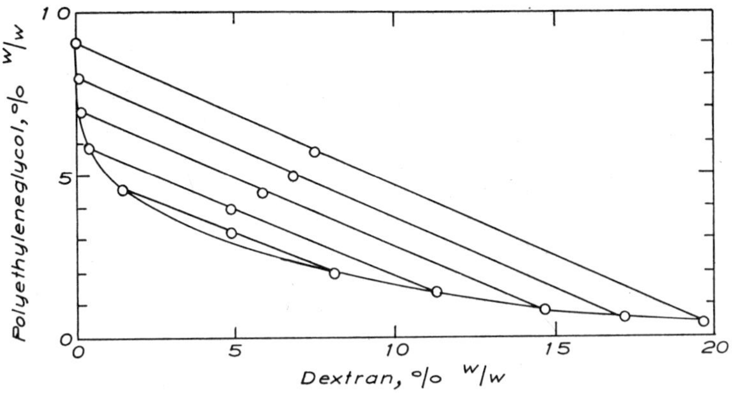

System	Total system Dextran % w/w	PEG % w/w	H_2O % w/w	Bottom phase Dextran % w/w	PEG % w/w	H_2O % w/w	Top phase Dextran % w/w	PEG % w/w	H_2O % w/w
A	4.84	3.27	91.89	8.12	1.99	89.89	1.45	4.65	93.90
B	4.89	4.00	91.11	11.32	1.45	87.23	0.47	5.87	93.66
C	5.86	4.50	89.64	14.60	0.82	84.58	0.13	6.96	92.91
D	6.84	5.00	89.16	17.07	0.57	82.36	0.05	7.98	91.97
E	7.50	5.76	86.74	19.56	0.41	80.03	0.02	9.10	90.88

FIGURE 12.7*a*. The composition of the phases in the systems of D48–PEG 6000 at 4°C.

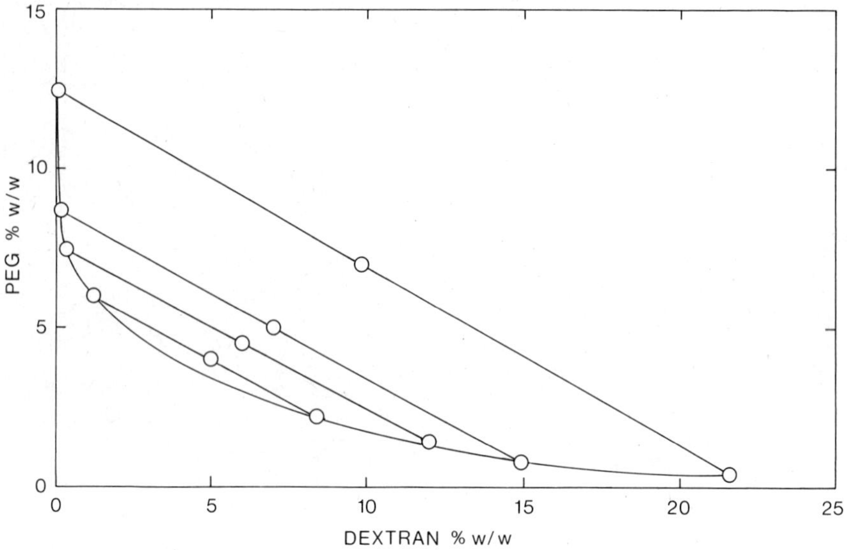

FIGURE 12.7*b*. Phase diagram of the dextran–polyethylene glycol system D500 (D48)–PEG 6000 at 37°C.

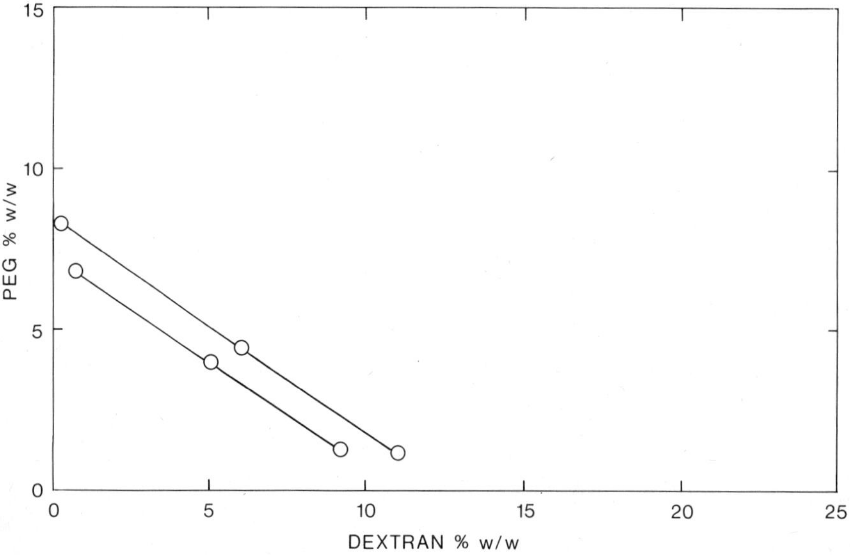

FIGURE 12.7*c*. The same as Figure 12.7*b* at 75°C.

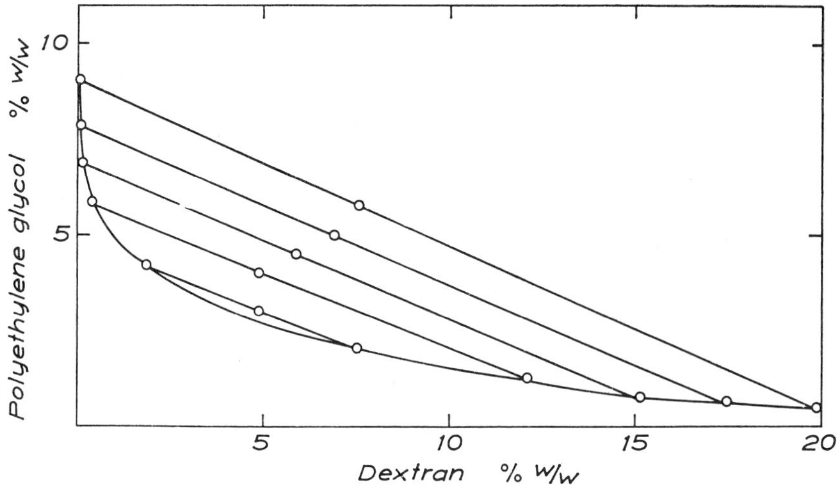

Sys-tem	Total system: Dextran % w/w	Total system: Poly-ethylene glycol % w/w	Total system: H_2O % w/w	Bottom phase: Dextran % w/w	Bottom phase: Poly-ethylene glycol % w/w	Bottom phase: H_2O % w/w	Top phase: Dextran % w/w	Top phase: Poly-ethylene glycol % w/w	Top phase: H_2O % w/w
A	4.89	3.00	92.11	7.50	2.03	90.47	1.87	4.20	93.93
B	4.89	4.00	91.11	12.13	1.28	86.59	0.36	5.84	93.80
C	5.86	4.50	89.64	15.13	0.79	84.08	0.13	6.90	92.97
D	6.84	5.00	88.16	17.48	0.67	81.85	0.06	7.89	92.05
E	7.52	5.77	86.71	19.93	0.55	79.52	0.03	9.03	90.94

FIGURE 12.8. Phase diagram and phase compositions of the dextran–polyethylene glycol system D48–PEG 6000 at 0°C.

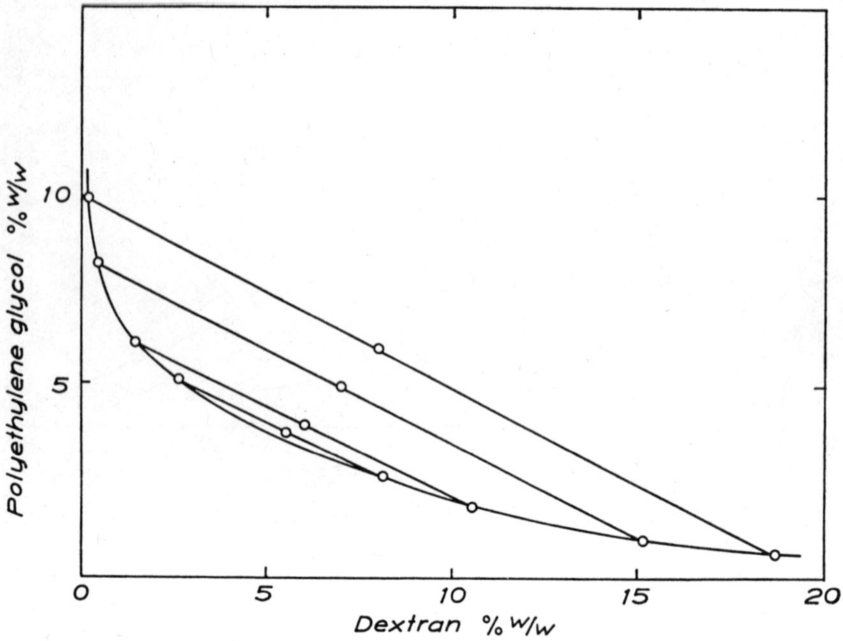

Sys-tem	Total system: Dextran % w/w	Total system: Polyethylene glycol % w/w	Total system: H_2O % w/w	Bottom phase: Dextran % w/w	Bottom phase: Polyethylene glycol % w/w	Bottom phase: H_2O % w/w	Top phase: Dextran % w/w	Top phase: Polyethylene glycol % w/w	Top phase: H_2O % w/w
A	5.50	3.80	90.70	8.10	2.67	89.23	2.62	5.20	92.18
B	6.00	4.00	90.00	10.50	1.85	87.65	1.43	6.18	92.39
C	7.00	5.00	88.00	15.17	0.96	83.87	0.41	8.24	91.35
D	8.00	6.00	86.00	18.68	0.63	80.69	0.16	9.96	89.88

FIGURE 12.9. Phase diagram and phase compositions of the dextran–polyethylene glycol system D37–PEG 6000 at 20°C.

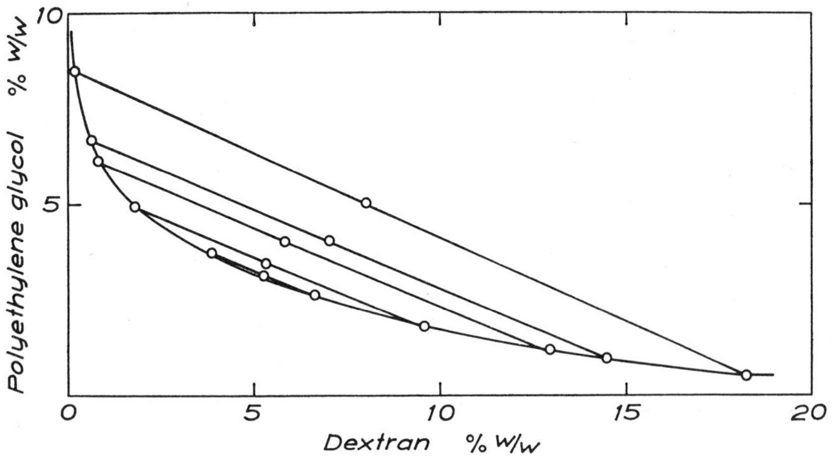

Sys-tem	Total system Dextran % w/w	Poly-ethylene glycol % w/w	H_2O % w/w	Bottom phase Dextran % w/w	Poly-ethylene glycol % w/w	H_2O % w/w	Top phase Dextran % w/w	Poly-ethylene glycol % w/w	H_2O % w/w
A	5.24	3.13	91.63	6.67	2.66	90.67	3.84	3.73	92.43
B	5.30	3.44	91.26	9.56	1.80	88.64	1.74	4.91	93.35
C	5.82	4.00	90.18	12.94	1.20	85.86	0.72	6.11	93.17
D	7.02	4.02	88.96	14.45	0.95	84.60	0.61	6.69	92.70
E	8.02	5.00	86.98	18.21	0.50	81.29	0.15	8.50	91.35

FIGURE 12.10. Phase diagram and phase compositions of the dextran–polyethylene glycol system D37–PEG 6000 at 0°C.

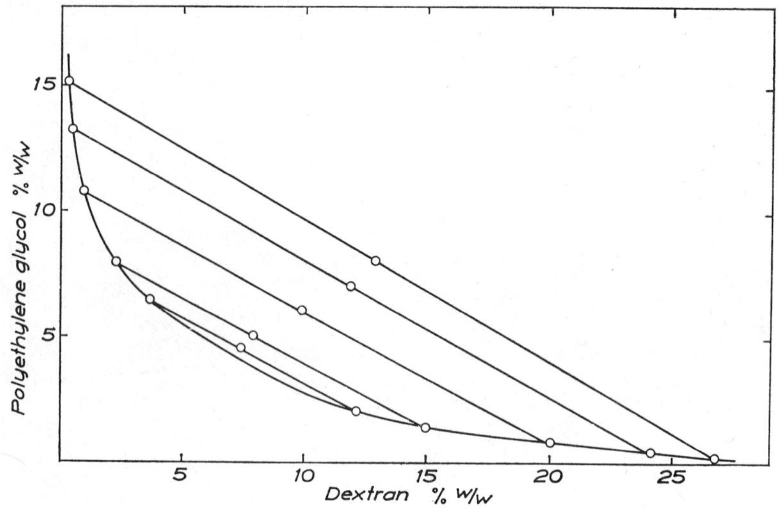

Sys-tem	Total system Dextran % w/w	Total system Poly-ethylene glycol % w/w	Total system H_2O % w/w	Bottom phase Dextran % w/w	Bottom phase Poly-ethylene glycol % w/w	Bottom phase H_2O % w/w	Top phase Dextran % w/w	Top phase Poly-ethylene glycol % w/w	Top phase H_2O % w/w
A	7.39	4.52	88.09	12.13	2.04	85.83	3.62	6.46	89.92
B	7.89	5.02	87.09	14.96	1.40	83.64	2.25	7.94	89.81
C	9.87	6.03	84.10	20.01	0.85	79.14	0.93	10.75	88.32
D	11.84	7.04	81.12	24.10	0.43	75.47	0.43	13.25	86.32
E	12.83	8.04	79.13	26.72	0.22	73.06	0.27	15.11	84.62

FIGURE 12.11. Phase diagram and phase compositions of the dextran–polyethylene glycol system D24–PEG 6000 at 20°C.

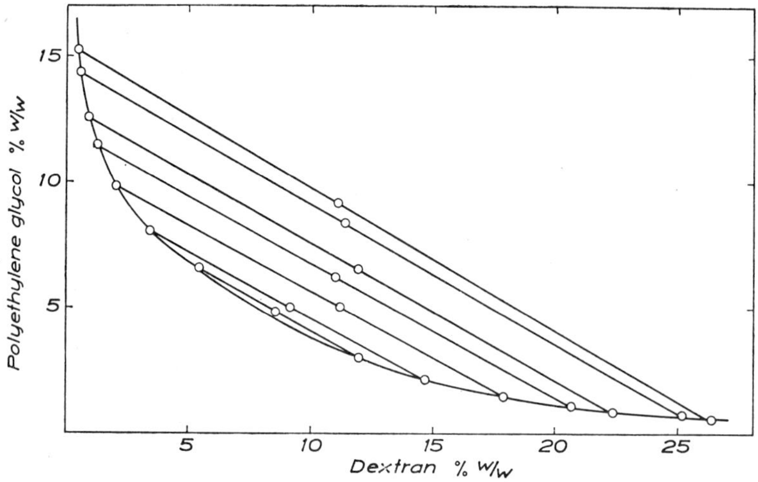

System	Total system: Dextran % w/w	Total system: Polyethylene glycol % w/w	Total system: H_2O % w/w	Bottom phase: Dextran % w/w	Bottom phase: Polyethylene glycol % w/w	Bottom phase: H_2O % w/w	Top phase: Dextran % w/w	Top phase: Polyethylene glycol % w/w	Top phase: H_2O % w/w
A	8.54	4.77	86.69	11.91	3.00	85.09	5.41	6.59	88.00
B	9.14	5.00	85.86	14.67	2.10	83.23	3.41	8.02	88.57
C	11.18	5.00	83.82	17.89	1.46	80.65	2.01	9.80	88.19
D	10.98	6.22	82.80	20.69	1.12	78.19	1.22	11.47	87.31
E	11.87	6.51	81.62	22.38	0.85	76.77	0.89	12.55	86.56
F	11.33	8.35	80.32	25.17	0.78	74.05	0.55	14.37	85.08
G	11.05	9.16	79.79	26.40	0.50	73.10	0.41	15.21	84.38

FIGURE 12.12. Phase diagram and phase compositions of the dextran–polyethylene glycol system D17–PEG 6000 at 20°C.

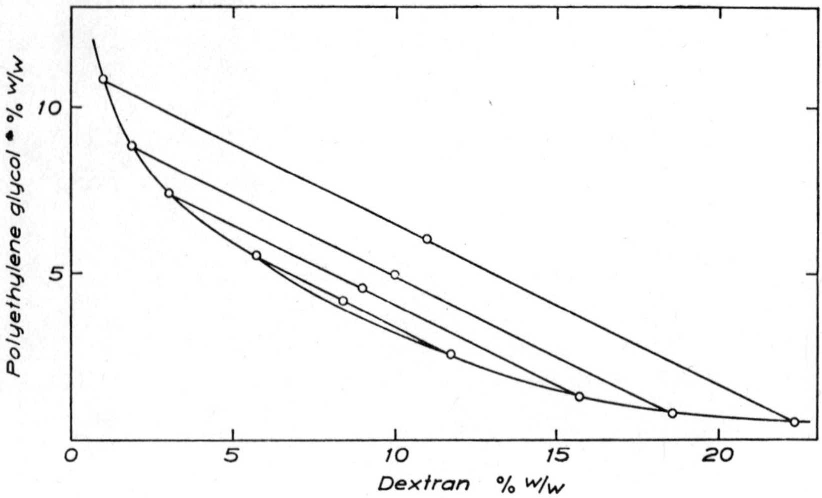

	Total system			Bottom phase			Top phase		
System	Dextran % w/w	Polyethylene glycol % w/w	H_2O % w/w	Dextran % w/w	Polyethylene glycol % w/w	H_2O % w/w	Dextran % w/w	Polyethylene glycol % w/w	H_2O % w/w
A	8.31	4.15	87.54	11.66	2.55	85.79	5.70	5.55	88.75
B	8.92	4.58	86.50	15.71	1.30	82.99	3.01	7.42	89.57
C	9.92	4.97	85.11	18.56	0.80	80.64	1.85	8.85	89.30
D	10.93	6.02	83.05	22.32	0.55	77.13	0.99	10.81	88.20

FIGURE 12.13. Phase diagram and phase compositions of the dextran–polyethylene glycol system D17–PEG 6000 at 0°C.

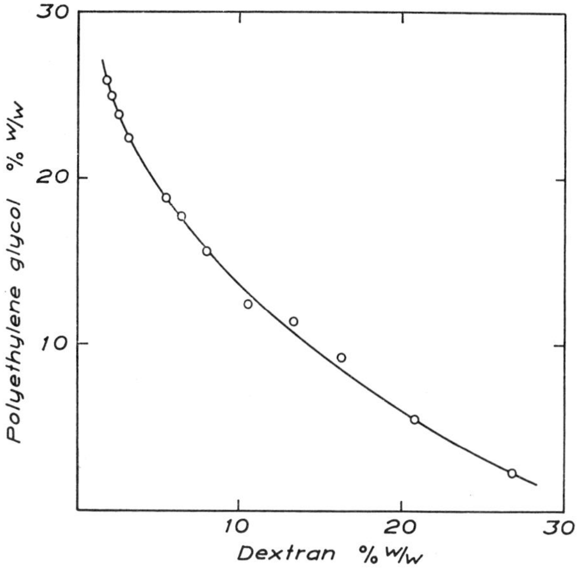

FIGURE 12.14. Binodial of the dextran–polyethylene glycol system D5–PEG 6000 at 20°C.

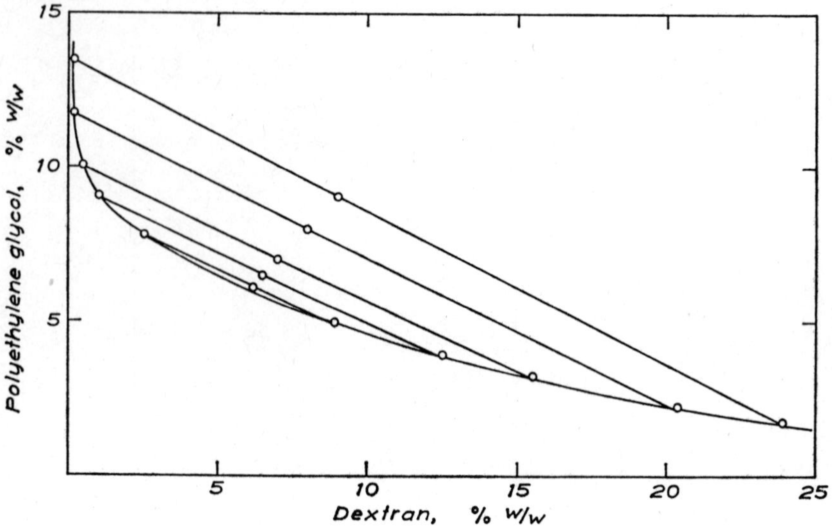

System	Total system Dextran % w/w	Total system Polyethylene glycol % w/w	Total system H_2O % w/w	Bottom phase Dextran % w/w	Bottom phase Polyethylene glycol % w/w	Bottom phase H_2O % w/w	Top phase Dextran % w/w	Top phase Polyethylene glycol % w/w	Top phase H_2O % w/w
A	6.14	6.09	87.77	8.91	4.99	86.10	2.52	7.82	89.66
B	6.50	6.50	87.00	12.48	3.93	83.59	1.00	9.09	89.91
C	7.00	7.00	86.00	15.50	3.25	81.25	0.44	10.07	89.49
D	8.00	8.00	84.00	20.34	2.28	77.38	0.15	11.80	88.05
E	9.00	9.00	82.00	23.81	1.90	74.29	0.13	13.46	86.41

FIGURE 12.15. Phase diagram and phase compositions of the dextran–polyethylene glycol system D48–PEG 4000 at 20°C.

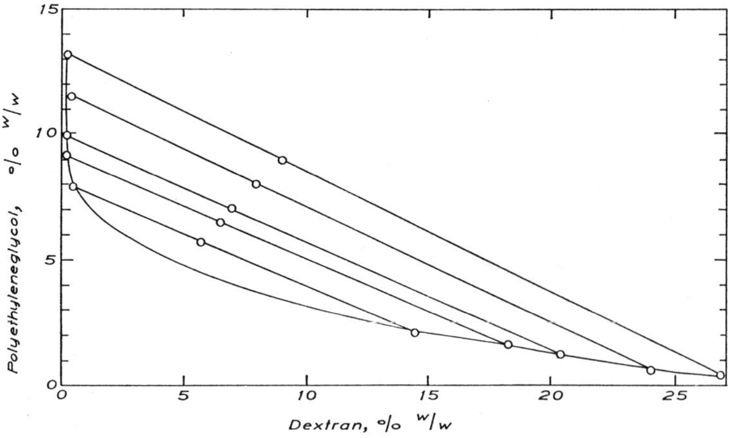

System	Total System Dextran % w/w	Total System Polyethylene glycol % w/w	Total System H_2O % w/w	Bottom phase Dextran % w/w	Bottom phase Polyethylene glycol % w/w	Bottom phase H_2O % w/w	Top phase Dextran % w/w	Top phase Polyethylene glycol % w/w	Top phase H_2O % w/v
A	5.70	5.70	88.00	14.38	2.57	83.05	0.40	7.92	91.68
B	6.50	6.50	87.00	18.21	2.05	79.74	0.18	9.13	90.69
C	7.00	7.00	86.00	20.31	1.81	77.88	0.20	9.96	89.84
D	8.00	8.00	84.00	23.96	1.49	74.55	0.35	11.53	88.12
E	9.00	9.00	82.00	26.80	1.26	71.94	0.26	13.20	86.54

FIGURE 12.16*a*. Phase diagram and phase composition of the dextran–polyethylene glycol system D48–PEG 4000 at 0°C.

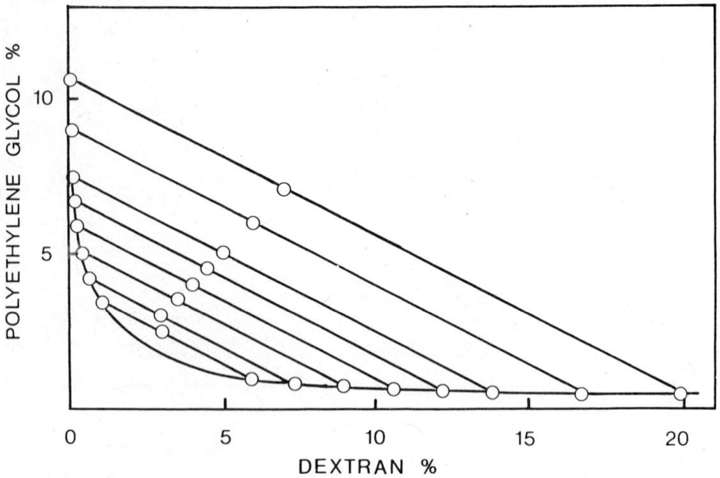

FIGURE 12.16*b*. Phase diagram of the dextran–polyethylene glycol system Dextran 500–PEG 20000 at 20°C (Schürch *et al*, 1981).

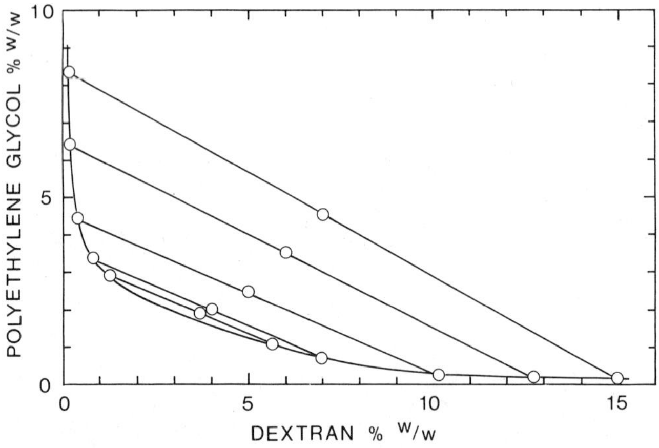

FIGURE 12.16c. Phase diagram of the dextran–polyethylene glycol system Dextran 500–PEG 40000 at 20°C.

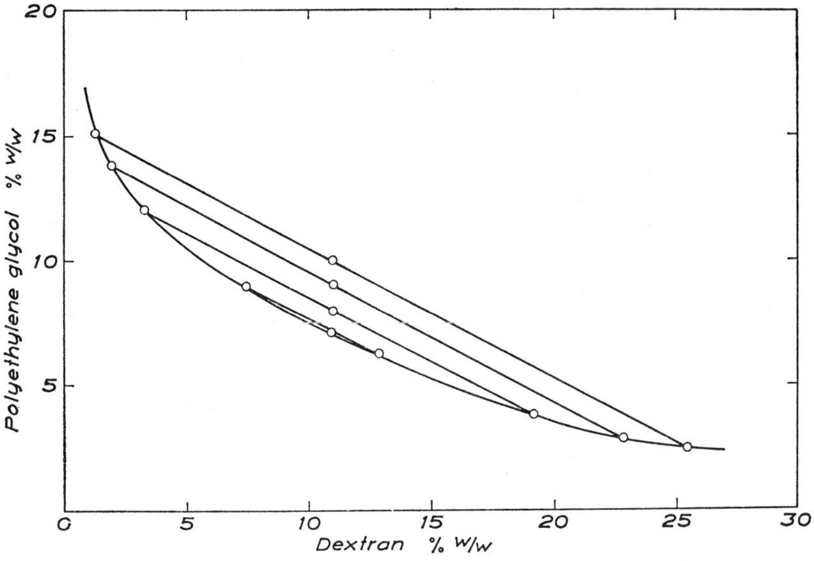

System	Total system: Dextran % w/w	Total system: Polyethylene glycol % w/w	Total system: H_2O % w/w	Bottom phase: Dextran % w/w	Bottom phase: Polyethylene glycol % w/w	Bottom phase: H_2O % w/w	Top phase: Dextran % w/w	Top phase: Polyethylene glycol % w/w	Top phase: H_2O % w/w
A	10.94	7.12	81.94	12.88	6.29	80.83	7.46	8.99	83.55
B	11.04	7.97	80.99	19.14	3.90	76.96	3.25	12.01	84.74
C	11.06	9.00	79.94	22.80	2.89	74.31	1.90	13.82	84.28
D	11.00	10.00	79.00	25.45	2.32	72.23	1.28	15.18	83.54

FIGURE 12.17. Phase diagram and phase compositions of the dextran–polyethylene glycol system D17–PEG 4000 at 20°C.

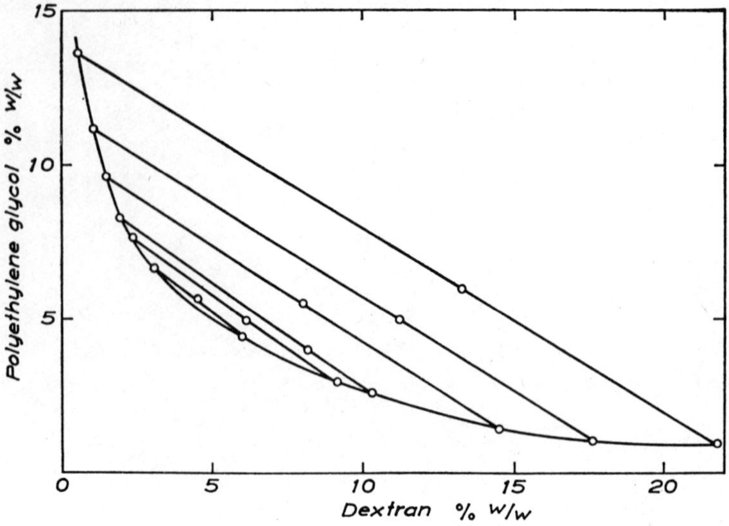

Sys-tem	Total system Dextran % w/w	Total system Poly-ethylene glycol % w/w	Total system H_2O % w/w	Bottom phase Dextran % w/w	Bottom phase Poly-ethylene glycol % w/w	Bottom phase H_2O % w/w	Top phase Dextran % w/w	Top phase Poly-ethylene glycol % w/w	Top phase H_2O % w/w
A	4.51	5.65	89.84	6.00	4.39	89.61	3.05	6.63	90.32
B	4.98	6.10	88.92	9.19	2.97	87.84	2.32	7.62	90.06
C	3.98	8.14	87.88	10.32	2.55	87.13	1.87	8.27	89.86
D	8.00	5.50	86.50	14.47	2.33	83.20	1.44	9.65	88.91
E	4.98	11.18	83.84	17.60	1.05	81.35	1.04	11.17	87.79
F	5.98	13.22	80.80	21.79	0.97	77.24	0.53	13.65	85.82

FIGURE 12.18*a*. Phase diagram and phase compositions of the dextran–polyethylene glycol system D17–PEG 20000 at 20°C.

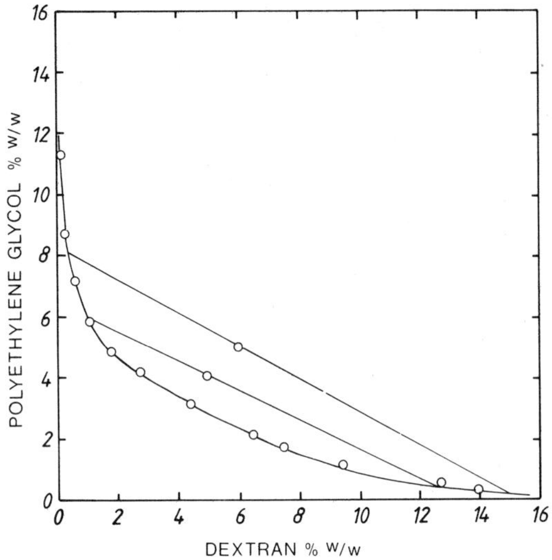

FIGURE 12.18*b*. Phase diagram of the dextran–polyethylene glycol system Dextran 40–PEG 40000 at 20°C.

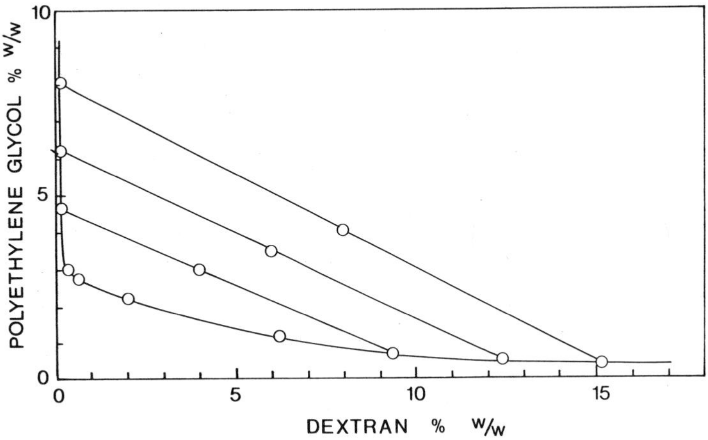

FIGURE 12.18c. Phase diagram of the dextran–polyethylene glycol system Crude dextran–PEG 20000.

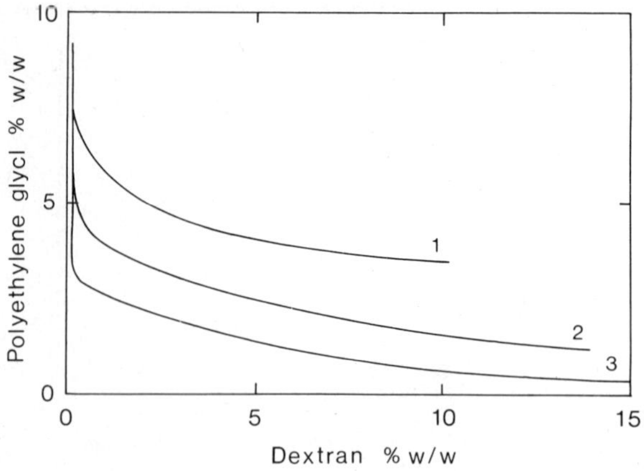

FIGURE 12.18*d*. Binodials of phase systems with crude dextran: 1, PEG 4000; 2, PEG 6000; 3, PEG 20000.

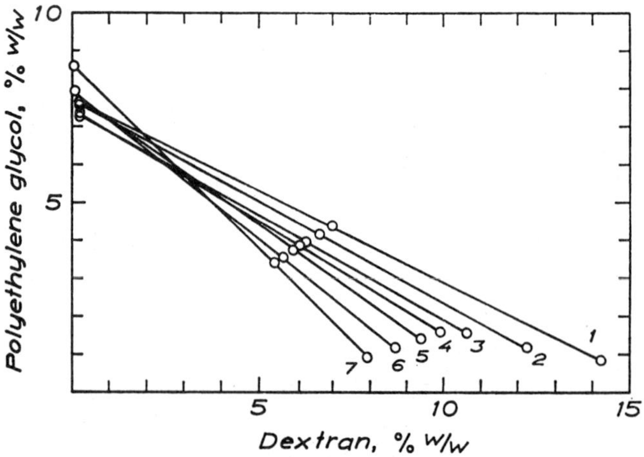

FIGURE 12.19. Tie lines of the dextran–polyethylene glycol system D48–PEG 6000 with the addition of various amounts of NaCl, at 20°C.

Line Number	Moles NaCl[a]	Line Number	Moles NaCl[a]
1	0.1	5	3.0
2	1.0	6	4.0
3	2.0	7	5.0
4	2.5		

[a]Added to 1 kg of the system 7% w/w D48; 4.4% w/w PEG 6000; 88.6% w/w H_2O.

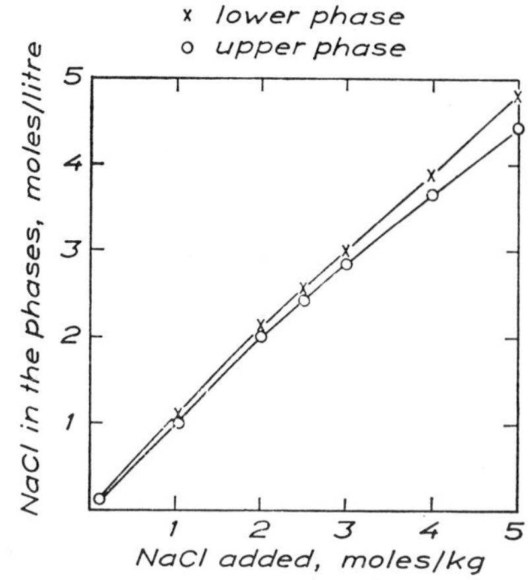

FIGURE 12.20. Partition of NaCl in the dextran–polyethylene glycol system D48–PEG 6000. *Abscissa:* molar concentration of NaCl in the phases. *Ordinate:* moles of NaCl added to 1 kg of the system 7% w/w D48; 4.4% w/w PEG 6000; 88.6% w/w H_2O, at 20°C.

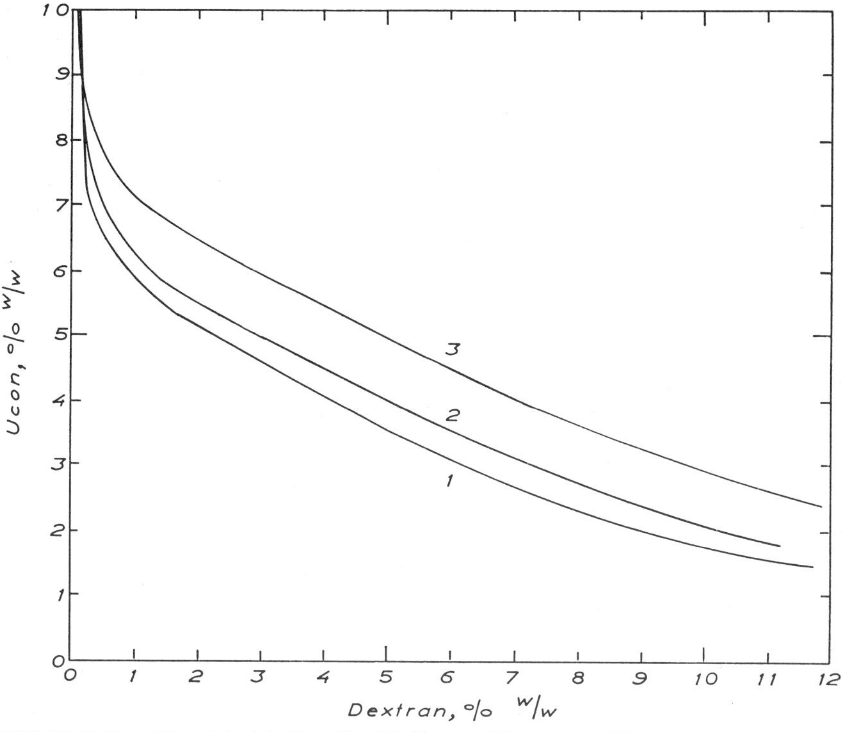

FIGURE 12.21. Bionodials of the Ucon 50 · HB–Dextran 500 system at 23°C: 1, Ucon 50 · HB · 5100; 2, Ucon 50 · HB · 3520; 3, Ucon 50 · HB · 2000.

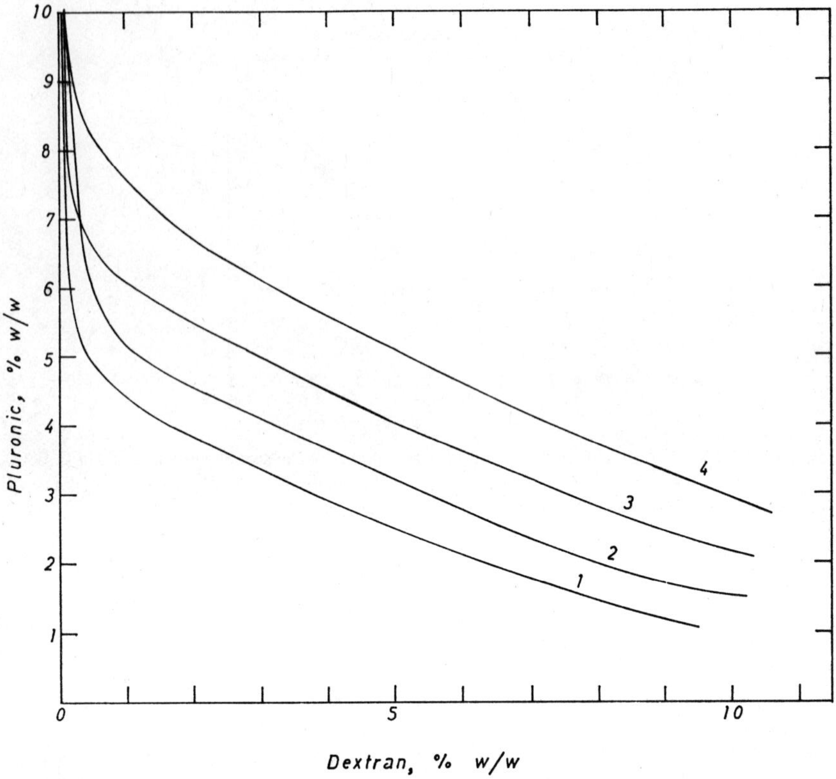

FIGURE 12.22. Binodials of dextran 500–Pluronic systems at 23°C: 1, Pluronic F-98; 2, Pluronic F-88; 3, Pluronic F-77; 4, Pluronic P-85.

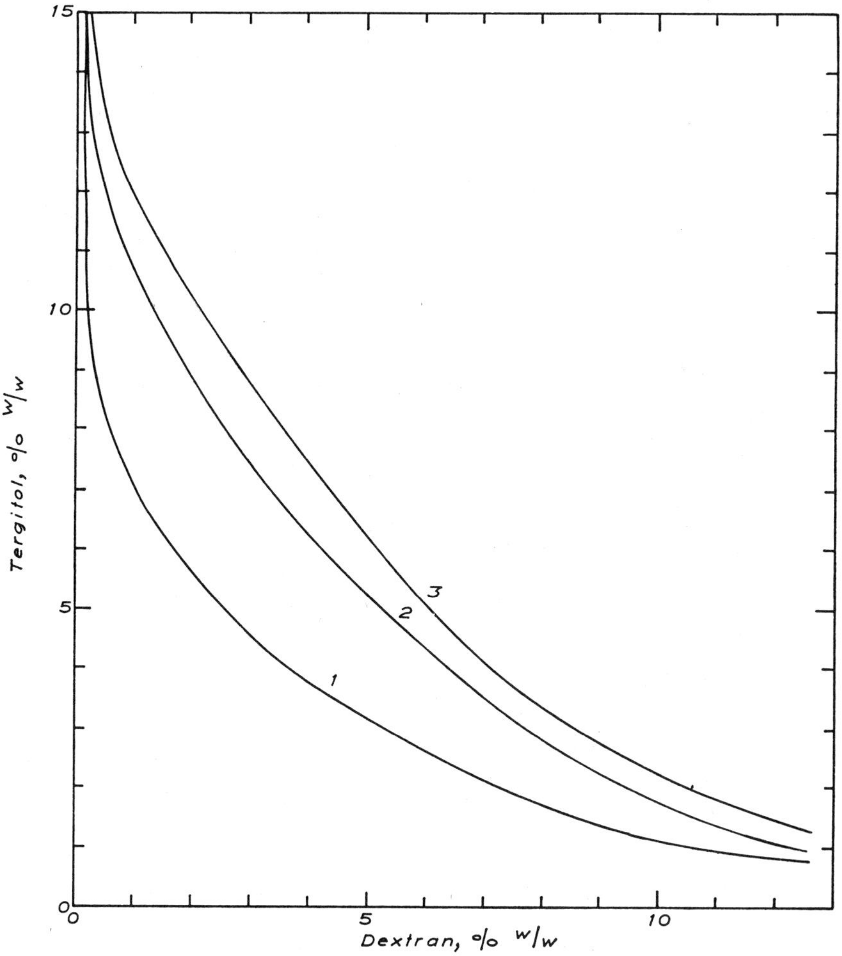

FIGURE 12.23. Binodials of the Tergitol NP–Dextran 500 system at 23°C: 1, Tergitol NP-44; 2, Tergitol NP-40; 3, Tergitol NP-35.

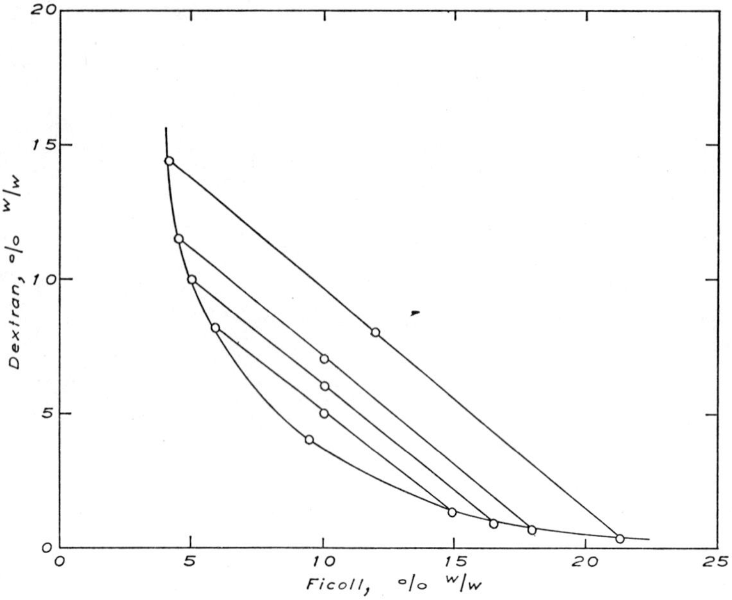

Sys-tem	Total system Ficoll % w/w	Total system Dextran % w/w	Total system H_2O % w/w	Bottom phase Ficoll % w/w	Bottom phase Dextran % w/w	Bottom phase H_2O % w/w	Top phase Ficoll % w/w	Top phase Dextran % w/w	Top phase H_2O % w/w
A	10.00	5.00	85.00	14.93	1.40	83.67	5.90	8.17	85.93
B	10.00	6.00	84.00	16.51	1.00	82.49	5.00	9.91	85.09
C	10.00	7.00	83.00	17.99	0.76	81.25	4.53	11.48	83.99
D	12.00	8.00	80.00	21.33	0.41	78.26	4.19	14.35	81.46

FIGURE 12.24*a*. Phase diagrams and phase compositions of the Ficoll–dextran system F–D500 at 23°C.

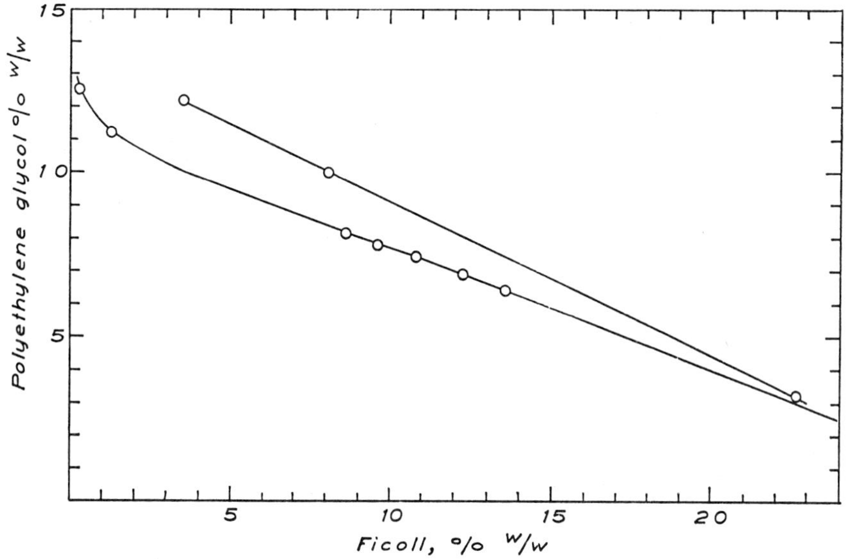

FIGURE 12.24*b*. Binodial of the Ficoll–polyethylene glycol system F–PEG 6000 at 23°C.

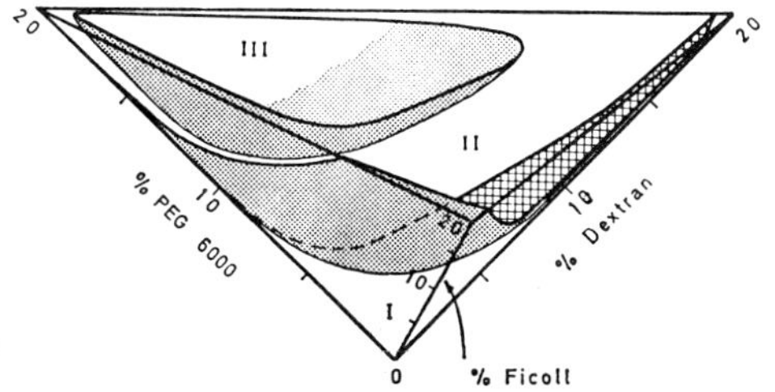

FIGURE 12.24c. Phase diagram of the aqueous Ficoll–Dextran 500–PEG 6000 system at 23°C. The three concentration axes are perpendicular to each other. The two curved surfaces separate the regions of one (I), two (II), and three (III) phases.

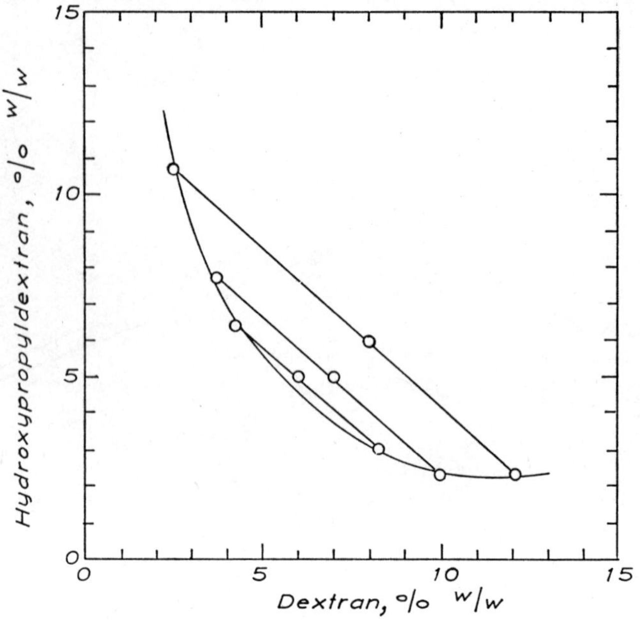

	Total system			Bottom phase			Top phase		
System	Dextran % w/w	HPD % w/w	H_2O % w/w	Dextran % w/w	HPD % w/w	H_2O % w/w	Dextran % w/w	HPD % w/w	H_2O % w/w
A	6.00	5.00	89.00	8.24	3.04	88.72	4.22	6.40	89.38
B	7.00	5.00	88.00	9.96	2.30	87.70	3.71	7.74	88.55
C	8.00	6.00	86.00	12.08	2.34	85.58	2.46	10.76	86.78

FIGURE 12.25. Phase diagrams and phase compositions of the dextran–hydroxypropyl dextran system D500–HDP 500 A at 23°C.

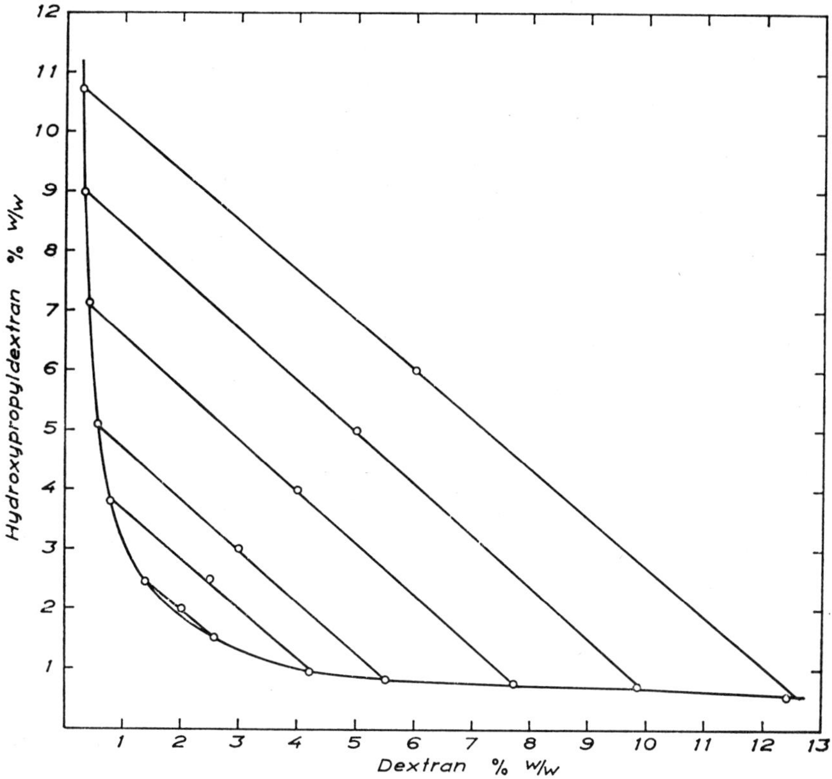

Sys-tem	Total system: Dextran % w/w	Total system: Hydroxy-propyl-dextran % w/w	Total system: H_2O % w/w	Bottom phase: Dextran % w/w	Bottom phase: Hydroxy-propyl-dextran % w/w	Bottom phase: H_2O % w/w	Top phase: Dextran % w/w	Top phase: Hyproxy-propyl-dextran % w/w	Top phase: H_2O % w/w
A	2.00	2.00	96.00	2.56	1.52	95.92	1.37	2.47	96.16
B	2.50	2.50	95.00	4.20	0.95	94.85	0.77	3.78	95.45
C	3.00	3.00	94.00	5.51	0.82	93.67	0.56	5.19	94.25
D	4.00	4.00	92.00	7.72	0.75	91.53	0.39	7.13	92.48
E	5.00	5.00	90.00	9.88	0.72	89.40	0.29	8.96	90.75
F	6.00	6.00	88.00	12.38	0.51	87.11	0.24	10.67	89.09

FIGURE 12.26. Phase diagram and phase compositions of the dextran–hydroxypropyldextran system D68–HPD 70 at 4°C.

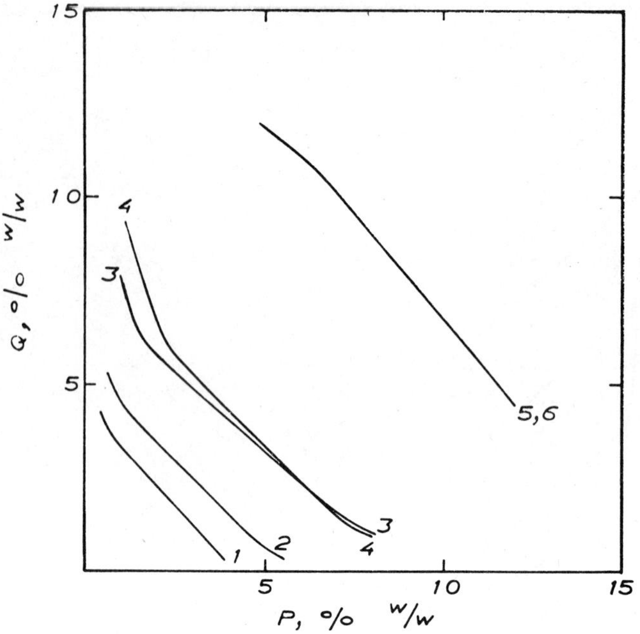

FIGURE 12.27. Binodials of the phase systems obtained for various pairs of the polymers Dextran 500 and hydroxypropyldextran 500 with different degrees of substitution (DS). 1, Dextran–HP Dextran C (DS = 1.50); 2, Dextran–HP Dextran B (DS = 0.76); 3, Dextran–HP Dextran A (DS = 0.39); 4, HP Dextran A–HP Dextran C; 5, HP Dextran B–HP Dextran C; 6, HP Dextran A–HP Dextran B.

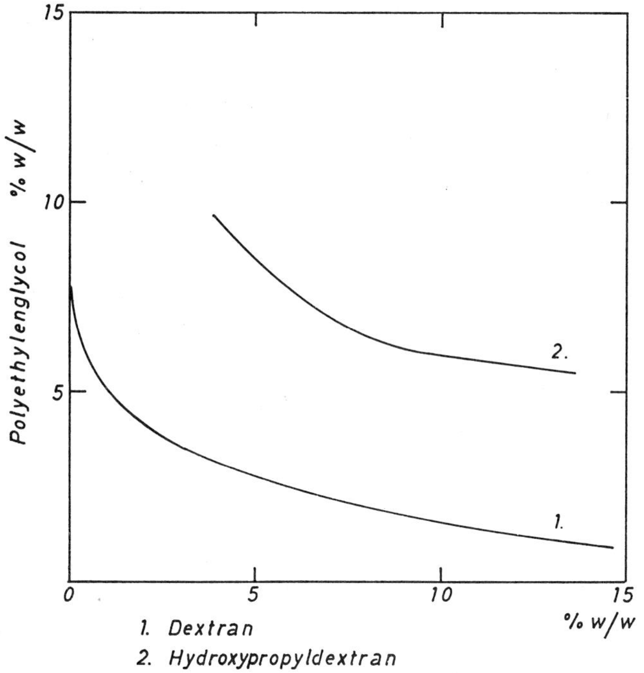

FIGURE 12.28. Binodials of the system dextran–polyethylene glycol and hydroxypropyl dextran–polyethylene glycol at 0–1°C. 1, Dextran 500–PEG 6000; 2, HP Dextran 500–PEG 6000.

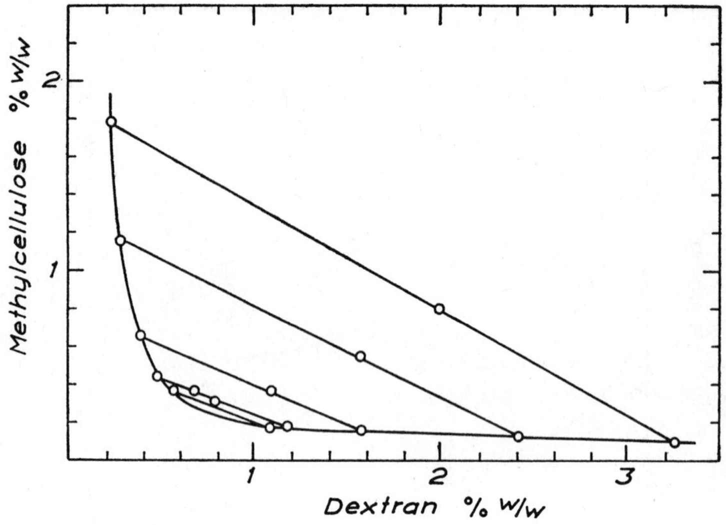

Sys-tem	Total system Dextran % w/w	Methyl-cellu-lose % w/w	H_2O % w/w	Bottom phase Dextran % w/w	Methyl-cellu-lose % w/w	H_2O % w/w	Top phase Dextran % w/w	Methyl-cellu-lose % w/w	H_2O % w/w
A	0.79	0.30	98.91	1.09	0.17	98.74	0.57	0.36	99.07
A 1	0.68	0.36	98.96	1.18	0.17	98.65	0.47	0.43	99.10
B	1.10	0.36	98.54	1.58	0.15	98.27	0.39	0.65	98.96
C	1.58	0.54	97.88	2.42	0.12	97.46	0.28	1.15	98.57
D	2.00	0.80	97.20	3.27	0.10	96.63	0.23	1.78	97.99

FIGURE 12.29. Phase diagram and phase compositions of the dextran–methylcellulose system D68–MC 4000 at 20°C.

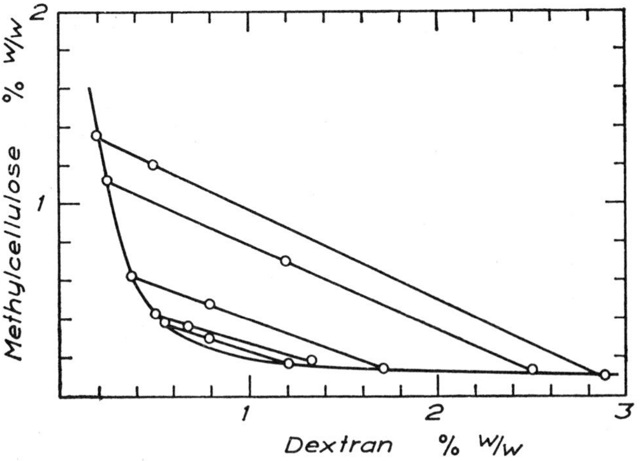

Sys-tem	Total system Dextran % w/w	Methyl-cellu-lose % w/w	H_2O % w/w	Bottom phase Dextran % w/w	Methyl-cellu-lose % w/w	H_2O % w/w	Top phase Dextran % w/w	Methyl-cellu-lose % w/w	H_2O % w/w
A	0.80	0.30	98.90	1.21	0.17	98.62	0.56	0.38	99.06
A 1	0.68	0.36	98.96	1.33	0.18	98.49	0.51	0.43	99.06
B	0.80	0.48	98.72	1.72	0.14	98.14	0.38	0.63	98.99
C	1.20	0.70	98.10	2.51	0.14	97.35	0.25	1.12	98.63
D	0.50	1.20	98.30	2.88	0.09	97.03	0.18	1.37	98.45

FIGURE 12.30. Phase diagram and phase compositions of the dextran–methylcellulose system D68–MC 4000 at 4°C.

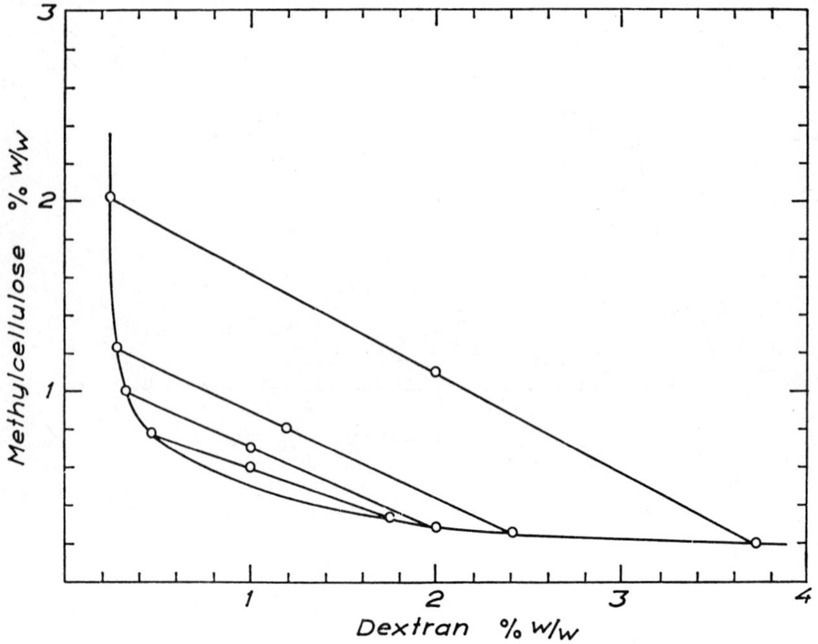

Sys-tem	Total system: Dextran % w/w	Total system: Methyl-cellu-lose % w/w	Total system: H_2O % w/w	Bottom phase: Dextran % w/w	Bottom phase: Methyl-cellu-lose % w/w	Bottom phase: H_2O % w/w	Top phase: Dextran % w/w	Top phase: Methyl-cellu-lose % w/w	Top phase: H_2O % w/w
A	1.00	0.60	98.40	1.76	0.33	97.91	0.47	0.78	98.75
B	1.00	0.70	98.30	2.01	0.29	97.70	0.33	1.00	98.67
C	1.20	0.80	98.00	2.42	0.26	97.32	0.28	1.23	98.49
D	2.00	1.10	96.90	3.74	0.20	96.06	0.24	2.02	97.74

FIGURE 12.31. Phase diagram and phase compositions of the dextran–methylcellulose system D68–MC 400 at 20°C.

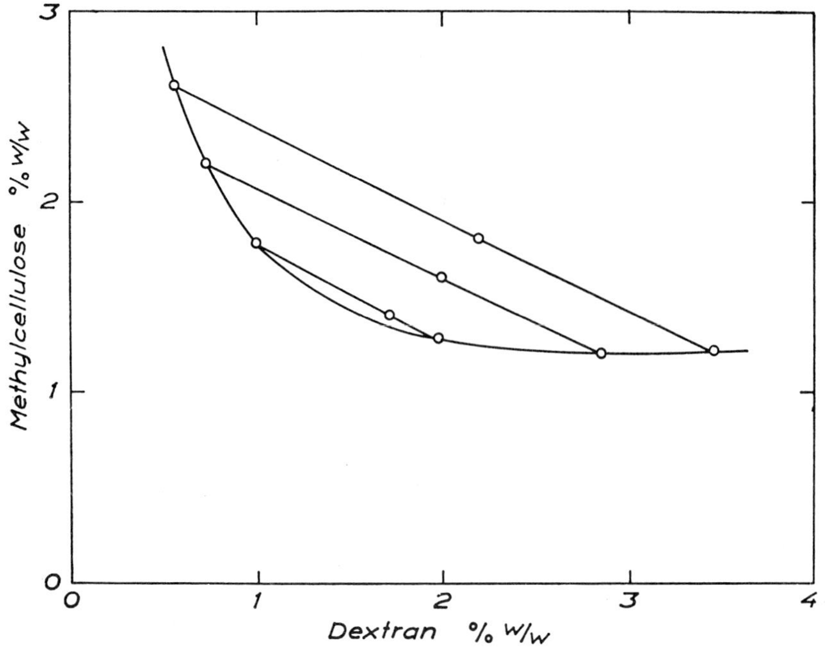

Sys-tem	Total system: Dextran % w/w	Total system: Methyl-cellu-lose % w/w	Total system: H_2O % w/w	Bottom phase: Dextran % w/w	Bottom phase: Methyl-cellu-lose % w/w	Bottom phase: H_2O % w/w	Top phase: Dextran % w/w	Top phase: Methyl-cellu-lose % w/w	Top phase: H_2O % w/w
A	1.72	1.40	96.88	1.98	1.29	96.73	1.00	1.78	97.22
B	2.00	1.60	96.40	2.85	1.20	95.95	0.73	2.20	97.07
C	2.20	1.80	96.00	3.45	1.22	95.33	0.55	2.61	96.84

FIGURE 12.32. Phase diagram and phase compositions of the dextran–methylcellulose system D68–MC 10 at 20°C.

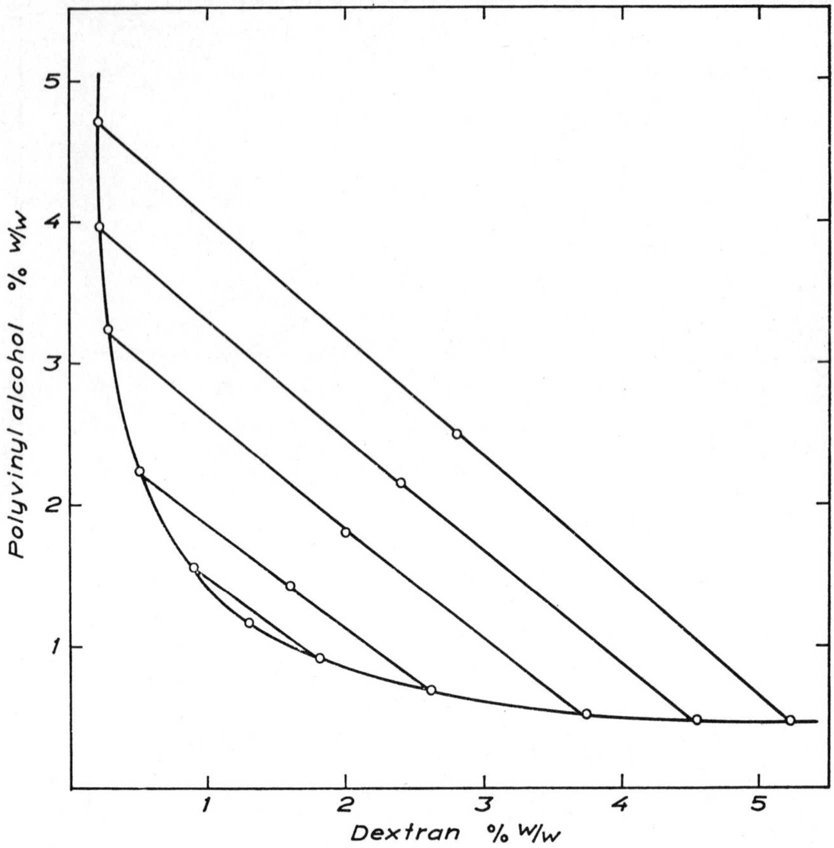

System	Total system Dextran % w/w	Total system Poly-vinyl-alcohol % w/w	Total system H_2O % w/w	Bottom phase Dextran % w/w	Bottom phase Poly-vinyl-alcohol % w/w	Bottom phase H_2O % w/w	Top phase Dextran % w/w	Top phase Poly-vinyl-alcohol % w/w	Top phase H_2O % w/w
A	1.30	1.16	97.54	1.81	0.92	97.27	0.91	1.56	97.53
B	1.60	1.42	96.98	2.61	0.68	96.71	0.50	2.23	97.27
C	2.00	1.78	96.22	3.74	0.53	95.73	0.28	3.25	96.47
D	2.40	2.14	95.46	4.55	0.48	94.97	0.21	3.99	95.80
E	2.80	2.49	94.71	5.23	0.47	94.30	0.20	4.69	95.11

FIGURE 12.33. Phase diagram and phase compositions of the dextran–polyvinyl alcohol system D68–PVA 48/20 at 20°C.

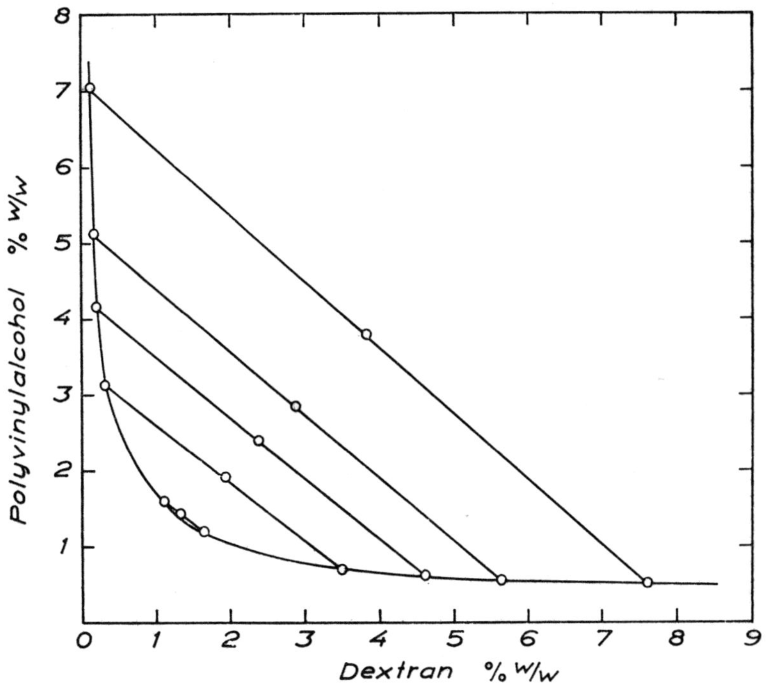

System	Total system Dextran % w/w	Total system Polyvinyl-alcohol % w/w	Total system H_2O % w/w	Bottom phase Dextran % w/w	Bottom phase Polyvinyl-alcohol % w/w	Bottom phase H_2O % w/w	Top phase Dextran % w/w	Top phase Polyvinyl-alcohol % w/w	Top phase H_2O % w/w
A	1.34	1.44	97.22	1.65	1.19	97.16	1.11	1.59	97.30
B	1.91	1.90	96.19	3.49	0.69	95.82	0.31	3.13	96.56
C	2.39	2.38	95.23	4.60	0.61	94.79	0.18	4.16	95.66
D	2.87	2.85	94.28	5.64	0.55	93.81	0.16	5.11	94.73
E	3.82	3.80	92.38	7.60	0.53	91.87	0.12	7.06	92.82

FIGURE 12.34. Phase diagram and phase compositions of the dextran–polyvinyl alcohol system D68–PVA 28/20 at 20°C.

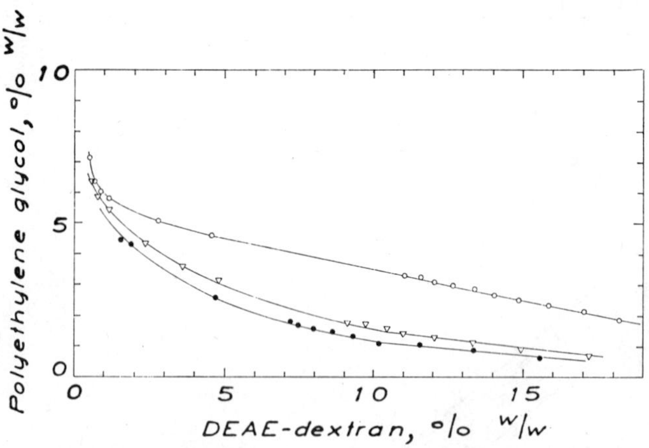

FIGURE 12.35. Binodial of the system DEAE-dextran 500–polyethylene glycol, PEG 6000–$LiSO_4$, at different concentrations of $LiSO_4$. ○, 0.1 *M*; ▽, 0.3 *M*; ●, 0.5*M*.

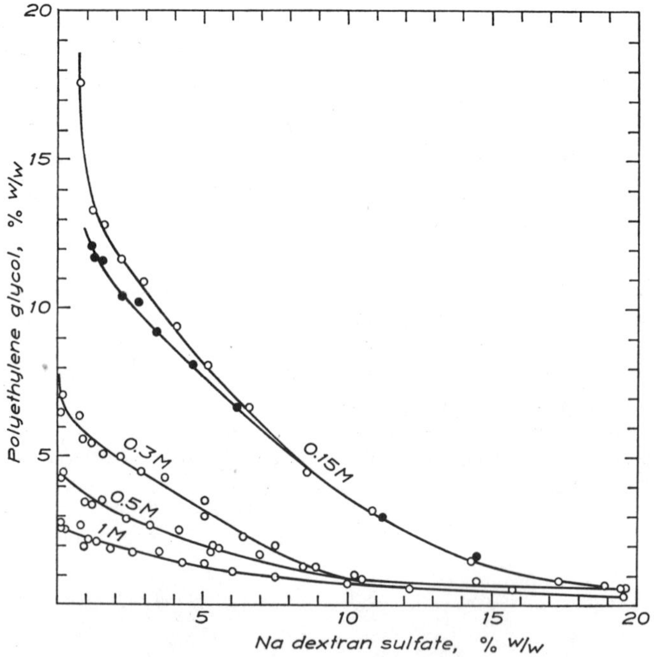

FIGURE 12.36. Binodials of the Na dextran sulfate–polyethylene glycol system NaDS68–PEG 6000 at various NaCl concentrations (expressed as moles per liter phase system). ○, 20°C; ●, 4°C.

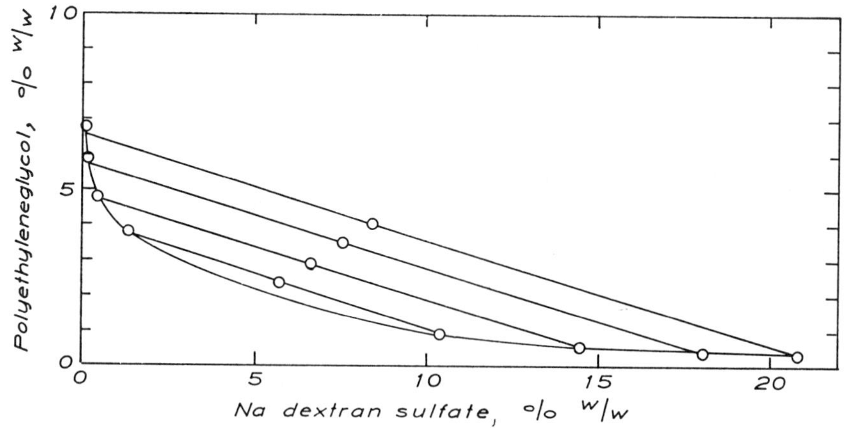

Sys-tem	Total system: Na dextran sulfate % w/w	Total system: Poly-ethylene glycol % w/w	Bottom phase: Na dextran sulfate % w/w	Bottom phase: Poly-ethylene glycol % w/w	Top phase: Na dextran sulfate % w/w	Top phase: Poly-ethylene glycol % w/w
A	5.71	2.38	10.40	—	1.24	3.80
B	6.63	2.84	14.46	—	0.40	4.79
C	7.54	3.48	18.07	—	0.15	5.88
D	8.43	4.03	20.81	—	0.09	6.78

FIGURE 12.37. Phase diagram and polymer composition of the Na dextran sulfate–polyethylene glycol system NaDS48–PEG 6000 in 0.3 *M* NaCl at 4°C. (NaDS48 = batch To 255 of dextran sulfate 500).

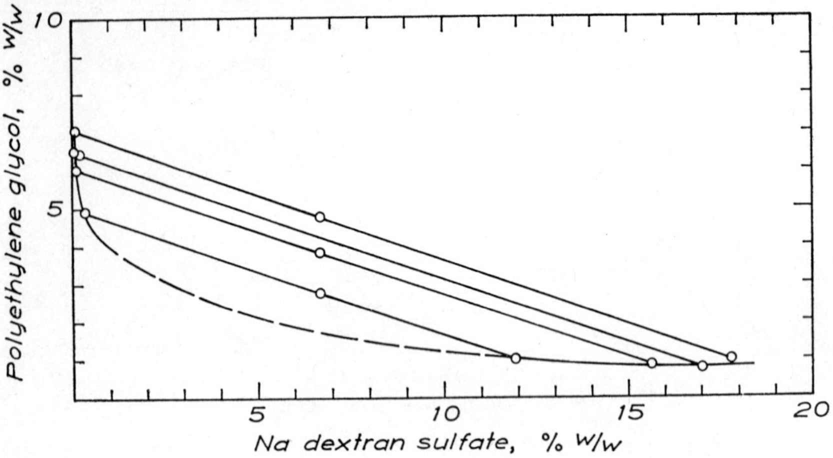

System	Total system: Na dextran sulfate % w/w	Total system: Polyethylene glycol % w/w	Bottom phase: Na dextran sulfate % w/w	Bottom phase: Polyethylene glycol % w/w	Top phase: Na dextran sulfate % w/w	Top phase: Polyethylene glycol % w/w
A	6.69	2.87	11.88	1.0	0.31	4.91
B	6.67	3.81	15.59	0.8	0.08	6.02
C	0.20	6.45	17.01	0.8	0.01	6.5
D	6.67	4.76	17.79	1.0	0.03	7.03

FIGURE 12.38. Phase diagram and polymer compositions of the Na dextran sulfate–polyethylene glycol system NaDS68–PEG 6000 in 0.3 *M* NaCl at 4°C.

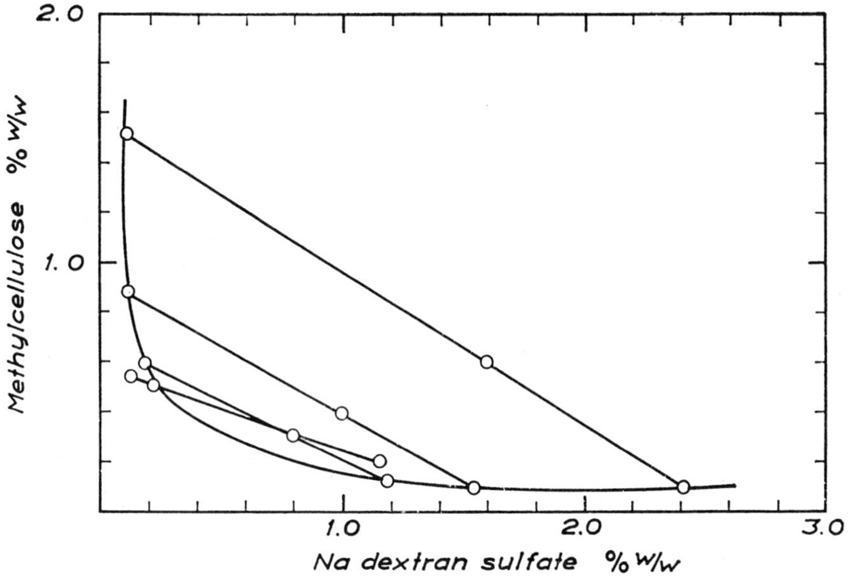

System	Total system: Na dextran sulfate % w/w	Total system: Methyl-cellu-lose % w/w	Bottom phase: Na dextran sulfate % w/w	Bottom phase: Methyl-cellu-lose % w/w	Top phase: Na dextran sulfate % w/w	Top phase: Methyl-cellu-lose % w/w
A	0.22	0.51	1.15	0.20	0.13	0.54
B	0.80	0.30	1.19	0.13	0.18	0.59
C	1.00	0.40	1.55	0.10	0.11	0.88
D	1.60	0.60	2.42	0.09	0.11	1.52

FIGURE 12.39. Phase diagram and polymer compositions of the Na dextran sulfate–methylcellulose system NaDS68–MC 4000 in 0.15 *M* NaCl at 4°C.

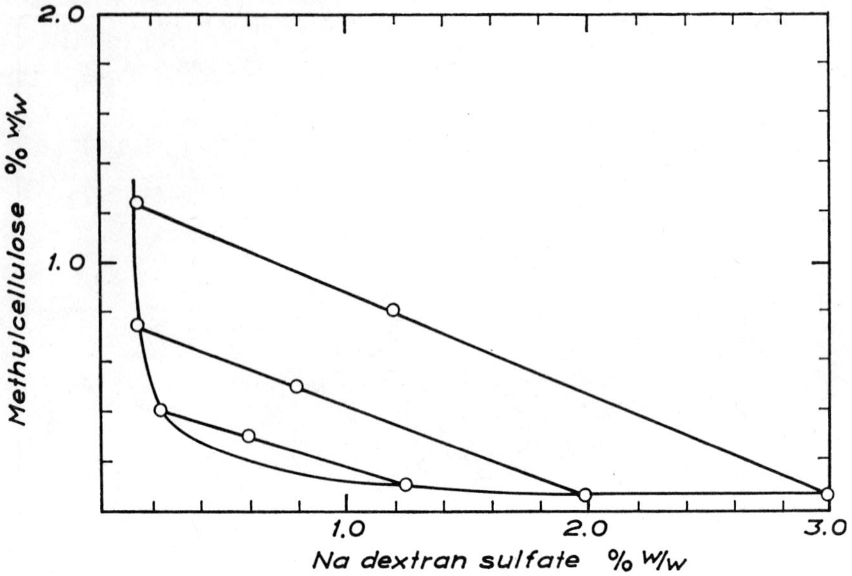

System	Total system: Na dextran sulfate % w/w	Total system: Methyl-cellu-lose % w/w	Bottom phase: Na dextran sulfate % w/w	Bottom phase: Methyl-cellu-lose % w/w	Top phase: Na dextran sulfate % w/w	Top phase: Methyl-cellu-lose % w/w
A	0.60	0.30	1.25	0.11	0.24	0.41
B	0.80	0.50	1.99	0.06	0.14	0.75
C	1.20	0.80	3.00	0.06	0.15	1.24

FIGURE 12.40. Phase diagram and polymer compositions of the Na dextran sulfate–methylcellulose system NaDS68–MC 4000 in 0.3 *M* NaCl at 4°C.

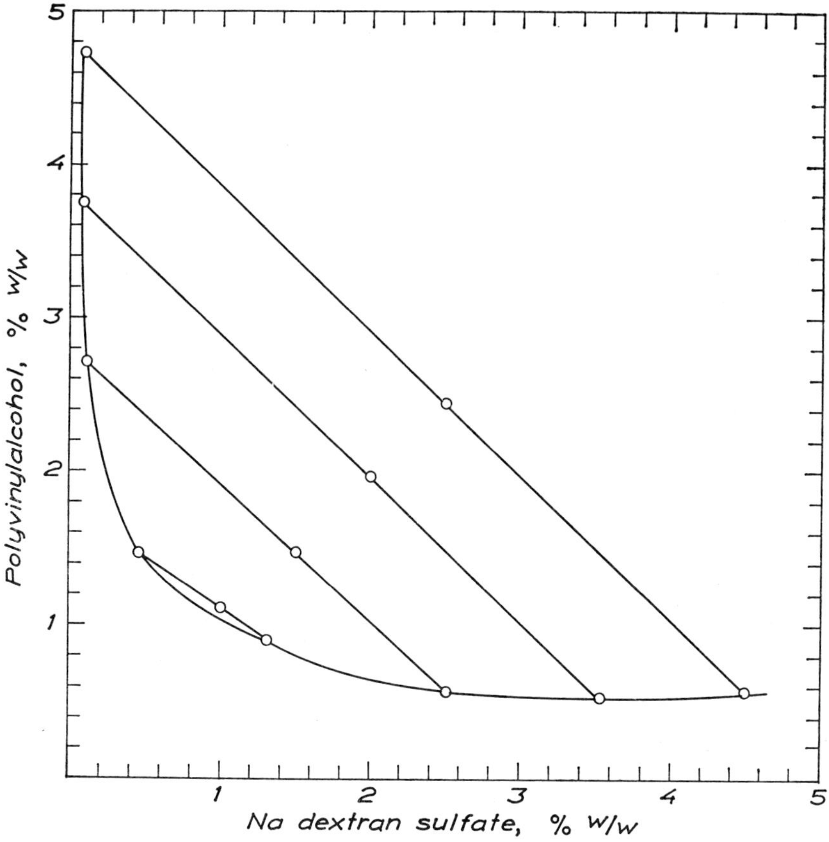

System	Total system: Na dextran sulfate % w/w	Total system: Poly-vinyl-alcohol % w/w	Bottom phase: Na dextran sulfate % w/w	Bottom phase: Poly-vinyl-alcohol % w/w	Top phase: Na dextran sulfate % w/w	Top phase: Poly-vinyl-alcohol % w/w
A	1.00	1.11	1.31	0.90	0.45	1.47
B	1.50	1.47	2.51	0.57	0.10	2.71
C	2.00	1.96	3.53	0.54	0.08	3.75
D	2.50	2.44	4.49	0.57	0.08	4.73

FIGURE 12.41. Phase diagram and polymer compositions of the phases of the Na dextran sulfate–polyvinyl alcohol system NaDS68–PVA 48/20 in 0.15 *M* NaCl at 4°C.

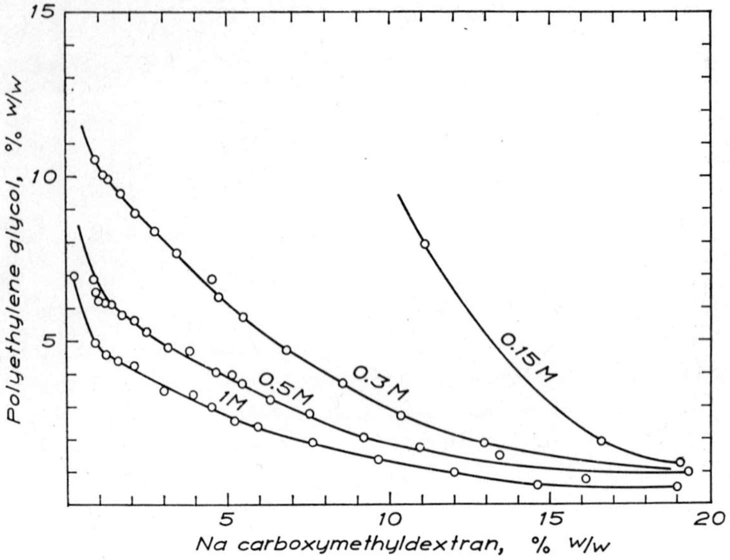

FIGURE 12.42*a*. Binodials of the Na carboxymethyldextran–polyethylene glycol system Na-CMD68–PEG 6000 at various NaCl concentrations at 20°C.

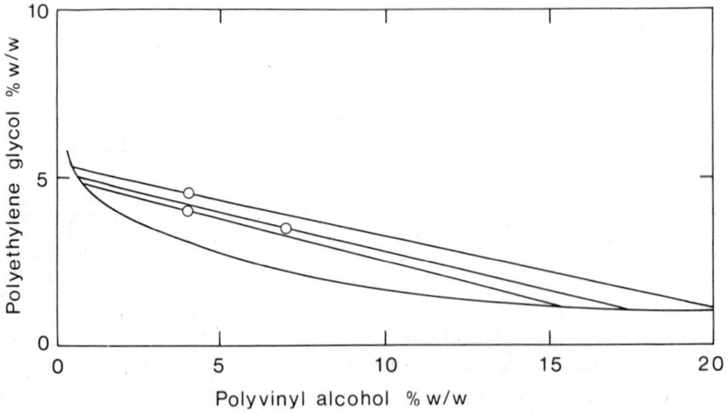

FIGURE 12.42*b*. Phase diagram of the system polyvinyl alcohol–polyethylene glycol PVA 13000–PEG 20000 at 20°C.

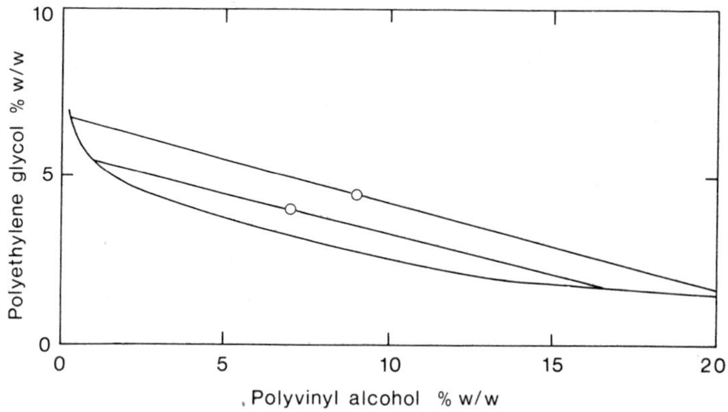

FIGURE 12.42c. Phase diagram of the system polyvinyl alcohol–polyethylene glycol PVA 22000–PEG 6000 at 20°C.

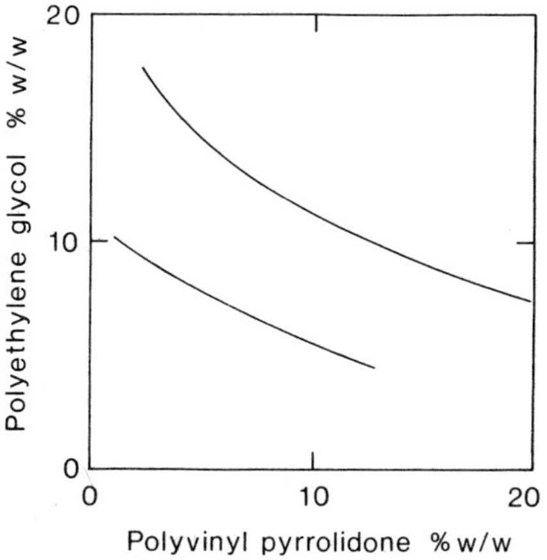

FIGURE 12.42*d*. Binodials of the phase system polyvinylpyrrolidone PVP 40000–PEG 20000 (*upper*) and PVP 350000–PEG 20000 (*lower*) at 20°C.

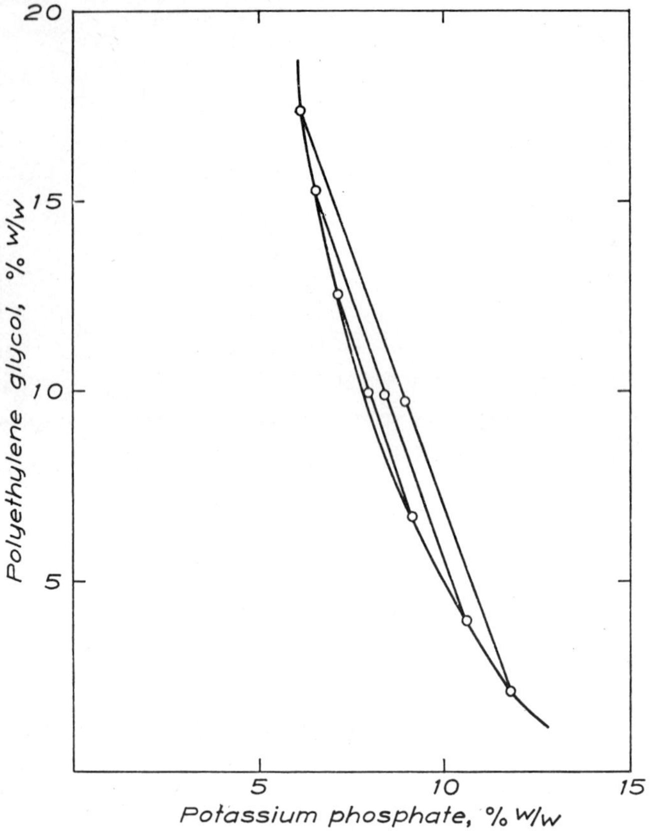

System	Total system: Potassium phosphate % w/w	Total system: Polyethylene glycol % w/w	Total system: H_2O % w/w	Bottom phase: Potassium phosphate % w/w	Bottom phase: Polyethylene glycol % w/w	Bottom phase: H_2O % w/w	Top phase: Potassium phosphate % w/w	Top phase: Polyethylene glycol % w/w	Top phase: H_2O % w/w
A	7.92	10.04	82.04	9.13	6.68	84.19	7.13	12.57	80.30
B	8.36	9.91	81.73	10.57	3.95	85.48	6.52	15.25	78.23
C	8.94	9.69	81.37	11.74	2.08	86.18	6.11	17.36	76.53

FIGURE 12.43. Phase diagram and phase compositions of the potassium phosphate–PEG 20000 system at 20°C.

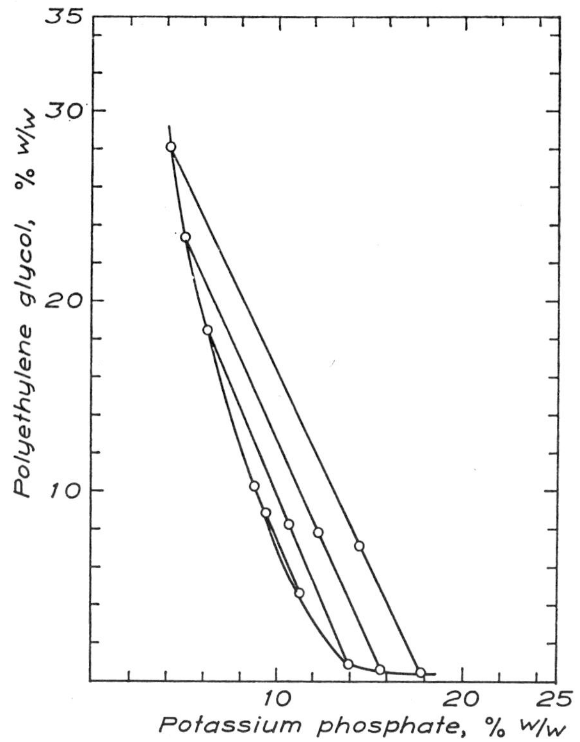

	Total system			Bottom phase			Top phase		
System	Potassium phosphate % w/w	Polyethylene glycol % w/w	H_2O % w/w	Potassium phosphate % w/w	Polyethylene glycol % w/w	H_2O % w/w	Potassium phosphate % w/w	Polyethylene glycol % w/w	H_2O % w/w
A	9.37	8.79	81.84	11.17	4.65	84.18	8.76	10.20	81.04
B	10.65	8.26	81.09	13.84	0.90	85.26	6.13	18.45	75.42
C	12.17	7.79	80.04	15.53	0.55	83.92	5.03	23.28	71.69
D	14.37	7.04	78.59	17.66	0.37	81.97	4.13	28.00	67.87

FIGURE 12.44. Phase diagram and phase compositions of the potassium phosphate–PEG 6000 system at 20°C.

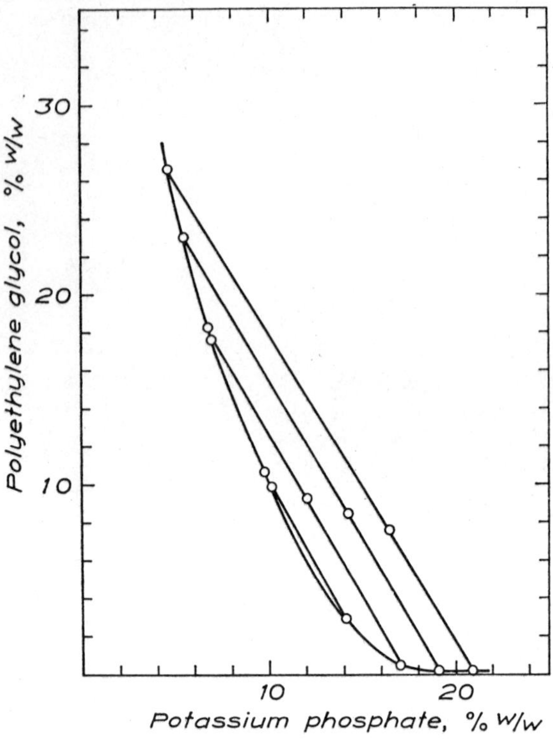

Sys-tem	Total system: Potassium phosphate % w/w	Total system: Polyethylene glycol % w/w	Total system: H_2O % w/w	Bottom phase: Potassium phosphate % w/w	Bottom phase: Polyethylene glycol % w/w	Bottom phase: H_2O % w/w	Top phase: Potassium phosphate % w/w	Top phase: Polyethylene glycol % w/w	Top phase: H_2O % w/w
A	10.16	9.92	79.92	13.99	2.94	83.07	9.74	10.63	79.63
B	12.02	9.20	78.78	17.01	0.50	82.49	6.70	18.26	75.04
C	14.17	8.39	77.44	19.00	0.19	80.81	5.46	23.03	71.51
D	16.38	7.55	76.07	20.87	0.18	78.95	4.60	26.57	68.83

FIGURE 12.45. Phase diagram and phase compositions of the potassium phosphate–PEG 6000 system at 0°C.

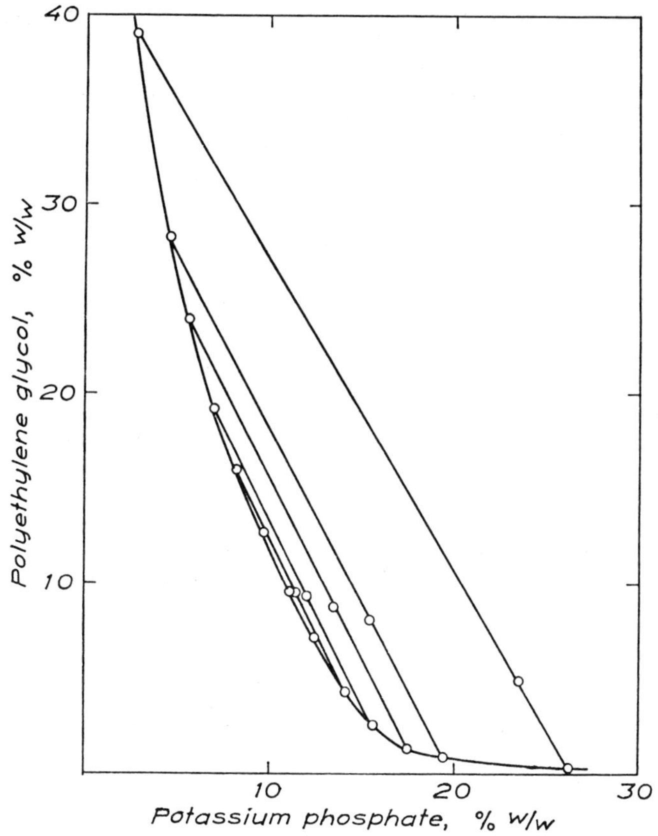

	Total system			Bottom phase			Top phase		
System	Potassium phosphate % w/w	Polyethylene glycol % w/w	H_2O % w/w	Potassium phosphate % w/w	Polyethylene glycol % w/w	H_2O % w/w	Potassium phosphate % w/w	Polyethylene glycol % w/w	H_2O % w/w
A	11.04	9.55	79.41	12.35	7.09	80.56	9.67	12.65	77.68
B	11.37	9.42	79.21	14.06	4.23	81.71	8.19	15.96	75.85
C	11.97	9.19	78.84	15.46	2.54	82.00	7.01	19.16	73.83
D	13.36	8.67	77.97	17.41	1.30	81.29	5.56	23.90	70.54
E	15.30	7.93	76.77	19.41	0.78	79.81	4.55	28.15	67.30
F	23.37	4.87	71.76	26.26	1.01	72.73	2.68	38.92	58.40

FIGURE 12.46. Phase diagram and phase compositions of the potassium phosphate–PEG 4000 system at 20°C.

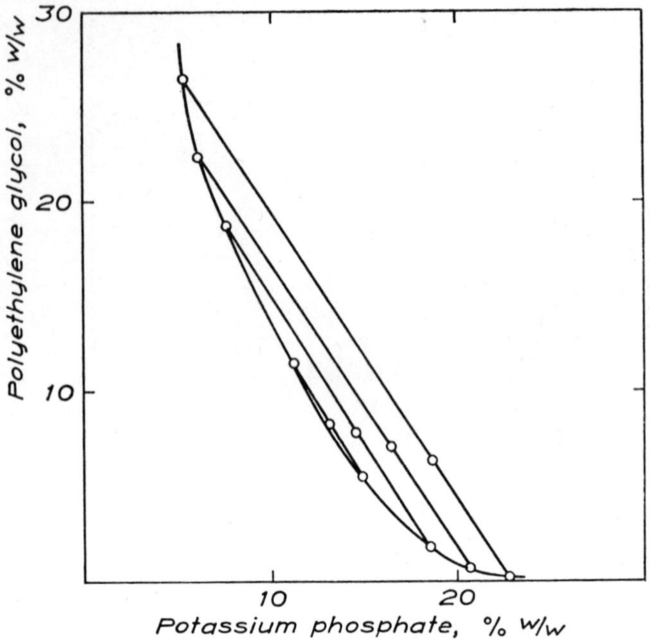

	Total system			Bottom phase			Top phase		
System	Potassium phosphate % w/w	Polyethylene glycol % w/w	H_2O % w/w	Potassium phosphate % w/w	Polyethylene glycol % w/w	H_2O % w/w	Potassium phosphate % w/w	Polyethylene glycol % w/w	H_2O % w/w
A	13.11	8.23	78.66	14.87	5.48	79.65	11.13	11.44	77.43
B	14.54	7.74	77.72	18.57	1.72	79.71	7.63	18.66	73.71
C	16.51	7.02	76.47	20.67	0.63	78.70	5.98	22.29	71.73
D	18.63	6.30	75.07	22.73	0.14	77.13	5.29	26.33	68.38

FIGURE 12.47. Phase diagram and phase compositions of the potassium phosphate–PEG 4000 system at 0°C.

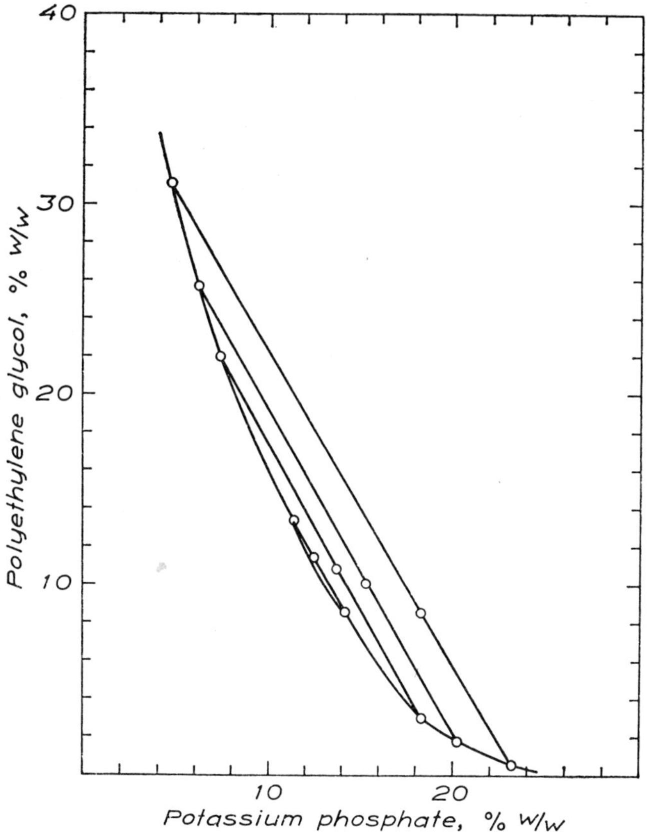

System	Total system: Potassium phosphate % w/w	Total system: Polyethylene glycol % w/w	Total system: H_2O % w/w	Bottom phase: Potassium phosphate % w/w	Bottom phase: Polyethylene glycol % w/w	Bottom phase: H_2O % w/w	Top phase: Potassium phosphate % w/w	Top phase: Polyethylene glycol % w/w	Top phase: H_2O % w/w
A	12.36	11.37	76.27	14.12	8.57	77.31	11.35	13.38	75.27
B	13.66	10.74	75.60	18.34	2.94	78.72	7.31	21.86	70.83
C	15.21	9.99	74.80	20.13	1.68	78.19	6.11	25.64	68.25
D	18.13	8.56	73.31	23.17	0.48	76.35	4.62	31.00	64.38

FIGURE 12.48. Phase diagram and phase compositions of the potassium phosphate–PEG 1540 system at 20°C.

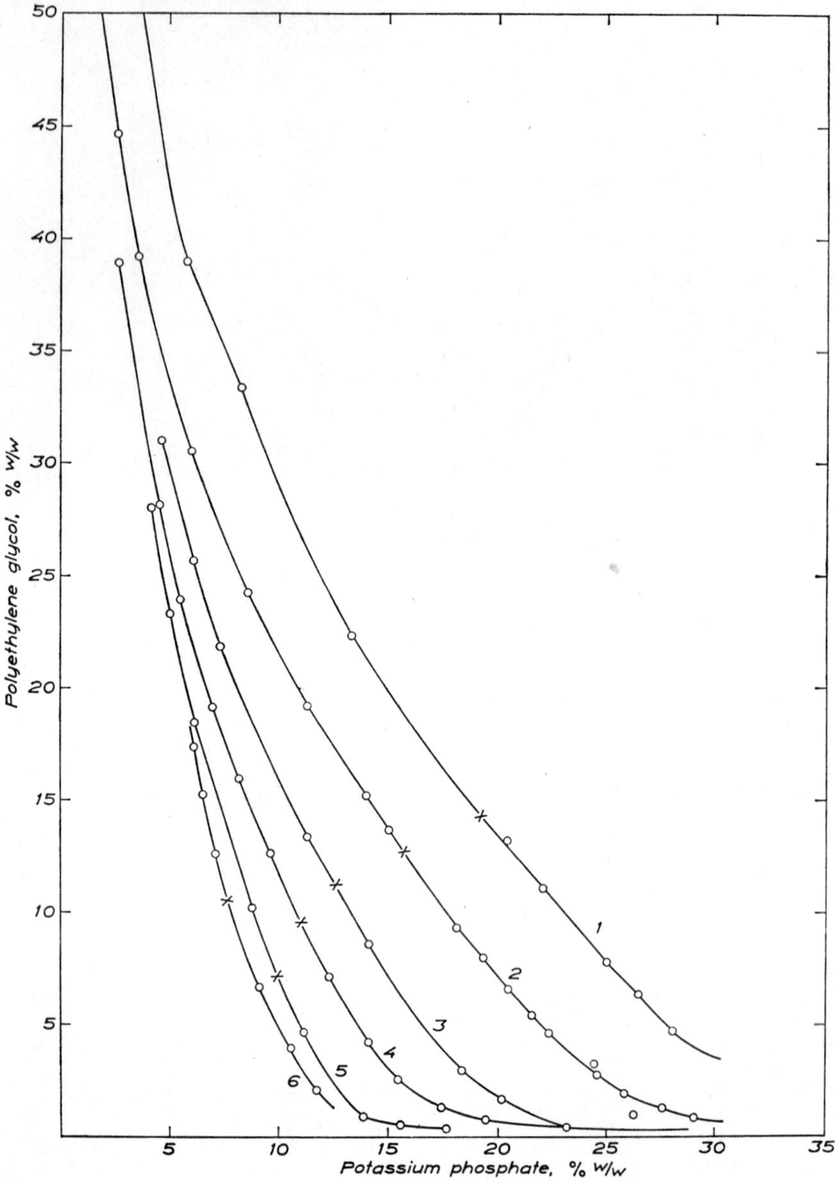

FIGURE 12.49. Binodials and critical points (×) of potassium phosphate–polyethylene glycol systems at 20°C with the following polyethylene glycol fractions: 1, PEG 300; 2, PEG 600; 3, PEG 1540; 4, PEG 4000; 5, PEG 6000; 6, PEG 20000.

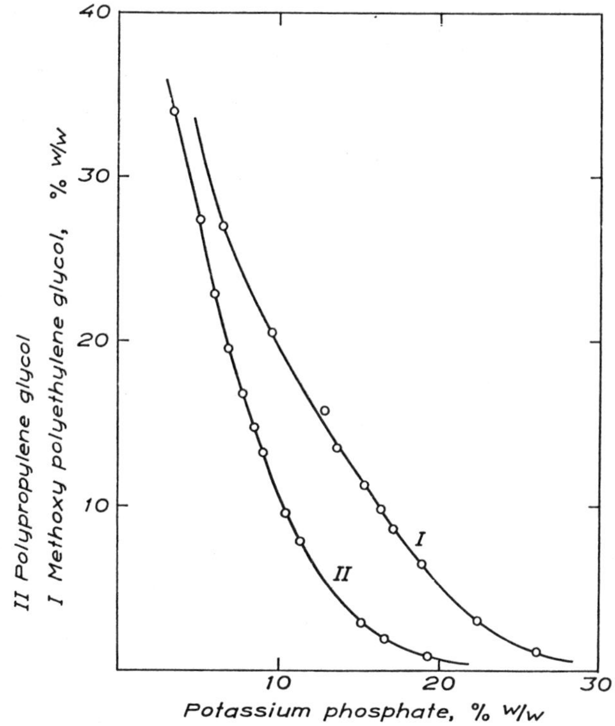

FIGURE 12.50. Binodials of the systems potassium phosphate–methoxypolyethylene glycol (I), and potassium phosphate–polypropylene glycol (II) at 20°C.

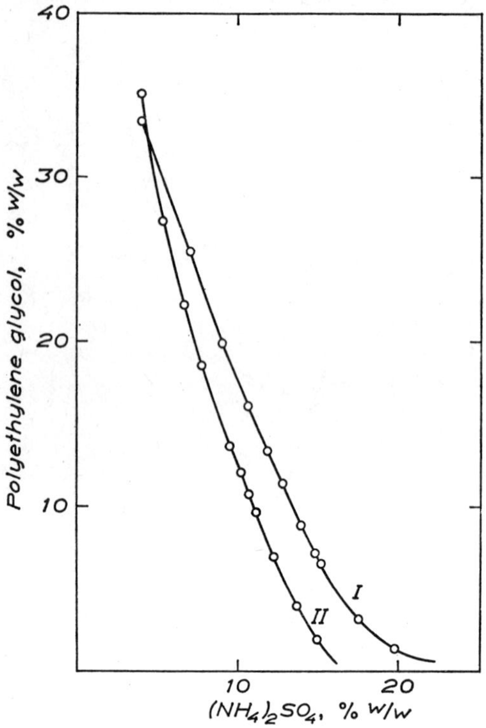

FIGURE 12.51. Binodials of the ammonium sulfate–polyethylene glycol systems $(NH_4)_2SO_4$–PEG 1540 (I) and $(NH_4)_2SO_4$–PEG 4000 (II) at 20°C.

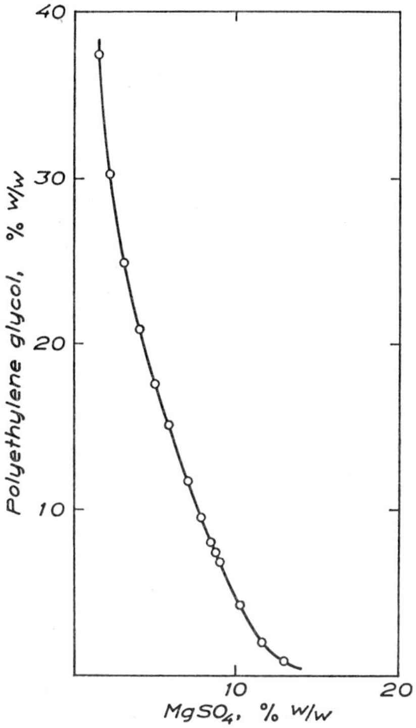

FIGURE 12.52. Binodial of the system $MgSO_4$–PEG 4000 at 20°C.

REFERENCES

Granath, K.A. (1958). *J. Colloid Sci.*, **13,** 308.

Lindberg, B., and Svensson, S. (1968). *Acta Chem. Scand.*, **22,** 1907.

Lindenfors, S., and Jullander, J. (1973). In R.L. Whistler (Ed.), *Industrial Gums,* Academic, New York, pp. 673–694.

Murphy, P.T., and Whistler, R.L. (1973). In R.L. Whistler (Ed.), *Industrial Gums,* Academic, New York, pp. 513–542.

Schürch, S., Gerson, D.F., and McIver, D.J.L. (1981). *Biochim. Biophys. Acta,* **640,** 557–571.

Sloan, J.W., Alexander, B.H., Lohmar, R.L., Wolff, I.A., and Rist, C.E. (1954). *J. Am. Chem. Soc.*, **76,** 4429.

Whistler, R.L. (Ed.) (1973). *Industrial Gums,* Academic, New York.

APPENDIX Experimental Procedures for Partition Experiments, Countercurrent Distribution, and Removal of Polymers

PARTITION EXPERIMENTS

Polymer solutions:
- 20% w/w Dextran 500 (Pharmacia, Uppsala)
- 40% w/w PEG 4000 (now renamed PEG 3400, Union Carbide, New York)
- 40% w/w PEG 6000 (now renamed PEG 8000, Union Carbide, New York)

1. ***Cell Organelles and Membrane Vesicles.*** In order to study the effect of polymer concentration the protocol in Table A.1 may be used, and to study the effect of salt the protocol of Table A.2 may be used.

Each tube is thoroughly shaken and the phases are allowed to settle for 30 min. Samples are taken from the upper phase and the concentration of cell organelles or membrane vesicles is determined using turbidity at 500 nm, protein content, or an appropriate enzyme activity. The amount in the upper phase expressed as percentage of total added to the tube is plotted against polymer concentration (see Fig. 4.1). One may also take a sample from the lower phase and then calculate the amount of material at the interface.

2. ***Cells.*** Cells require a certain tonicity in order to be stable. In the protocol of Table A.3 the partition of cells is studied at 100 m*M* Na_3PO_4, pH 7, with different polymer concentrations.

In the protocol of Table A.4 cells are studied in phase systems with constant polymer composition and varying ionic composition at roughly the same tonicity.

TABLE A.1 Protocol for Partition of Cell Organelles or Membrane Vesicles in Phase Systems Differing in Polymer Concentration at Constant Ionic Composition[a]

Tube No.	Dextran 500, 20%	PEG 4000, 40%	Sucrose, 1 *M*	Na Phosphate, 0.1 *M*, pH 8	H_2O	Sample, mL
1	1.38	0.69	1.0	0.5	0.93	0.5
2	1.43	0.71	1.0	0.5	0.86	0.5
3	1.5	0.75	1.0	0.5	0.75	0.5
4	1.63	0.81	1.0	0.5	0.56	0.5
5	1.75	0.88	1.0	0.5	0.37	0.5

[a] 10 m*M* sodium phosphate pH 8. All quantities in grams except the sample.

TABLE A.2 Protocol for Partition of Cell Organelles or Membrane Vesicles in a Phase System of 6% Dextran 500 and 6% PEG 4000 with Different Ionic Compositions[a]

Tube No.	Dextran 500, 20%	PEG 4000, 40%	Sucrose, 1 *M*	NaPB, 0.1 *M*, pH 8	LiPB, 0.1 *M*, pH 8	LiCl, 0.1 *M*	NaCl, 0.1 *M*	KCl, 0.1 *M*	H_2O	Sample, mL
1	1.5	0.75	1.0	0.5					0.75	0.5
2	1.5	0.75	1.0	—	0.5				0.75	0.5
3	1.5	0.75	1.0	—	—	0.5			0.75	0.5
4	1.5	0.75	1.0	—	—	—	0.5		0.75	0.5
5	1.5	0.75	1.0	—	—	—	—	0.5	0.75	0.5

[a]PB = phosphate buffer. All quantities in grams except sample.

TABLE A.3 Protocol for Partition of Cells in Phase Systems Differing in Polymer Concentration[a]

Tube No.	Dextran 500, 20%	PEG 4000, 40%	Na Phosphate, 0.5 *M*, pH 7	H_2O	Sample, mL
1	1.38	0.69	1.0	0.93	1.0
2	1.43	0.71	1.0	0.86	1.0
3	1.5	0.75	1.0	0.75	1.0
4	1.63	0.81	1.0	0.56	1.0
5	1.75	0.88	1.0	0.37	1.0

[a] At 100 m*M* sodium phosphate, pH 7. All quantities in grams except the sample.

TABLE A.4 Protocol for Partition of Cells (Erythrocytes) in a Phase System with Constant Polymer Concentration but with Different Ionic Compositions[a]

Tube No.	Dextran 500, 20%	PEG 6000, 20%	Na Phosphate, 0.4 *M*, pH 6.8	NaCl, 0.4 *M*	H_2O	Packed Erythrocytes, mL
1	2	1.6	2.2	—	2.2	0.1
2	2	1.6	1.8	0.6	2.0	0.1
3	2	1.6	1.2	1.5	1.7	0.1
4	2	1.6	0.6	2.4	1.4	0.1
5	2	1.6	0.2	3.0	1.2	0.1

[a] All quantities in grams except samples.

TABLE A.5 Publications Where Detailed Descriptions on Countercurrent Distribution Can be Found

Material	Reference
Membrane vesicles	Andersson et al. (1976)
Chloroplasts	Karlstam and Albertsson (1972); Albertsson (1974)
Cell Membranes	Widell et al. (1982)
Chromosomes	Pinaev et al. (1979)
Cells	Albertsson and Baird (1962); Walter (1977)

Countercurrent Distribution

For experimental procedures involving countercurrent distribution the reader is referred to the original articles listed in Table A.5.

Protein Determination

For protein determination on samples from phase partition and countercurrent distribution experiments the method based on Coomassie Brilliant Blue G (Bradford, 1976) is recommended. The Lowry method (Lowry et al., 1951) can be used if it is modified to remove the polymers by a precipitation step using trichloroacetic acid (TCA). The following procedure has been used in our laboratory:

Solutions:

A. 2% w/v Na_2CO_3.

B. 0.5% w/v $CuSO_4 \cdot 5H_2O$ in 1% w/v sodium citrate.

C. Mix 50 mL A and 1 mL B. Prepare fresh every day.

D. Folin reagent.

To 0.2 mL of the sample containing polymers, is added 0.2 mL ice-cold 10% TCA and 1.0 mL ice-cold 5% TCA and the mixture is allowed to stand for 30 min in ice. The precipitated proteins are centrifuged at 20,000 *g* for 20 min (Sorvall SE-12 rotor). The supernatant is sucked off and the pellet washed twice by centrifugation, as above, with 1.5 mL ice-cold 5% TCA. The pellet is then dissolved by treatment with 0.2 mL 0.5 *M* NaOH at room temperature for at least 2 hr. Then 1.0 mL of reagent C is added and after 10 min 0.1 mL Folin reagent is added and the contents of the tube are quickly mixed.

The following guide is useful in selecting a suitable system.
I. To make the partition higher:

(a) Use PEG 4000 instead of PEG 6000.

(b) Use a different ion. For negatively charged material the partition increases in the order $K^+ < Na^+ < Li^+$, while for positively charged material the opposite relation holds. For negatively charged material the partition coefficient is higher with $HPO^{2-}{}_4$, $SO^{2-}{}_4$, or citrate than with $H_2PO^-{}_4$, Cl^-, or Br^-. The opposite holds for positively charged material.

(c) Use a higher pH in the system.

(d) Use a phase system with a polymer composition closer to the critical point in the phase diagram.

(e) Use a charged polymer. A positively charged PEG will pull up negatively charged material into the top phase.

(f) Try a different ionic strength.

II. To make the partition lower:

(a) Use PEG 6000 instead of PEG 4000.

(b) Use a system further away from the critical point.

(c) Use a different ion. See I(b).

(d) Add extra salt, usually a chloride. Notice the relationship stated in I(b).

(e) Use a lower pH.

(f) Use a charged polymer.

REMOVAL OF POLYMERS

After a fractionation experiment is finished it is often desirable to remove the phase polymers from the substance under study. For labile material, however, it may sometimes be advantageous to retain the material in the polymer solutions since the polymers can have a stabilizing effect. Many different techniques have been used for removal of the polymers, including centrifugation, electrophoresis, and precipitation.

1. ***Centrifugation.*** In the case of large particles such as cells and cell fragments, the polymers can be washed away by repeated centrifugations. Also virus particles and DNA can be recovered from the polymers by high-speed centrifugations. Alberts (1967) showed that a polyethylene glycol top phase containing DNA could be layered on CsCl and the DNA banded directly by density gradient centrifugation. The polyethylene glycol stayed as a separate phase on top of the gradient.

2. ***Chromatography.*** Proteins usually can be adsorbed on hydroxylapatite or ion exchange columns. Since dextran and polyethylene glycol are uncharged they can be washed away. The proteins can then be eluted from the column. DNA has been freed from dextran and polyethylene glycol by similar chromatography on hydroxylapatite and methylated albumin.

3. ***Electrophoresis.*** Since dextran or polyethylene glycol are not charged, but proteins and nucleic acids are, electrophoresis can also be applied. Ordinary column methods may be used, but since only a group separation between charged and uncharged material is required, very simple electrophoresis apparatus can be applied. A number of different electrophoresis units have been constructed for the special purpose of recovering proteins, nucleic acids or other charged substances from non ionic polymers like dextran or polyethylene glycol. Figure A.1 shows a simple arrangement which is a modification of an apparatus designed by Hjertén (1971).

The mixture to be separated, for example dextran or polyethylene glycol and a protein, is layered above a number of sucrose layers with increasing density. When current passes through the tube the protein molecules will travel down through the sucrose layers. In 60% w/w sucrose the mobility is very low (Hjertén, 1971) and the protein is stopped when it enters this layer and is concentrated into a narrow band. After the electrophoresis is finished, the contents of the tube are collected by punching a hole in the dialysis membrane at the bottom of the tube and collecting drops in the same manner as one does from centrifugation tubes after gradient centrifugation. We have found this procedure useful for removing dextran and polyethylene glycol from proteins, nucleic acids, and also virus (experiments made by Birgitta Ericson in the author's laboratory). The procedure is simple and quick. It can be used only with fairly small samples and the polymer concentration should not be more than 3%; otherwise, droplet effects and convection will occur at the boundary between the polymer layer and the sucrose layer below.

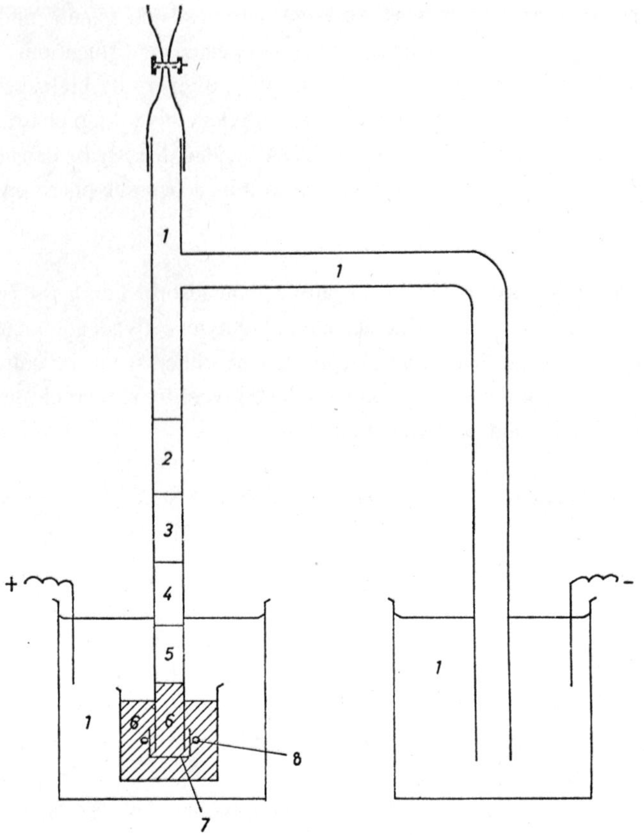

FIGURE A.1. Electrophoresis setup for separating proteins from neutral polymers. (Modified after Hjertén, 1971). (1) Buffer; (2) mixture of protein and polymer at a concentration of preferably less than about 3%; (3) 20% sucrose in buffer; (4) 30% sucrose in buffer; (5) 40% sucrose in buffer; (6) 60% sucrose in buffer; (7) dialysis membrane; (8) rubber band to keep the dialysis membrane fixed to the glass tube.

Figure A.2 shows another arrangement. It consists essentially of a U tube with side arms. The polymer–protein mixture is first added at 1. When current is applied the protein (negatively charged) will move up the tube at 9 while the uncharged polymer stays at the bottom. At 10 the proteins move over to the side arm and down at 11. At 3 there is a high concentration of buffer or salt so that the movement of the protein stops at the boundary at 12. All the protein is concentrated there and may be collected through 5. Gel plugs, 4, of polyacrylamide prevent liquid flow from one electrode compartment to the other. The gel plugs contain high buffer concentrations to prevent electroosmotic flow. At the

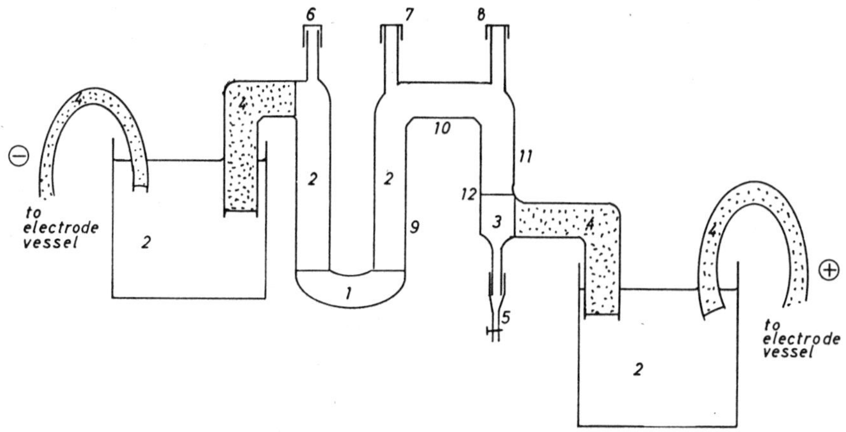

FIGURE A.2. Electrophoresis apparatus for removal of polymers without any anticonvection media. The polymer–protein mixture is added at 1. When voltage is applied, the protein (negative in the case shown) will move through the buffer (2) and concentrate at the boundary at 12 between a high ionic strength solution (3) and the buffer. The protein is collected through 5. Procedure: The entire length of the s-shaped tube between the two gel plugs (4) is first filled with buffer (for example Trix-HCl, 0.01 *M*). A solution with high ionic strength (1 *M* $(NH_4)_2SO_4$) is then added slowly with a syringe through 8 into the bottom part at 3. The polymer mixture is then added slowly with a syringe through 6 into the bottom part at 1. The current is put on. After a suitable time the current is switched off and the solution 3 together with a small layer above 12 is collected through 5. The gel plug at 4 contains both buffer and a high ionic strength solution as in 3.

boundaries between 1 and 2 there is stability against convection during the entire run because of the high density of the polymer solution. However, at the end of the experiment when the protein solution has left the polymer layer there will be convection in the tail of the protein zone due to the density difference between protein solution and the buffer. Therefore, a spreading will occur for the last protein portion which moves up the tube at 9. This will delay the transport and dilute the protein, but this will, in any case, eventually arrive at the boundary at 12 and be concentrated there. Experiments with solutions containing 20% dextran or polyethylene glycol and 1% protein have shown that 99.9% of the polymers are removed from the protein by this procedure. This apparatus has the advantage over the apparatus shown in Figure A.1 in that larger volumes and higher concentrations of polymer solutions can be handled. In fact, the higher the polymer concentration in the polymer–protein mixture at 1 in Figure A.2 the more stable is this layer during the run. The apparatus of Figures A.1 and A.2 thus complement each other. That of Figure A.1 is suitable for mixtures with low polymer concentration in the sample while the one in Figure A.2 is suitable for concentrated, viscous polymer mixtures.

4. *Transfer into a Polymer-Free Phase.*

A. Proteins. As may be seen in Figure 5.3 a number of proteins are transferred to the top phase of the dextran–polyethylene glycol system when a high concentration of NaCl is present. This top phase contains polyethylene glycol as the only polymer. Polyethylene glycol forms a phase system with ammonium

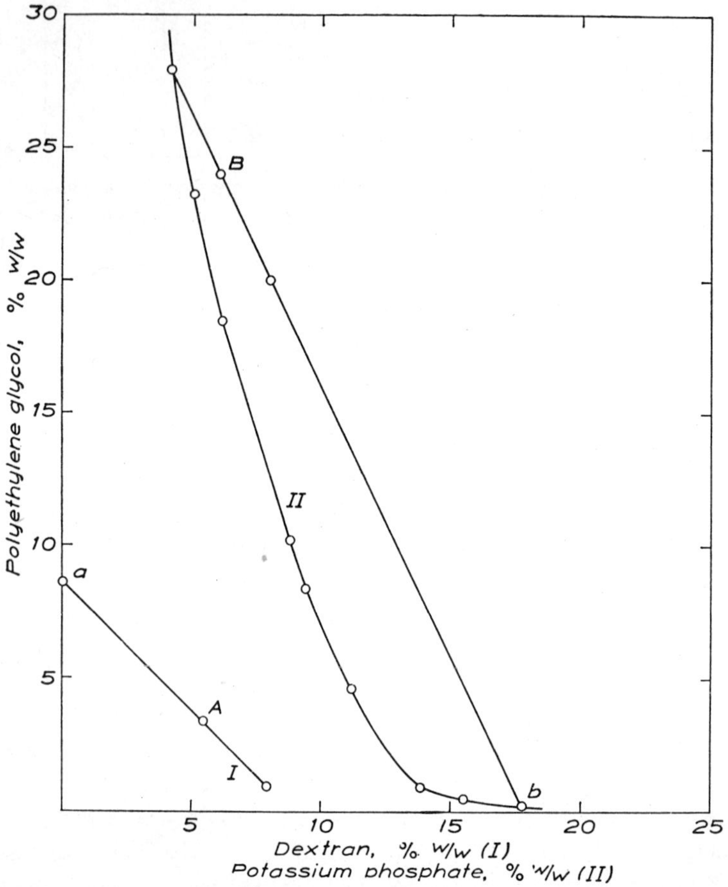

FIGURE A.3. Parts of the phase diagrams of the dextran–polyethylene glycol system (D 48–PEG 6000; system no. 7 of Fig. 12.19) and potassium phosphate–polyethylene glycol system (Fig. 12.44) used for the transfer of a protein into a polymer-free phase. The enzyme is first partitioned in system A, in which most of the enzyme activity is found in the top phase represented by point a. This top phase is collected and PEG 6000 and potassium phosphate are added to give a new phase system represented by point B. In this system the enzyme is transferred to the bottom phase represented by point b and containing practically no polyethylene glycol.

sulfate or potassium phosphate, the bottom phase of which contains almost only salt (see Fig. 12.45). If a protein is transferred to this salt phase, we have separated it from the polymers. An example of such a procedure follows.

Fractions, containing upper and lower phase, from a countercurrent distribution experiment with ceruloplasmin were combined and NaCl added up to 5 *M*. A phase system similar to No. 7 of Figure 12.19 is obtained. The top phase contains most of the enzyme and about 8.5% w/w polyethylene glycol but no dextran. In Figure A.3, the top phase is represented by point a. It is dialyzed to remove the major part of the NaCl and then added to PEG 6000 and potassium phosphate (equimolar parts of KH_2PO_4 and K_2HPO_4) in such proportions that a system containing 24% w/w PEG 6000 and 6% potassium phosphate is obtained (system B in Figure A.3). The enzyme is thereby transferred to the bottom phase represented by point b in Figure A.3 and containing almost no polyethylene glycol.

In the procedures described the proteins are concentrated (10–100 times) concomitant with their recovery. This is an advantage, since after countercurrent distribution, the protein has been diluted by spreading over a number of tubes. To concentrate the protein after a run is often a desirable step prior to further purification or analysis.

This technique has been used by a number of workers to recover protein after phase partition. As an example, the procedure by Okazaki and Kornberg (1964) may be mentioned.

B. Nucleic Acids. The same technique may also be applied to nucleic acids. If the phase system is sufficiently removed from the critical point, the upper phase of the dextran–polyethylene glycol system consists only of polyethylene glycol. By adding $(NH_4)_2SO_4$ or potassium-phosphate to the separated top phase a new phase system is obtained and the nucleic acids are found in the lower, salt phase, which is free from polyethylene glycol.

5. *Transfer of the Protein to a Phase from which It Can be Precipitated.* When ammonium sulfate or potassium phosphate is added to a dextran–polyethylene glycol system the polyethylene glycol concentration in the lower phase diminishes and, if enough salt is added, the dextran phase becomes free of polyethylene glycol. Then a protein that partitions to the dextran phase may be separated from the upper phase, and the protein precipitated with further additions of $(NH_4)_2SO_4$.

6. *Ultrafiltration.* Polyethylene glycol can also be separated from proteins by ultrafiltration (Ingham and Busby, 1980).

REFERENCES

Alberts, B. (1967). *Meth. Enzymol.*, **12A,** 566.

Albertsson, P.-Å. (1974). *Meth. Enzymol.*, **31,** 761–769.

Albertsson, P.-Å. and Baird, D. (1962). *Exptl. Cell Res.*, **28,** 296–322.

Andersson, B., Åkerlund, H.-E., and Albertsson, P.-Å. (1976). *Biochim. Biophys. Acta,* **423,** 122–132.

Bollum, F.J. (1963). *Cold Spring Harbor Symp. Quant. Biol.*, **28,** 21.

Bradford, M.M. (1976). *Anal. Biochem.*, **72,** 248.

Hjerten, S. (1971). *Biochim. Biophys. Acta,* **237,** 395.

Ingham, K.C. and Busby, T.F. (1980). *Chem. Eng. Comm.*, **7,** 315–326.

Karlstam, B. and Albertsson, P.-Å. (1972). *Biochim. Biophys. Acta,* **255,** 539–552.

Lowry, O.H., Rosebrough, N.J., Farr, A.L., and Randall, R.J. (1951). *J. Biol. Chem.*, **193,** 265.

Okazaki, T. and Kornberg, A., (1964). *J. Biol. Chem.*, **239,** 259.

Pinaev, G., Bandyopadhyay, D., Glebov, O., Shanbhag, V., Johansson, G., and Albertsson, P.-Å. (1979). *Exptl. Cell Res.*, **124,** 191–203.

Walter, H. (1977). In N. Catsimpoolas (Ed.), *Methods of Cell Separation,* Vol. 1, Plenum, New York, pp. 307–354.

Widell, S., Lundborg, T., and Larsson, C. (1982). *Plant Physiol.*, **70,** 1429–1435.

INDEX